The ART *of* BOTANICAL PAINTING

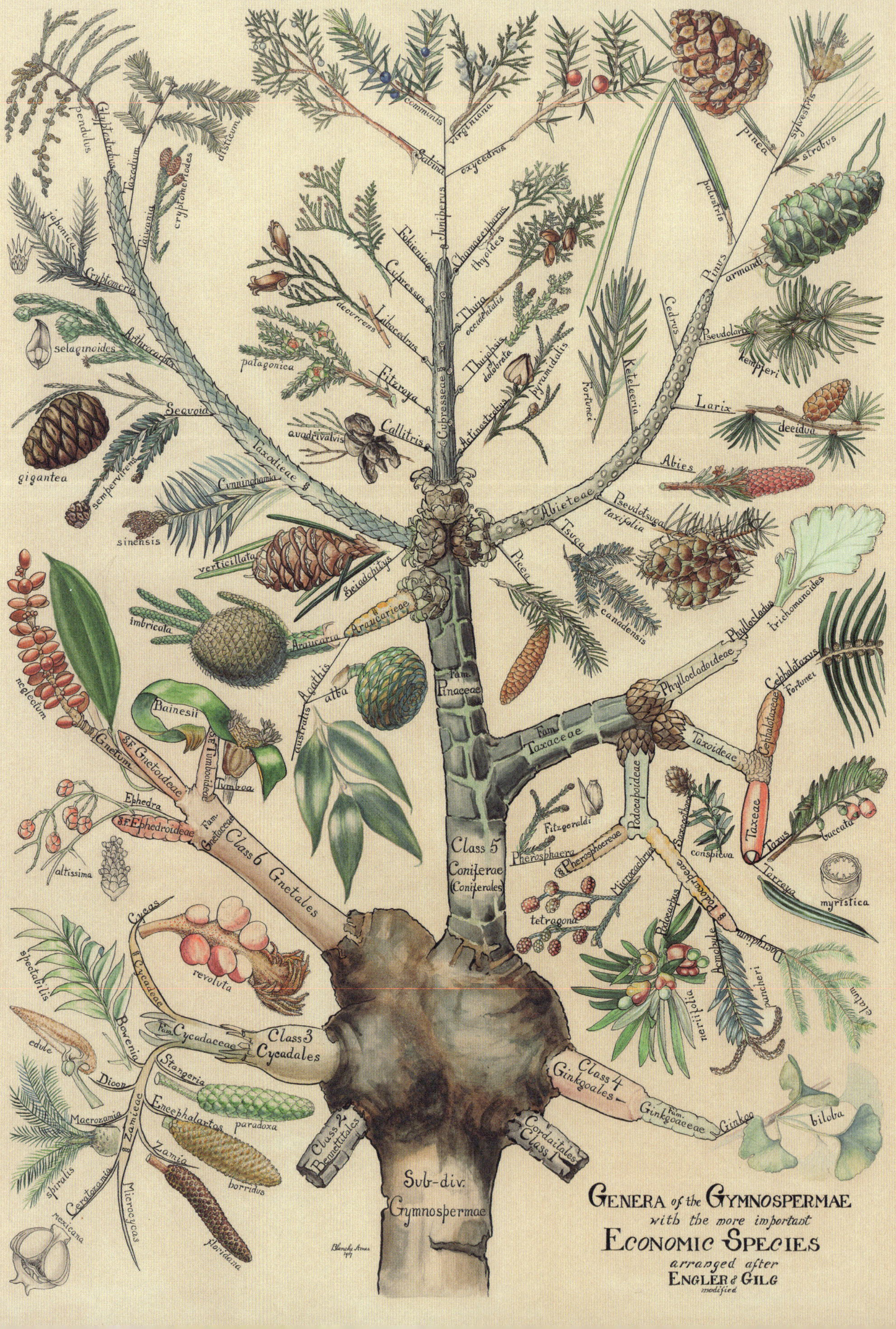

GENERA of the GYMNOSPERMAE
with the more important
ECONOMIC SPECIES
arranged after
ENGLER & GILG
modified
Sub-div. Gymnospermae
Class 1 Cordaitales
Class 2 Bennettitales
Class 3 Cycadales
Class 4 Ginkgoales
Class 5 Coniferae (Coniferales)
Class 6 Gnetales
Fam. Cycadaceae
Cycas
revoluta
Cycadeae
spectabilis
Bowenia
edule
Dioon
Stangeria
Zamieae
Macrozamia
spiralis
Encephalartos
paradoxa
horridus
Zamia
floridana
Ceratozamia
mexicana
Microcycas
Fam. Ginkgoaceae
Ginkgo
biloba
Fam. Pinaceae
Fam. Taxaceae
Fam. Gnetaceae
S.F. Gnetoideae
Gnetum
gnemon
S.F. Ephedroideae
Ephedra
altissima
S.F. Tumboideae
Tumboa
Bainesii
Araucarieae
Araucaria
imbricata
Agathis
australis
alba
Sciadopitys
verticillata
Taxodieae
Cunninghamia
sinensis
Sequoia
gigantea
sempervirens
Arthrotaxis
selaginoides
Cryptomeria
japonica
Taiwania
cryptomerioides
Taxodium
distichum
Glyptostrobus
pendulus
Cupresseae
Callitris
quadrivalvis
Fitzroya
patagonica
Libocedrus
decurrens
Cupressus
Fokienia
Juniperus
Sabina
communis
virginiana
excelsa
Chamaecyparis
thyoides
Thuja
occidentalis
Thujopsis
dolabrata
Actinostrobus
pyramidalis
Abieteae
Picea
Tsuga
canadensis
Pseudotsuga
taxifolia
Abies
Larix
decidua
Keteleeria
Fortunei
Pseudolarix
kaempferi
Cedrus
Pinus
palustris
armandi
pinea
sylvestris
strobus
Phyllocladoideae
Phyllocladus
trichomanoides
Taxoideae
Cephalotaxeae
Cephalotaxus
Fortunei
Taxeae
Taxus
baccata
Torreya
myristica
Podocarpoideae
Pherosphaereae
Pherosphaera
Fitzgeraldi
Microcachrys
tetragona
Saxegothaea
conspicua
Podocarpeae
Podocarpus
neriifolia
Acmopyle
pancheri
Dacrydium
elatum

The ART *of* BOTANICAL PAINTING

ICONIC BOTANICAL WALL CHARTS FROM THE GOLDEN AGE OF SCIENTIFIC DISCOVERY

Anna Laurent

ilex

First published in Great Britain in 2016 by Ilex,
an imprint of Octopus Publishing Group Ltd
Carmelite House, 50 Victoria Embankment
London, EC4Y 0DZ
www.octopusbooks.co.uk
www.octopusbooksusa.com

An Hachette UK company
www.hachette.co.uk

The authorized representative in the EEA is Hachette Ireland,
8 Castlecourt Centre, Dublin 15, D15 XTP3, Ireland
(email: info@hbgi.ie)

Distributed in the US by Hachette Book Group
1290 Avenue of the Americas, 4th & 5th Floors
New York, NY 10104

Distributed in Canada by Canadian Manda Group
664 Annette Street, Toronto, Ontario MS62C8

This material was previously published as *The Botanical Wall Chart*.
This revised edition published in 2025.

Cover image: Fondazione Museo Civico di Rovereto
Back cover image: Randolph College, Lynchburg

ISBN 978-1-84091-895-3

A CIP catalogue record for this book is available from the British Library

Printed and bound in China

10 9 8 7 6 5 4 3 2 1

Commissioned by Zara Larcombe

Publisher: Alison Starling
Consultant Editorial Director: Ellie Corbett
Managing Editor: Rachel Silverlight
Assistant Editor: Ellen Sleath
Art Director: Ben Gardiner
Picture Manager: Giulia Hetherington
Production Manager: Lisa Pinell

FSC www.fsc.org MIX Paper | Supporting responsible forestry FSC® C008047

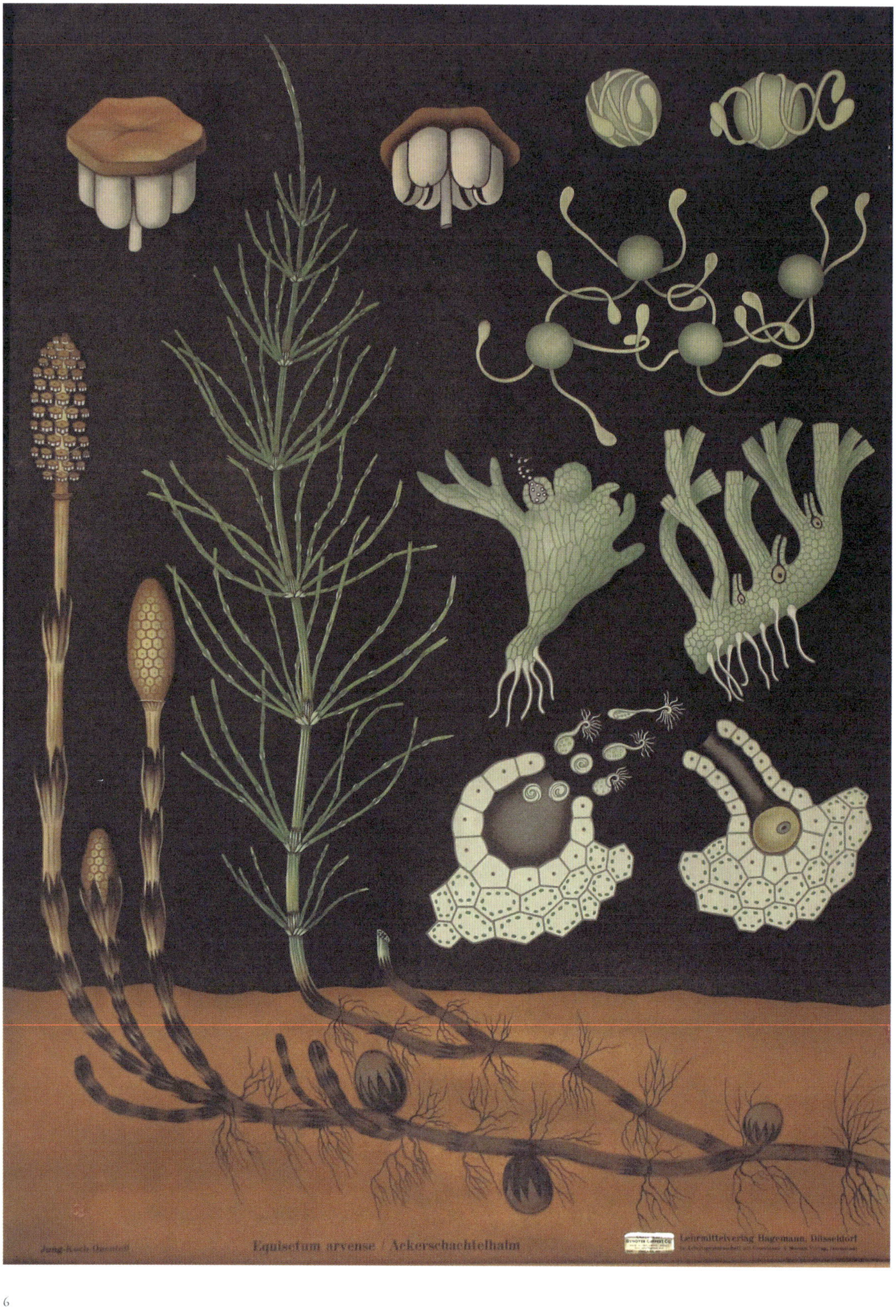
Equisetum arvense / Ackerschachtelhalm
Lehrmittelverlag Hagemann, Düsseldorf

INTRODUCTION: THE SCIENTIST, THE ILLUSTRATOR & THE EDUCATOR

'Natural, scientifically reliable wall charts can replace a natural object in classroom teaching and in lectures; they are more enlightening than the spoken word.'

DODEL-PORT

More than an archive of illustration and inquiry, this book documents an extraordinary convergence of disciplines that flourished in the late nineteenth and early twentieth centuries. Europe was enjoying a golden age of scientific discovery; naturalists were exploring the globe and there was a clamour for knowledge of the natural world. A pedagogical curiosity was no longer limited to elite salons and research; education was now considered a right afforded to all, in classrooms across the continent. And thus the botanical wall chart was born: a synthesis of art, science and education.

The first educational wall charts appeared in Germany in the 1820s. Subjects weren't limited to science – they included historic and religious lessons – but botanical wall charts were unique in the collection of dedicated and brilliant authors they brought together: professors, biologists, illustrators, writers and, of course, botanists. A confluence of emerging social trends – compulsory education across all social classes, and advancing printing technologies – would sow the seeds for a revolution in scientific education, disseminated through gorgeous drawings of anthers, petals, rhizomes, stamens, and seeds.

Two aspects characterized the ideal wall chart: a large format, which allowed for easy viewing across a large classroom, and a comprehensive view of an organism in its entirety. With magnified pollen grains and deconstructed ovaries, wall charts could replace dissection labs and microscopes. Explanatory text was usually absent (sometimes it appeared in a short key at the bottom, or on the back), prompting students to determine the scientific narrative themselves. Educators embraced pedagogical inquiry as the best way to learn, prompting the student to explore visual details and leverage their own knowledge to narrate a pollination sequence, a morphological diagram, or a survey of indehiscent fruits. This is not to say that wall charts themselves were enough; while applauding their achievements, educators were aware of the need to examine a living specimen. A review of Otto Schmeil's series articulates, 'The execution in drawing and colour is quite excellent, the images are so large that everything can be seen clearly at long distances, and you have to give them praise therefore. Nevertheless, they carry some risk in itself, even if only in the hands of inept or lazy teacher who bases all his teaching solely on them. Because when studying plants, nature must be the teacher itself; that is, the students must hold and dissect the fresh copies by hand. Once this is done, so the panels can be returned to in the classroom with great benefits.'

This collection includes charts from some of the most notable series, including ecology-founder Otto Schmeil's *Botanische Wandtafeln* (*Botanical Wallcharts*, 1913), Hermann Zippel and Carl Bollmann's *Ausländische Kulturpflanzen in farbigen Wandtafeln* (*Foreign Crops in Coloured Wall Panels*, 1889), and Jung, Koch, Quentell's *Neuen Wandtafeln* (*New Wallcharts*, 1902–3), with their iconic deep black backgrounds. It includes charts by Swiss botanists Arnold and Carolina Dodel-Port, who were careful to note the date a specimen was drawn from nature, and by André and Madeleine Rossignol, who emphasized simplicity and clarity.

It includes examples from countries across the world, providing a wealth of comparisons and a testament to the timeless sway of the botanical wall chart. The charts appear just as we've found them, with the wear and tear that testifies to their function as teaching tools. While they were certainly considered works of fine botanical illustration (now, and then), created by superlative painters and lithographers, their utility and intended context should not be overlooked. So from time to time you'll see staples, tape, hanging rods, hand-written notes, creases and translation labels, all paying testimony to their place in the story of education.

OPPOSITE:
Title: *Equisetum arvense/ Aekerschaehtelhalm (Field horsetail)*
Authors: Heinrich Jung, Dr. Friedrich Quentell
Illustrator: Dr. Gottlieb von Koch
Language: N/A
Country: Germany
Series/Book: *Neue Botanische Wandtafeln*
Plate: 42
Publisher: Fromann & Morian (Darmstadt, Germany); Hagemann (Düsseldorf, Germany)
Year: 1928; 1951–63

With the bold colours against an iconic black background and their dynamic compositions, it is no wonder that the wall charts of Jung, Koch, Quentell have enjoyed a second life as covetable artefacts in the design world. This book contains a wonderful variety of art, from the microscopic landscapes depicted in the *Dodel-Port Atlas* to the delicate artistry of Henriette Schilthuis' hand-signed watercolour plates. As much as the accompanying text, the charts also tell their own stories, expressed through the unique style and vision of the artists who created them.

The book's chapters are organized by plant family, allowing readers to compare how illustrators from different countries and teaching backgrounds would represent the same species or family, and to learn about basic taxonomy (a summary of the family's characteristics is included at the beginning of each chapter). In most cases, nineteenth-century botanists grouped species according to a classification system that persists today. In other cases, a species has since been renamed or reclassified – taxonomy is a subject that has long vexed botanists, striving to categorize plant species into groups that reflect their behaviours and evolution. Although Linneaus developed a useful framework of binomial nomenclature, it was only a beginning. Botanists today are still reclassifying and weighing morphological features. Just as species evolve, so does science. By grouping charts according to plant family, the book offers some insight into how taxonomy was understood then, and now. Some families are surprising. The *Araceae* family, for example, includes the smallest flowering plant (*Wolffia* spp.) and the largest single inflorescence (*Amorphophallus titanum*). Why are the potato and tomato in the same family as poisonous plants known for their flowers (*Atropa belladonna*)? The authors of these wall charts addressed variation within a family very differently. Zippel and Bollmann, for example, seemed to fill their charts with as many species as would fit, while the Rossignols focused on a single species, with no mention of its family.

OPPOSITE:
Title: *Aspidium filix* (top); *Pinus sylvestris* (bottom)
Author: L. Kny
Language: German
Country: Germany
Series/Book: *Botanische Wandtafeln*
Plate: 97 (top); 60 (bottom)
Publisher: Paul Parey (Berlin, Germany)
Year: 1874

OPPOSITE:
The development of the microscope opened fantastic new vistas to scientists and students, but equipment was expensive and limited to one pair of eyes at a time. The botanical wall chart offered an easy way to disseminate the knowledge gathered under the microscope throughout a whole classroom at once. And in the hands of artists and illustrators, science became a thing of beauty, as in these charts from Leopold Kny's *Botanische Wandtafeln*, that illustrate the microscopic anatomy of a *Aspidium filix* (now *Dryopteris filix-mas*; the male fern), top, and a *Pinus sylvestris* (Scots pine), bottom.

When I began researching this book, I expected that I'd be building on existing academia. Botanical wall charts are at the intersection of so many perennially relevant subjects – horticulture, fine art, history, education, agriculture, politics – I assumed my book wasn't the first. But it was. I wrote an outline for my book and found myself forging research – a few research papers that offered speculative conclusions, a few more university exhibits with promising collections and little history on the illustrators and authors, and the lonely curricula that included botanic wall charts as teaching tools, acknowledging the understudied discipline of wall charts, but lacking the funding to study their collections. The more I researched, the fewer answers I found.

When I visited the Czech University of Agriculture in Prague, assistant professor Milan Skalický introduced me to charts that filled hanging cabinets in classrooms, adjacent to magnified wax models and herbarium specimens. I was thrilled to touch the artifacts, after researching them for so many months. Zippel and Bollmann, A. Peter, Leopold Kny... And while the state of the charts, with their torn edges and worn lettering, was poignantly disappointing, I was excited to see the chalk markings and shorthand notes on many of the charts. 'These charts are still in circulation in your university,' I said to Milan. 'Some are faded and worn. Why haven't you replaced them with newer, more durable, visuals?' I was surprised that Milan hadn't anticipated my question. 'Why would we? Who could improve on these charts?' The quality produced has been unparalleled.

With rigorous research and meticulous illustrations, the wall chart ushered in a new way of thinking about an individual's relationship to the natural world. This book includes annotated captions that describe the illustrations, thereby emulating the experience of a schoolchild in the nineteenth century. Moreover, this collection is unique in its comprehensive survey of styles and information. By the time wall charts were more or less rolled up and shelved, a diverse group of educators and illustrators had produced a stunning selection of very different charts, all with the same purpose: to make botanical subjects accessible, relevant and memorable.

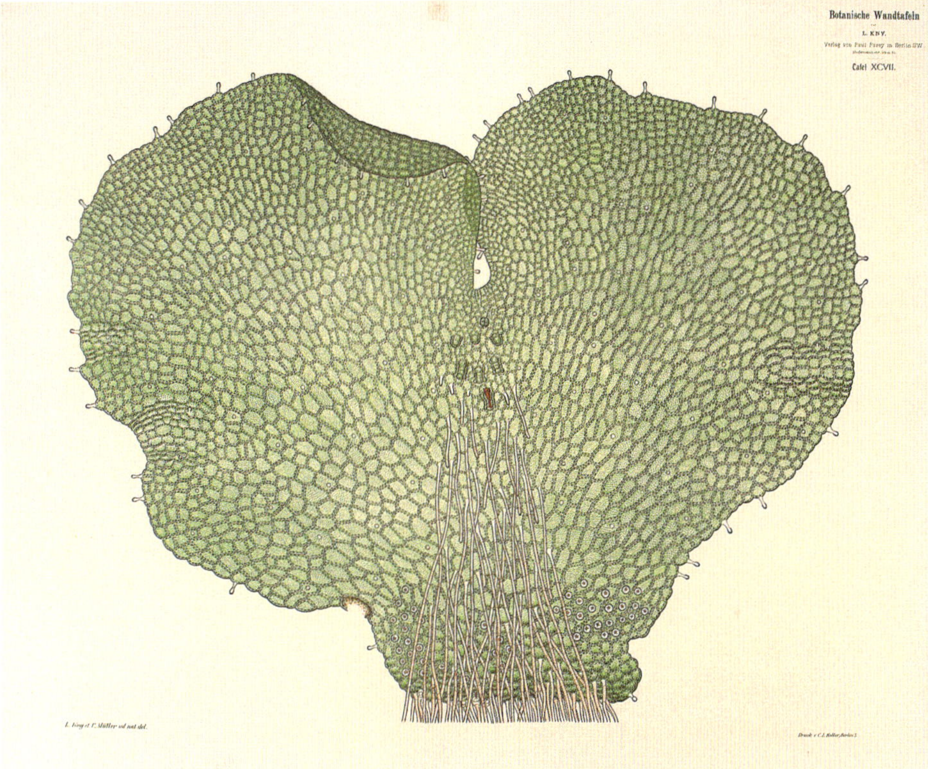
Botanische Wandtafeln
L. KNY.
Verlag von Paul Parey in Berlin SW
Tafel XCVII.
L. Kny et C. Müller ad nat. del.

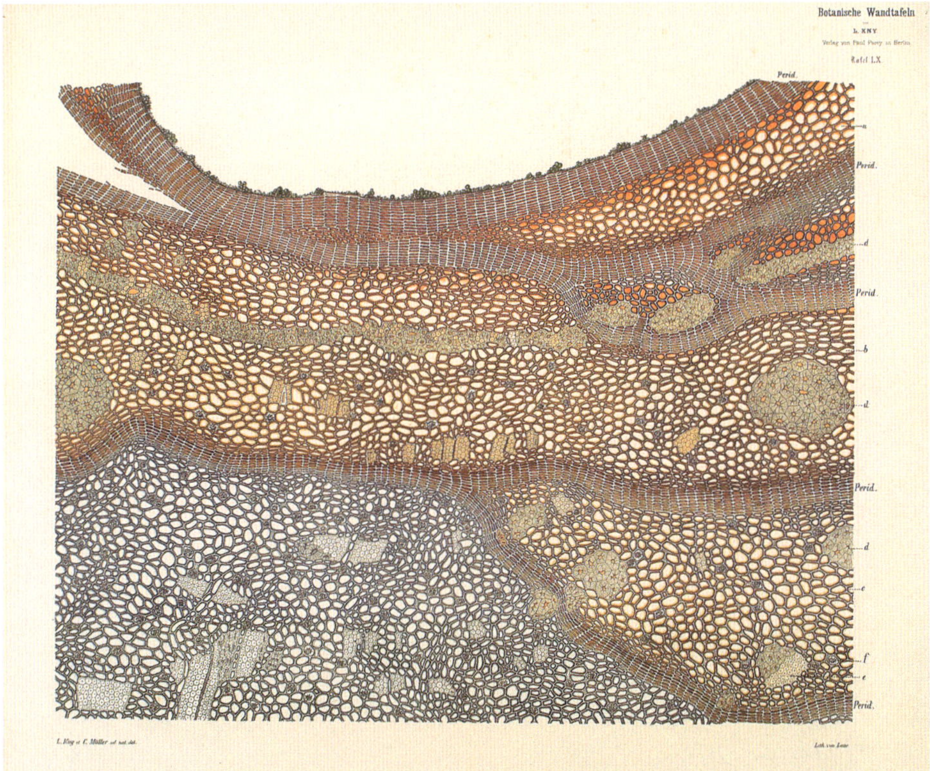
Botanische Wandtafeln
L. KNY
Verlag von Paul Parey in Berlin
Tafel LX
Perid.
a
Perid.
d
Perid.
b
d
Perid.
d
c
f
e
Perid.
L. Kny et C. Müller ad nat. del.

I AMARYLLIDACEAE / AMARYLLIS FAMILY

This family of 71 genera and about 1,600 species has become larger because of the inclusion of *Agapanthaceae* and *Alliaceae*. The plants originate mainly from tropical and subtropical regions, including South Africa, but are also found globally, especially in the Andes. They are herbaceous perennials, usually bulbous, occasionally rhizomatous, and include many favourite garden plants, such as *Agapanthus*, *Crinum*, *Galanthus*, *Ipheion*, *Leucojum*, *Narcissus*, *Nerine* and *Sternbergia*. *Narcissus pseudonarcissus* grows wild in parts of Britain and was famously feted by Wordsworth in his poem 'The Daffodils.'

In frost-prone areas, some of the more tender genera with showy flowers make good container or greenhouse displays, such as *Amaryllis*, *Hippeastrum*, *Lycoris*, *Pamianthe*, *Pancratium*, *Sprekelia*, *Tulbaghia* and *Zephyranthes*. The alliums include popular ornamental species as well as staples of the vegetable garden such as onions, chives, leeks, and garlic. The pungent smell of alliums derives from aliphatic disulfides, which have anti-bacterial properties.

Characteristics: linear, usually basal leaves; six tepals, often fused to form hypanthium or floral tube, sometimes with corona; six stamens; flower is solitary or in umbel, on a scape or leafless flower stem; fruit a capsule or berry.

OPPOSITE:
Title: *Schneeglöckchen* (*Galanthus nivalis*)
Author: Quirin Haslinger; Illustrator: Hans Pertlwieser
Language: German
Country: Austria
Series/Book: *Schoolplaat: Haslinger Botanische Wandtafeln*
Plate: 1
Publisher: teNeues & Co. (Kempen, Germany)
Year: 1950

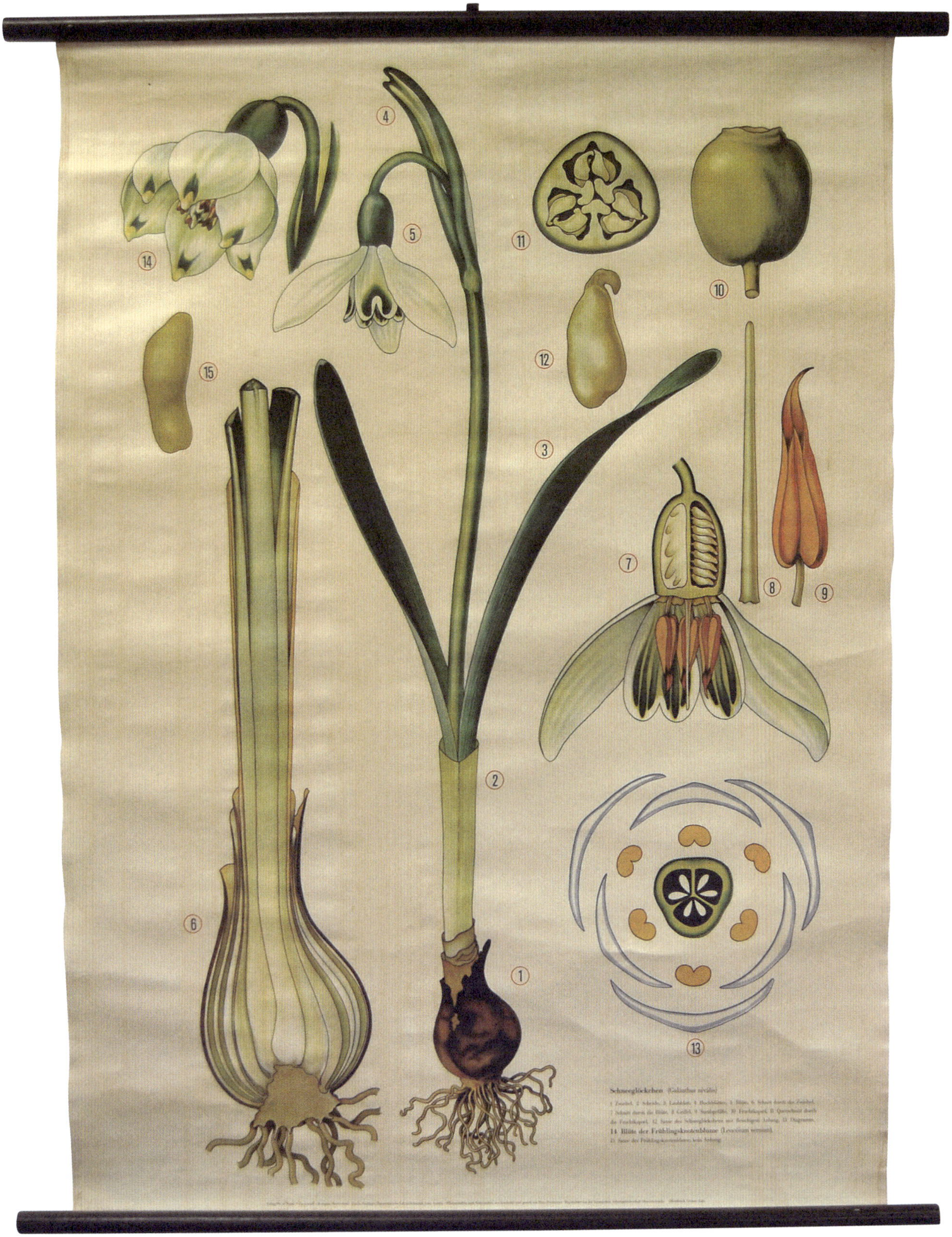
14
4
5
11
10
15
12
3
7
8
9
2
6
1
13

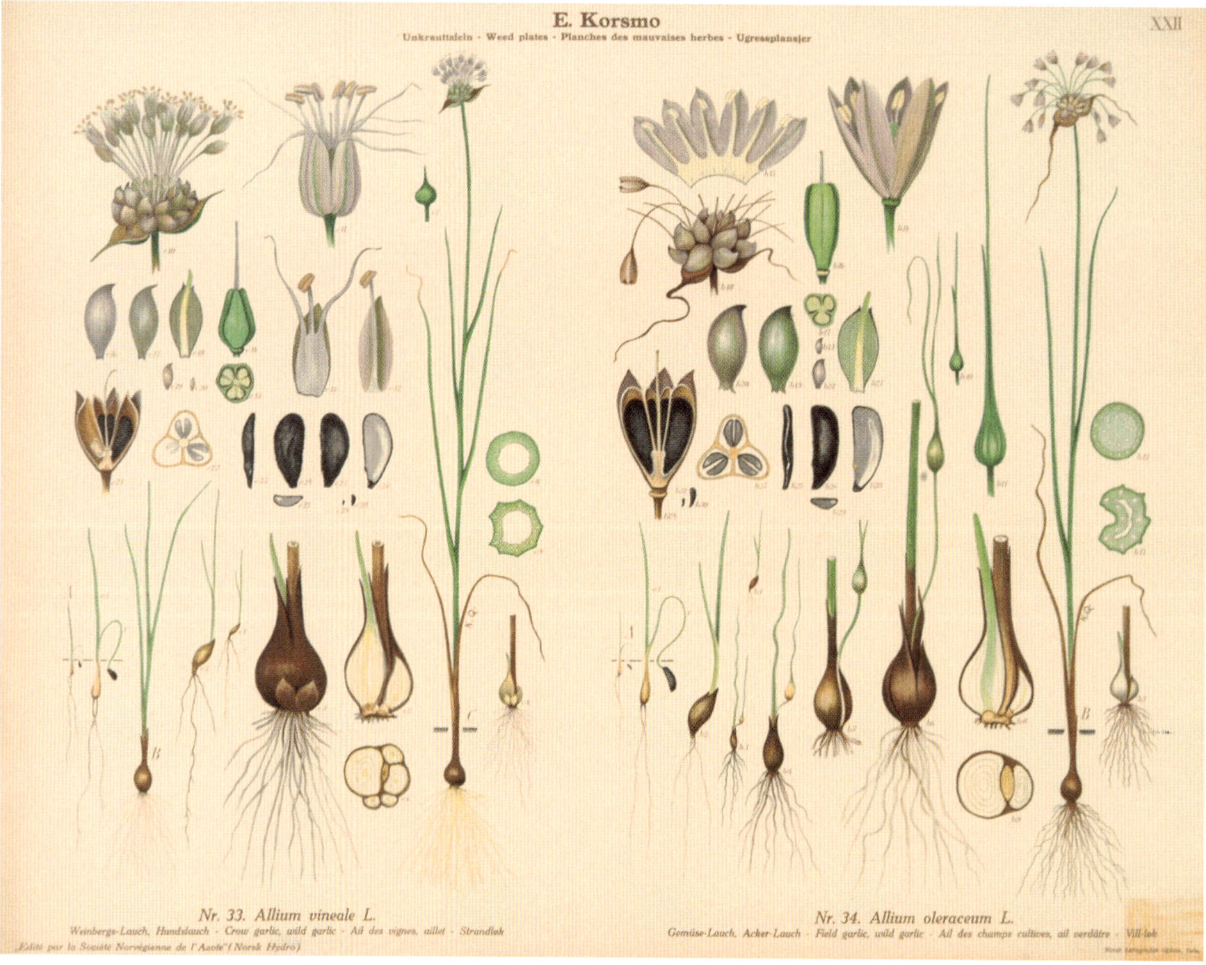

OPENING PAGE: Botanically accurate and poetically appropriate, a demure *Galanthus nivalis* bows its head as though greeting spring's carpet of green. Heralding the end of winter, a snowdrop can sometimes be found peeking through a late-January snow cover. Quirin Haslinger has included fourteen parts of the flower, numbered in sequence.

A classic *Amaryllidaceae*, from bulb (fig. 1), to scape (fig. 3), to a single inflorescence (fig. 5), *G. nivalis* has a pendant blossom, which serves the double purpose of protecting pollen and forming a curved glabrous peduncle that prevents unwanted insects from reaching the blossom. Green markings on the inner whorl of petals will guide desired pollinators, such as bees, toward honey and nectaries. A magnified anther (fig. 9) illustrates the small fissures through which pollen must be enticed. Haslinger opts not to illustrate a dehiscent form, rather using two cross sections (figs. 7, 11) as shorthand to reveal a three-chambered, syncarpous fruit with many ovaries. From this, the student can predict the form of a mature fruit. Similar to many *Amaryllidaceae*, a snowdrop's seeds are produced in capsules.

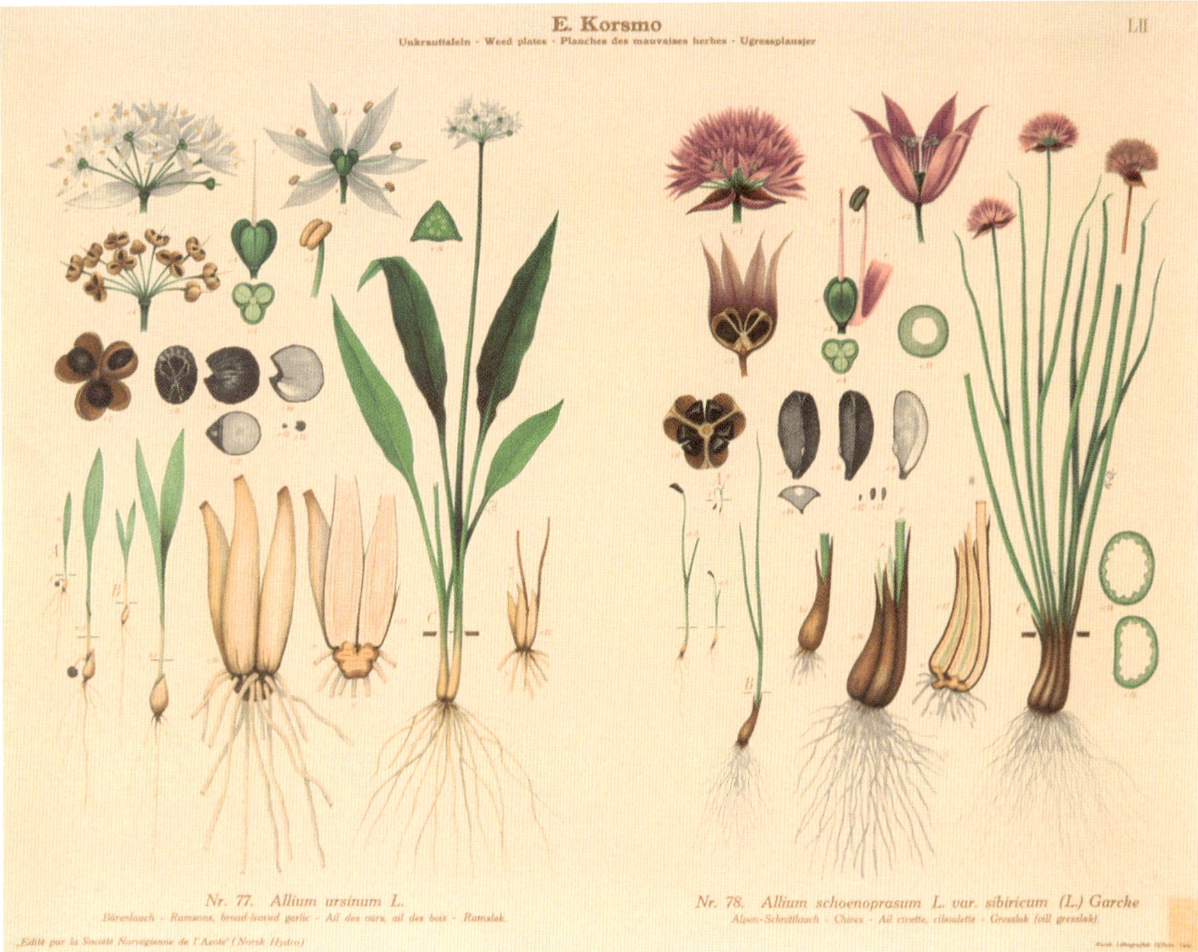

THIS SPREAD: If it seems curious that Emil Korsmo illustrated four species of alliums, yet omitted the onion, garlic, scallion, and shallot, consider the title of his series: *Weed Plates*. All his subjects were considered weeds in early-nineteenth century Europe; none had been cultivated, and all grew more or less wild. While some authors didn't seem to adhere to curation rules and named their series ambiguously (such as Jung, Koch, Quentell's *Botanische Wandtafeln* – or *Botanical Wallcharts*), for others, instruction began the collection's title: Dr. P. Esser's *Poisonous Plants of Germany* and Zippel and Bollmann's *Native Plants of Germany*, for example.

On the two boards shown, Korsmo has illustrated four species of alliums likely to be found in Europe's wood plains, roadsides, or sunny banks: on one chart (opposite), a pistil-to-pistil comparison of *A. vineale* (stag's garlic) and *A. oleraceum* (field garlic); on the second (above), he further details the genus with *A. ursinum* (wild garlic) and *A. schoenoprasum* (chives). Both charts are included in order to illustrate Korsmo's diligent coverage of a genus, his methodological comparisons, and his meticulous dissection of this oft-neglected category – a category that is socially constructed and not botanically based – namely, weeds.

OPPOSITE & ABOVE:

Title, opposite: *Nr. 33 Allium vineale*; *Nr. 34 Allium oleraceum*; Above: *Nr. 77 Allium ursinum*; *Nr. 78 Allium schoenoprasum var. sibiricum*
Author: Emil Korsmo; Illustrator: Knut Quelprud
Language: German, English, French, Norwegian
Country: Norway
Series/Book: *Unkrauttafeln* (*Weed Plates*)
Plates: 22; 52
Publisher: Norsk Hydro (Oslo, Norway)
Year: 1934

OPPOSITE: Throughout their *Botanic Atlas* series, Carolina and Arnold Dodel-Port more or less maintain a focus on plant reproduction, specifically the fertilization of the ovary, which occurs in the union of male and female plasma.

They chose *Narcissus poeticus* (pheasant's eye or poet's daffodil) to represent this because, as they say in the companion text to the illustrations, 'the conditions can probably be seen as typical of the vast majority of flowering plants and, moreover, they are easily accessible.' Because their concern is the ovule, that's what they've emphasized; the gorgeous daffodil blossom is included almost cursorily in the top left corner in order to give the student 'at least some idea of the appearance of the whole,' along with a brief enumeration of the pistils, petals, stigma, etc. (figs. 1, 3).

Cross-sections of the ovary (figs. 4, 5), longitudinally and horizontally, respectively, bring us to the flower's inner sanctum. Beginning at the base of the style, the ovary's backbone is lined with placentas waiting for pollen.

As always, the companion text concludes with '(All figures are drawn from nature).'

OPPOSITE:
Title: *Narcissus poeticus*
Authors: Arnold and Carolonia Dodel-Port
Language: German
Country: Switzerland
Series/Book: *Anatomisch physiologische Atlas der Botanik* (*The Anatomical & Physiological Atlas of Botany*)
Plate: 35
Publisher: J. F. Schreiber (Esslingen, Germany)
Year: 1878–93

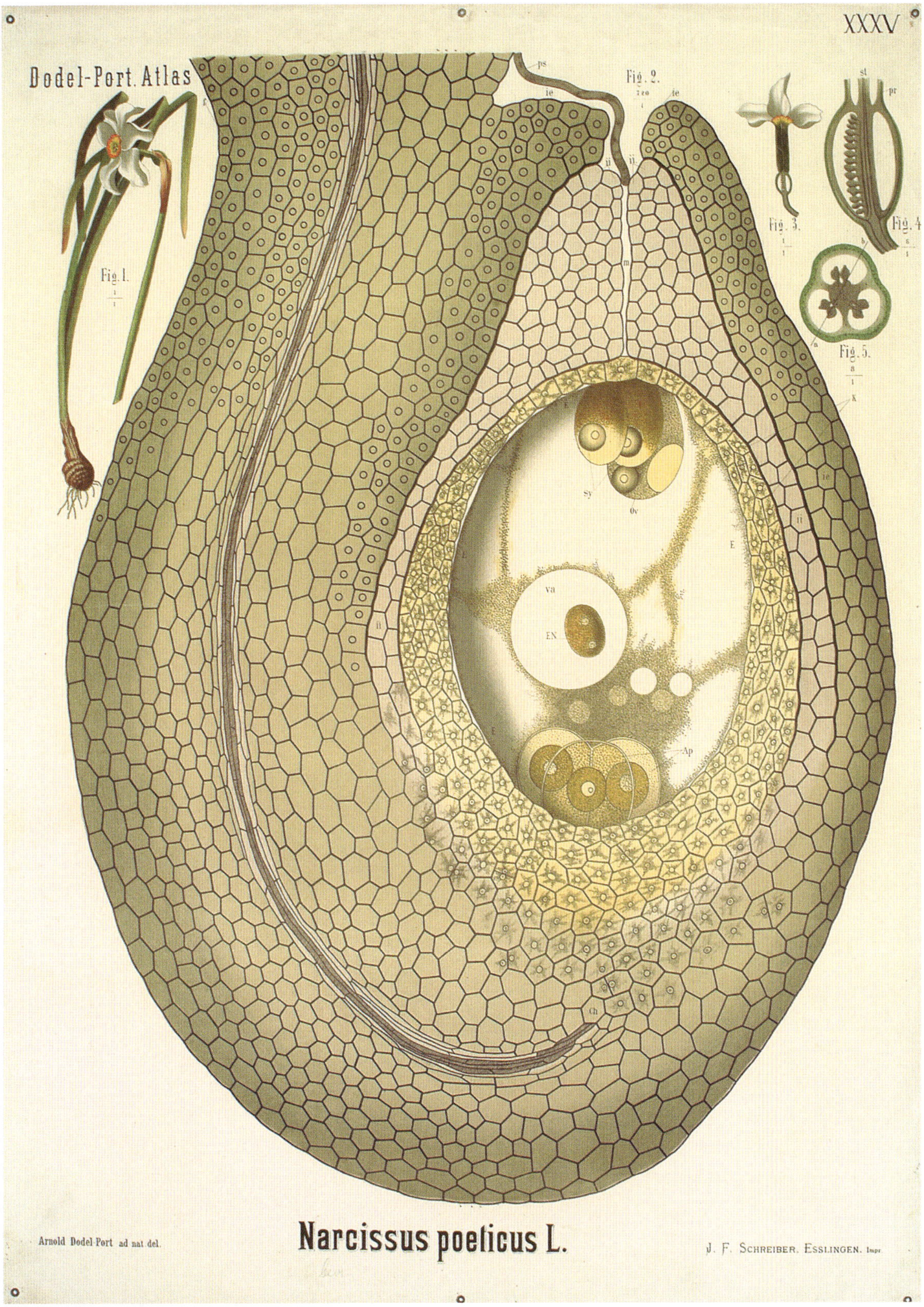
XXXV
Dodel-Port. Atlas
Fig. 1.
Fig. 2.
Fig. 3.
Fig. 4
Fig. 5.
Arnold Dodel-Port ad nat del.
Narcissus poeticus L.
J. F. Schreiber, Esslingen. Impr.

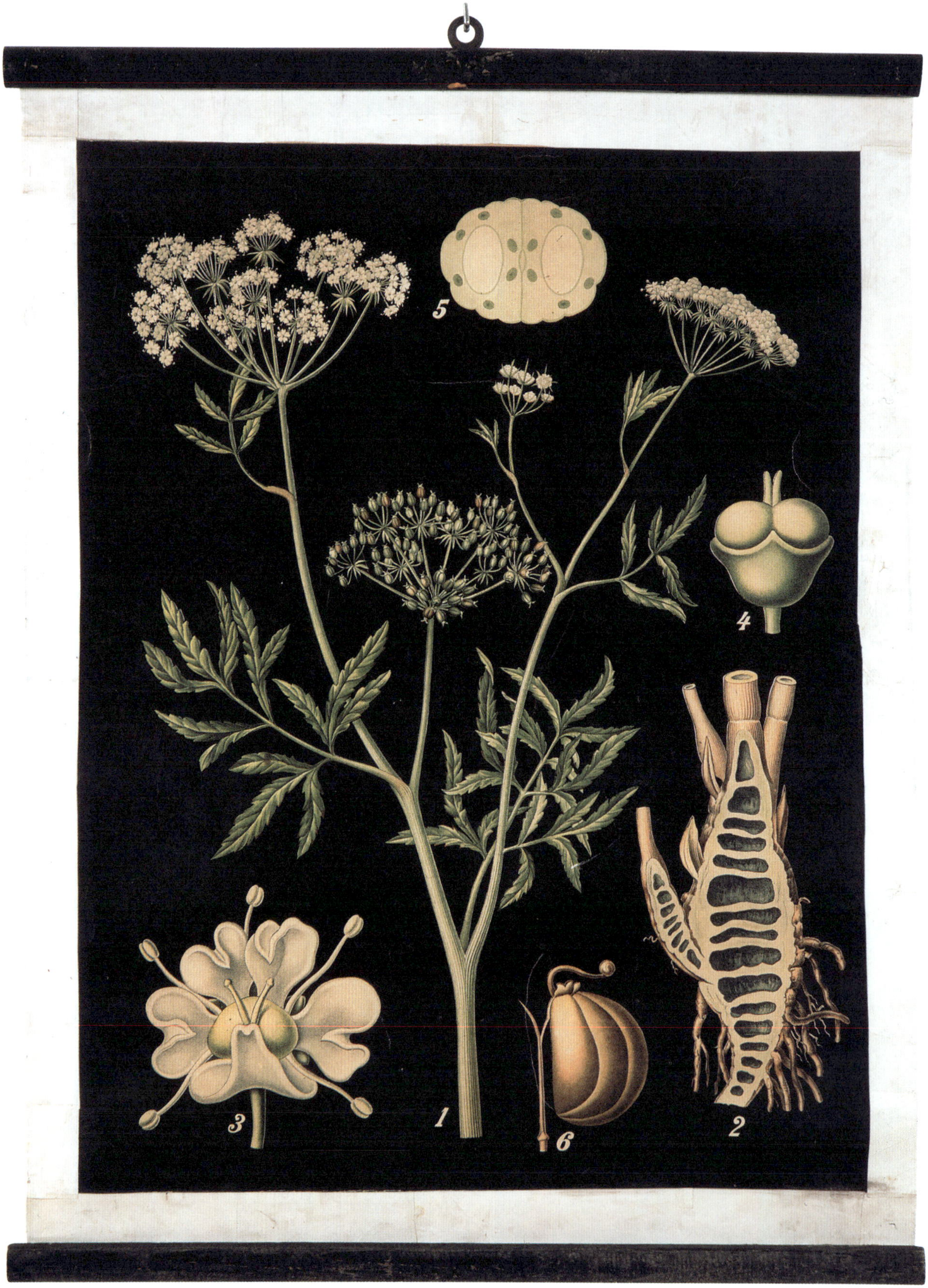
5
4
2
3
1
6

II APIACEAE / CARROT FAMILY

This is a uniform family of 449 genera and over 3,800 species. They are also known as the *Umbelliferae*, because of the typical flat-topped flowerheads – ideal landing pads for pollinating insects. *Apiaceae* are herbaceous annuals, biennials, and perennials – rarely shrubs – that occur mainly in northern temperate regions around the globe. Many are grown as culinary herbs and spices, such as *Anethum graveolens* (dill), *Anthriscus cerefolium* (chervil), *Carum carvi* (caraway), *Coriandrum sativum* (coriander), *Cuminum* (cumin), *Foeniculum vulgare* (fennel), *Levisticum officinale* (lovage), *Petroselinum crispum* (parsley) and *Pimpinella aromatica* (anise). Others are vegetables, for example carrot, celery and parsnip.

Common garden plants include *Astrantia*, *Eryngium*, *Trachyspermum*, while garden weeds such as the dreaded *Aegopodium podagraria* (ground elder) and *Heracleum sphondylium* (common hogweed) also feature in this family. Some are very poisonous: the ancient Greek philosopher Socrates, when sentenced to death, chose to drink an infusion of *Conium maculatum* (poison hemlock).

Characteristics: many aromatic, with aniseed or celery scent; large leaves deeply divided or dissected or pinnately compound, sometimes bristly; umbel of tiny flowers, subtended by bracts or bracteoles; sepals reduced or absent; five free petals, sometimes overlapping or inflexed; five minute stamens; fruit a dry schizocarp, some ridged, some winged.

OPPOSITE:
Title: *Cicuta virosa*
Author: Dr. P. Esser; Illustrator: Carl Bollmann
Language: German
Country: Germany
Series/Book: *Die Giftpflanzen Deutschlands (Poisonous Plants of Germany)*
Plate: 13
Publisher: Friedrich Vieweg & Sohn (Braunschweig, Germany)
Year: 1910

OPENING PAGE & RIGHT: The late nineteenth century was confronted with a peculiar botanic dilemma: while exotic species were arriving with increasing frequency, there was a concern that the effects of wars and industrialization would eclipse a pastoral heritage that sought to identify local plants – the good, the bad, and the poisonous. Here, we have two authors tackling the latter: on this page, Siegmund Schlitzberger's representation of *Aethusa cynapium* (fool's parsley or poison parsley) and *Conium maculatum* (hemlock or poison hemlock), from his series *Unsere verbreiteten Giftpflanzen* (*Our Common Toxic Plants*); and Dr. Peter Esser's careful deconstruction of the *Cicuta virosa* (water hemlock or cowbane), opening the chapter. Dr. Esser, director of the Botanic Garden at Cologne, was concerned that the German people were unaware of the snakes in their grass. His *Poisonous Plants of Germany* series profiles especially dangerous species, both native and introduced, that were literally a threat to his country's health.

Esser writes, 'Among poisonous plants, the genus *Cicuta* is of especial interest, as it is probably the most violently toxic of all the plants growing in temperate regions ... [and] the water hemlock is the most toxic of our native umbelliferous.' His chart entreats students to know the characteristics of the species – its lanceolate leaves, umbellate inflorescence, downward bracts, tiny chalices of fruits – but most of all, to memorize the rhizome. 'Most incidents of poisoning are caused by the confusion of the large rootstock for that of celery or parsley; the rootstock is particularly dangerous to children because it tastes sweet.' *Cicuta*'s rootstock may, however, be easily identified, Esser goes on to say: 'If the rootstock is cut longitudinally there will be seen, more or less clearly, a number of transverse chambers.'

As with most published series, Schlitzberger included companion text with his charts. Most authors were scientists, professors or experts for whom the illustration was a partial story. Accompanying books enabled teachers to understand the content before presenting the charts to students. They also allowed the charts to be disseminated wherever appropriate, regardless of existing material or language. Schlitzberger's charts could be used in a general elementary school classroom, a higher-level plant taxonomy lesson, or at a specialized agricultural school.

RIGHT:
Title: *Aethusia cynapium, Conium maculatum*
Author: Siegmund Schlitzberger
Language: German
Country: Germany
Series/Book: *Unsere verbreiteten Giftpflanzen (Our Common Toxic Plants)*
Plate: 7
Publisher: Theodor Fischer (Berlin, Germany)
Year: 1892

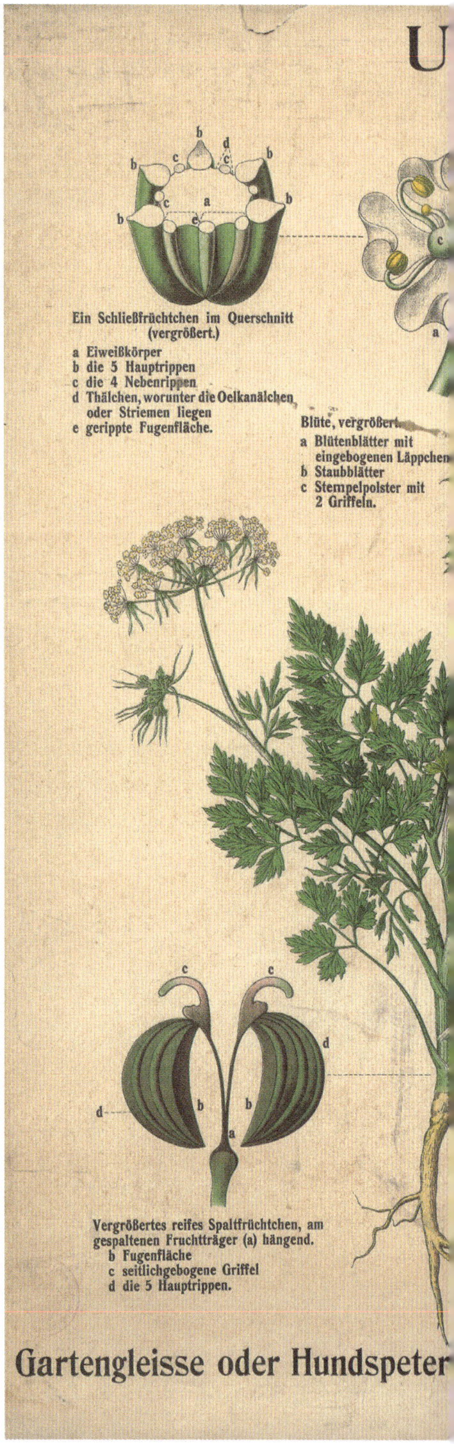

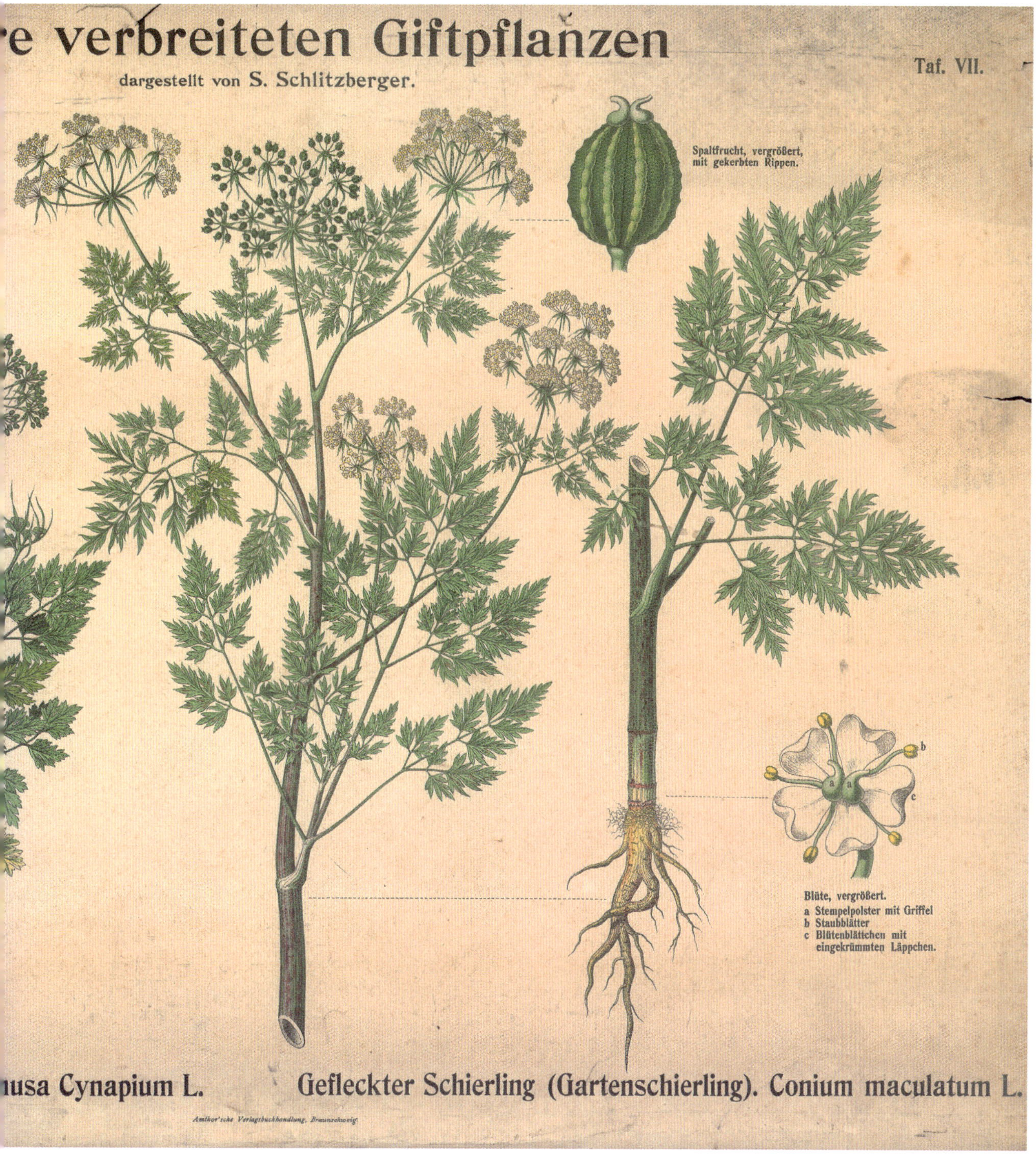
re verbreiteten Giftpflanzen
dargestellt von S. Schlitzberger.
Taf. VII.
Spaltfrucht, vergrößert, mit gekerbten Rippen.
Blüte, vergrößert.
a Stempelpolster mit Griffel
b Staubblätter
c Blütenblättchen mit eingekrümmten Läppchen.
a
b
c
usa Cynapium L.
Gefleckter Schierling (Gartenschierling). Conium maculatum L.
Amthor'sche Verlagsbuchhandlung, Braunschweig.

RIGHT: Wall charts were designed to facilitate a new kind of education based on inquisition. Rather than recite a lesson, students would learn to discern both the question and the answer. A. Peter's illustrations are the epitome of the pedagogical ideal. They provide a sliver of information, often only elucidating one perspective on a species, while providing a great deal of information for the inquisitive student to deduce. The chart shown here explores the reproductive development of the *Apiaceae* family.

With its domed inflorescence of tiny flowers *Heracleum sphondylium* (common hogweed) is a typical *Apiaceae*. Peter chose to abridge the inflorescence to four pedicels, showing four flowers at different stages of development (fig. 1). Flowers in the inflorescence mature progressively from the outside in, to attract passing pollinators. Here, the lower central flower's young anthers bow inward, while the flower above has already matured and shed its stamen. By its side (fig. 2), a mature schizocarp offers its seeds to the wind.

Peter also illustrates *Daucus carota* (wild carrot or Queen Anne's Lace), shown as a single flower in profile (fig. 4) and an artificially sparse inflorescence (fig. 3). Fluctuating the magnification among the umbels, he inflates the central flower. The function of this tiny red flower, coloured by an anthocyanin pigment, is to attract pollinating insects.

At the bottom of the chart, the bristled fruit of *Caucalis daucoides*, now known as *C. platycarpos* (burr parsley) angles defiantly in the corner, a barbed reminder that *Apiaceae*'s double-carpel fruits have adapted not to rely on the wind for dispersal, instead, opportunistically catching a ride on an unwitting passerby.

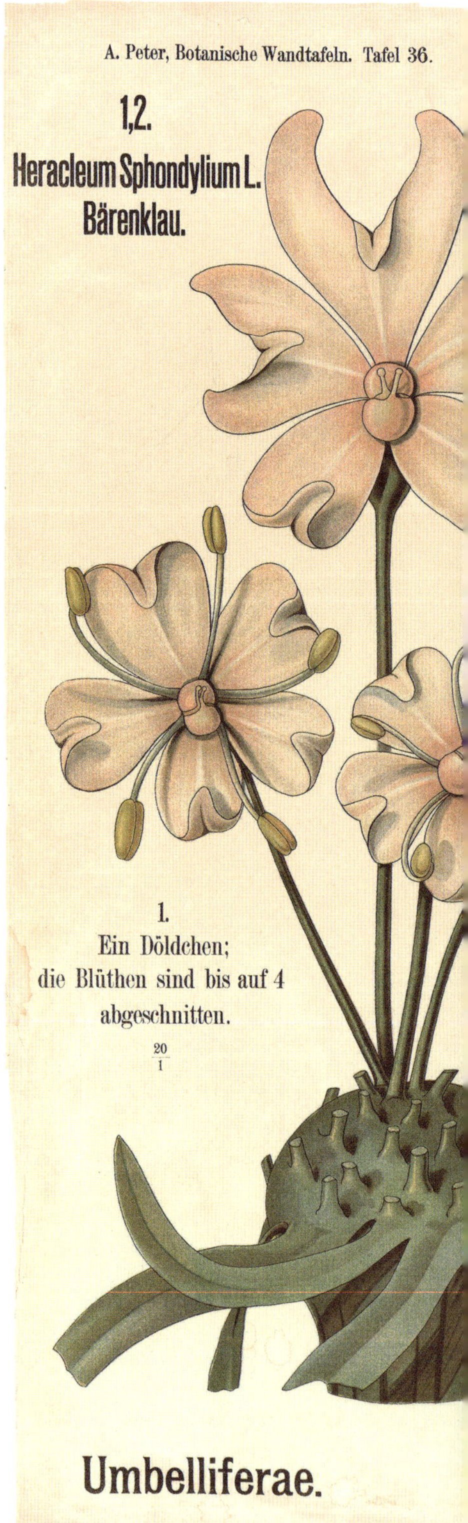

RIGHT:
Title: *Umbelliferae*
Author: A. Peter
Language: German
Country: Germany
Series/Book: *Botanische Wandtafeln*
Plate: 36
Publisher: Paul Parey (Berlin, Germany)
Year: 1901

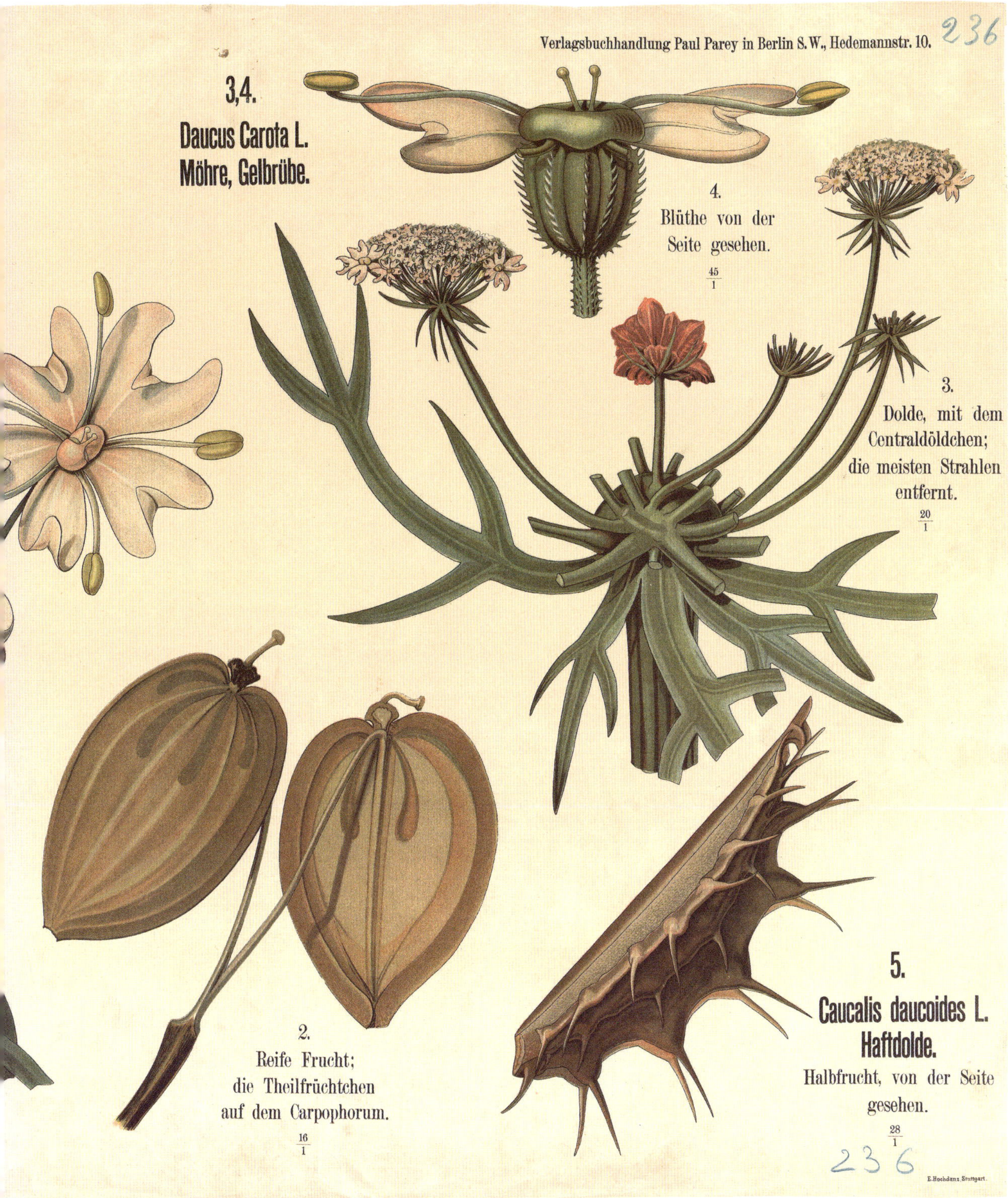
Verlagsbuchhandlung Paul Parey in Berlin S. W., Hedemannstr. 10.
3,4.
Daucus Carota L.
Möhre, Gelbrübe.
4.
Blüthe von der
Seite gesehen.
45/1
3.
Dolde, mit dem
Centraldöldchen;
die meisten Strahlen
entfernt.
20/1
2.
Reife Frucht;
die Theilfrüchtchen
auf dem Carpophorum.
16/1
5.
Caucalis daucoides L.
Haftdolde.
Halbfrucht, von der Seite
gesehen.
28/1
E. Hochdanz, Stuttgart.

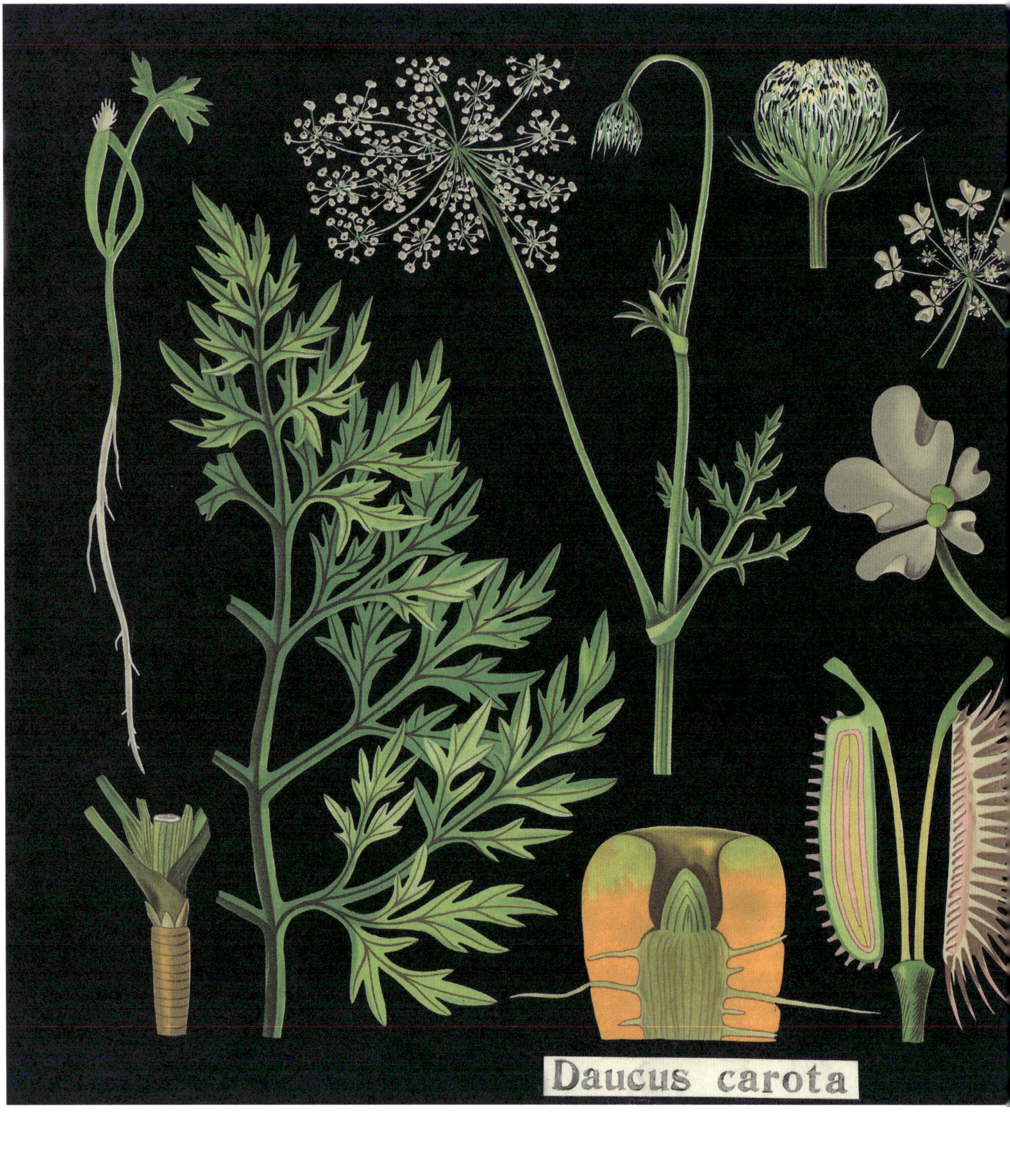
Daucus carota

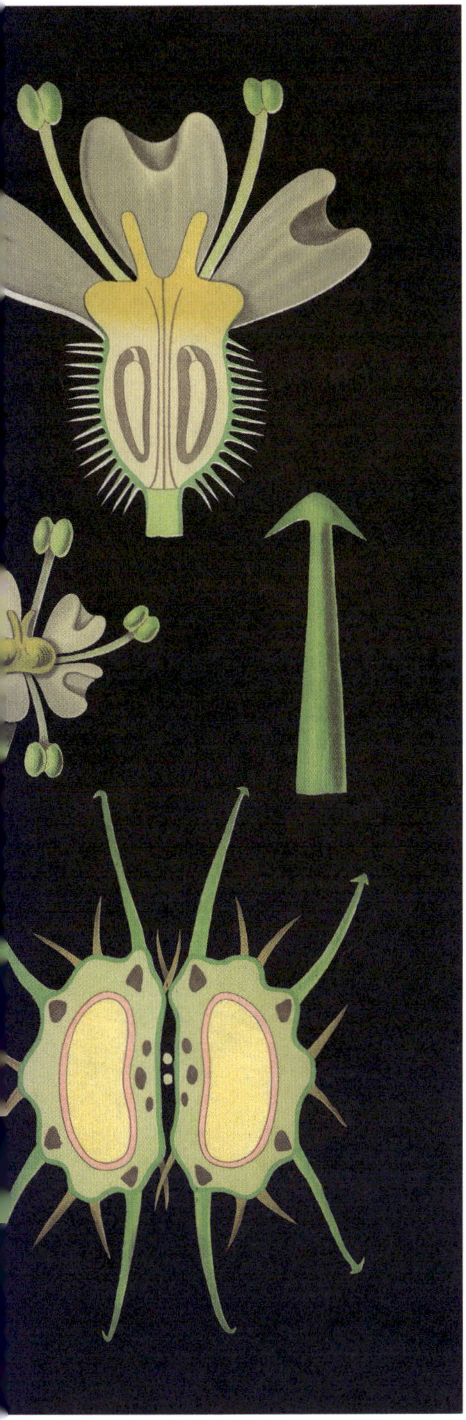

LEFT: While A. Peter introduced his species with a single discreet cross-section or organ, Jung, Koch, Quentell were among the illustrators whose boards offered an expanded portrait, meticulously rendered and comprehensively detailed. Here, they portray *Daucus carota* (wild carrot or Queen Anne's lace) as an unmistakeable *Apiaceae*, with features that include a feathery plume of foliage, a pointillist inflorescence of tiny flowers, and the schizocarp dehiscent seeds with many outstretched, delicate burrs. Upon a a whorl of delicate latticed bracts, five-petalled flowers curve slightly inward around five stamens and two styles, alongside a double ovary and young seeds with nascent bristles that will grow to disperse the seed.

The board illuminates every stage of the plant's blooming cycle, including a globular umbel after pollination. When pollinated, the umbrella shape of the inflorescence reverses and curls inward like a bird's nest. Appropriately, this mirrors the form that the plant had when it was young – a huddled inflorescence as the umbel's flowers matured from the outside in. Modestly shrouded by radiating pedicels, ovaries swell until the flower head glows a pale green, then the umbel's rays straighten out to introduce its young fruits to the world.

Jung, Koch, Quentell illustrate this stage in the top-centre position of the chart, in profile perspective, an aristocratic globe with a delicate green collar. Here we can distinguish *D. carota* from other *Apiaceae*, whose architecture doesn't change with maturity. All plant families have slipped in poisonous species that resemble their non-toxic relatives, and the *Apiaceae* is no different. Mistaking a hemlock for Queen Anne's lace can be fatal, so any defining characteristics are important to teach.

Yet the portrait is incomplete, neglecting a second identifying characteristic of *D. carota* – the little red flower exaggerated by A. Peter on the previous page. This sterile central flower will never bear fruit; its job is to attract pollinating insects to the umbel. It is surprising that Jung, Koch, Quentell fail to mention the red flower that gives the species its common name in North America, for the single drop of blood supposedly shed by Queen Anne when she pricked her finger while sewing lace.

LEFT:
Title: *Daucus carota*
Authors: Heinrich Jung, Dr. Friedrich Quentell
Illustrator: Dr. Gottlieb von Koch
Language: N/A
Country: Germany
Series/Book: *Neue Wandtafeln*
Plate: 19
Publisher: Fromann & Morian (Darmstadt, Germany)
Year: 1902–3

III ARACEAE / ARUM FAMILY

The arum or aroid family includes 143 genera and approximately 3,750 species, mostly from the wet tropics. They are herbaceous or woody perennials, usually with rhizomatous or tuberous roots. Some are climbers, such as *Monstera* and *Scindapsus*, some epiphytic, like *Anthurium*, and a few are aquatic or marsh plants, such as *Peltandra*. *Araceae* are fly-pollinated, so many, for instance *Arisarum* and *Dracunculus*, are unpleasantly scented. Some plants are also highly irritant. Cultivated species are grown for their showy inflorescences – *Arisaema*, *Arum*, *Calla*, *Lysichiton*, and *Zantedeschia* – or for their attractive foliage – *Caladium*, *Dieffenbachia* (dumb cane) and *Philodendron*. *Colocasia esculenta* (taro) is grown as a food crop for its tubers. There are several attractive wildflowers in this family, such as *Arum italicum* and *A. maculatum* (lords-and-ladies).

The largest plant in the family, the endangered *Amorphophallus titanum* (titan arum), is so huge and flowers so rarely in cultivation that it features in the *Guinness Book of Records*. It causes a sensation each time it blooms because of its extremely malodorous and huge inflorescence – up to 10 feet (3m) – and single leaf – up to 20 feet (6m).

Characteristics: leaf usually basal, broad, reticulate, with distinct petiole; stem often with bitter, milky sap; four to six sepaloid segments; tiny, sessile flowers borne on spadix, often with or surrounded by a petal- or leaf-like spathe. The fruit is usually a berry.

OPPOSITE:
Title: *Arum maculatum*
Author: Dr. P. Esser; Illustrator: Carl Bollmann
Language: German
Country: Germany
Series/Book: *Die Giftpflanzen Deutschlands (Poisonous Plants of Germany)*
Plate: 13
Publisher: Friedrich Vieweg & Sohn (Braunschweig, Germany)
Year: 1910

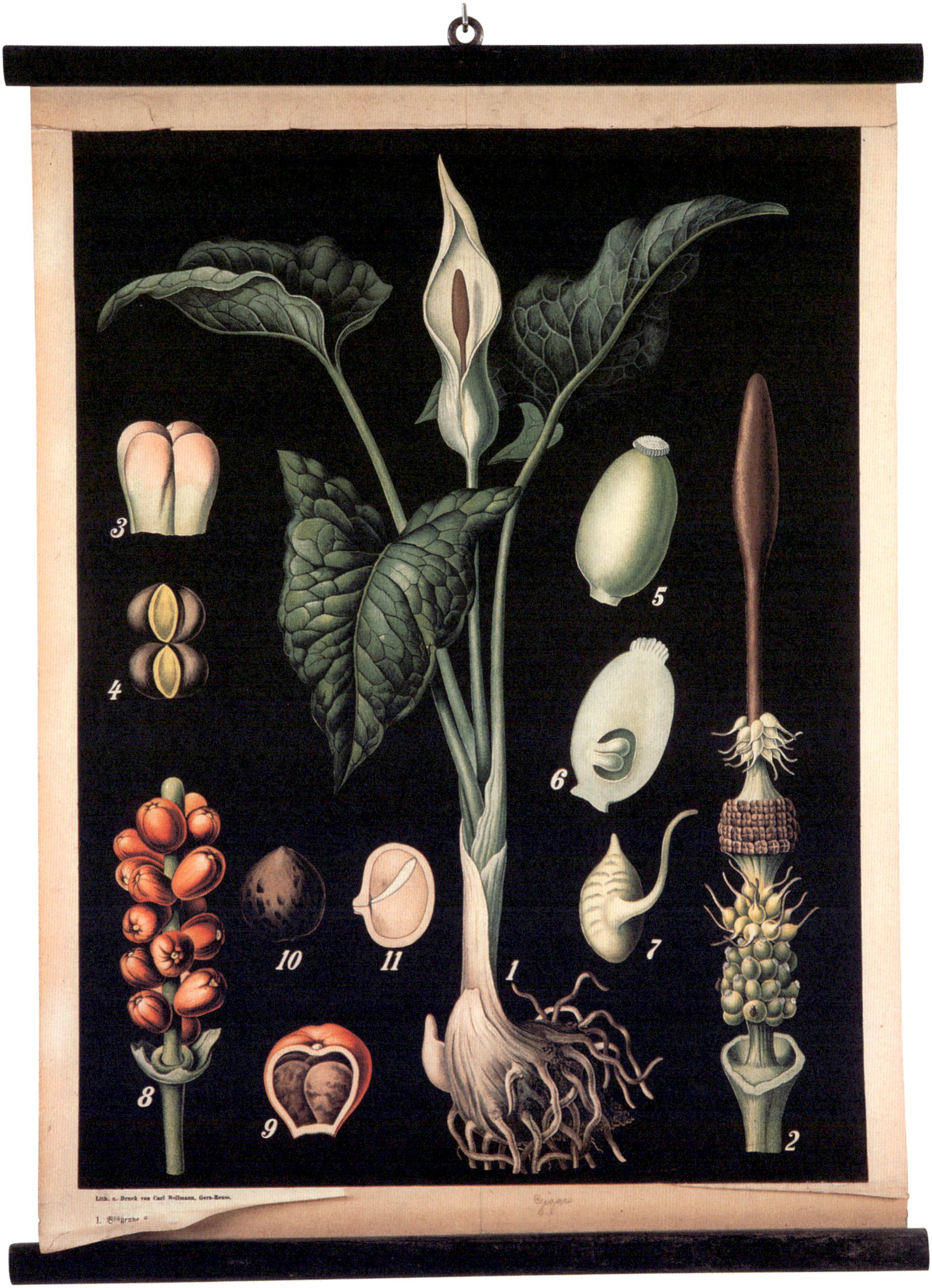
3
4
5
6
7
8
9
10
11
1
2
Lith. u. Druck von Carl Bollmann, Gera-Reuss.

OPENING PAGE: In his introduction to *Die Giftpflanzen Deutschlands* (*Poisonous Plants of Germany*) Dr. P. Esser, director of Cologne's Botanic Garden, articulated his reasons for publishing the series: 'There still occur frequent cases where people lose their lives by eating poisonous plants, the vast majority due to ignorance ... [and this] justifies the publication of a work whose purpose is the knowledge of poisonous plants.' Esser enlisted Carl Bollmann, who had previously worked with Hermann Zippel to publish *Ausländische Kulturpflanzen in Farbigen Wandtafeln* (*Foreign Crops on Coloured Wall Panels*) and *Repräsentanten einheimischer Pflanzenfamilien* (*Representatives of Indigenous Plant Families*). The collaboration resulted in comprehensive profiles with extensive anatomy and development, such that audiences could identify a species in any season. Distinguishing morphological details were emphasized with appropriate magnification (as was Bollmann's style), on the black background introduced by Jung, Koch, Quentell.

Of all species in the *Araceae* family, *Arum maculatum* (jack-in-the-pulpit or lords-and-ladies) was a particularly common subject in nineteenth-century material, due to its toxicity (high), ubiquity (widespread across temperate northern Europe) and structure (unique to the family). Esser's chart for *A. maculatum* emphasizes its tall spadix, looming veined leaves and the tightly clustered bright red berries that make this plant instantly recognizable when it is fruiting. At the centre of the chart, a mature plant (fig. 1) displays a rhizome, spear-shaped leaves, and a single inflorescence called the spadix. Shielded within a modified leaf called a spathe, the spadix keeps its anatomy hidden: at the base is a ring of carpellary flowers (fig. 5, enlarged) each with a one-chambered ovary containing several ovules, then a ring of staminate flowers (fig. 3, enlarged). Above both sets of flowers is a ring of hair-like structures (fig. 7) critical for pollination.

OPPOSITE: Jung, Koch, Quentell's version of *A. maculatum* emphasizes the large, glossy leaves with dark purple spots – a warning to any foragers attracted by the poisonous red berries, which glow in the chart. In spite of their inconspicuousness, jack-in-the-pulpit's flowers are insect-pollinated; the inflorescence emits a peculiar odor, attracting small flies that become entrapped beneath the hairs. The ovaries mature before the stamens, preventing self-pollination, so any pollen that reaches the stigmas is carried from the flies, which bear pollen from other individuals. Finally, the stigmas wither and secrete honey that is eaten by the trapped insects. Now the anthers ripen and pour out their pollen, dusting the captive insects – which are released only when the hairs wither, and may now go off to visit another arum chamber. The whole inflorescence has evolved to ensure cross-pollination, assisting the development of the plant's signature fruits: bright red berries that bear seeds. The illustration also includes the pollinators – tiny flies that become trapped in the spathe chamber thereby dusting the stigmas and fertilizing the ovaries.

OPPOSITE:
Title: *Arum maculatum*
Authors: Heinrich Jung, Dr. Friedrich Quentell
Illustrator: Dr. Gottlieb von Koch
Language: N/A
Country: Germany
Series/Book: *Neue Botanische Wandtafeln*
Plate: 32
Publisher: Fromann & Morian (Darmstadt, Germany); Hagemann (Düsseldorf, Germany)
Year: 1928; 1951–63

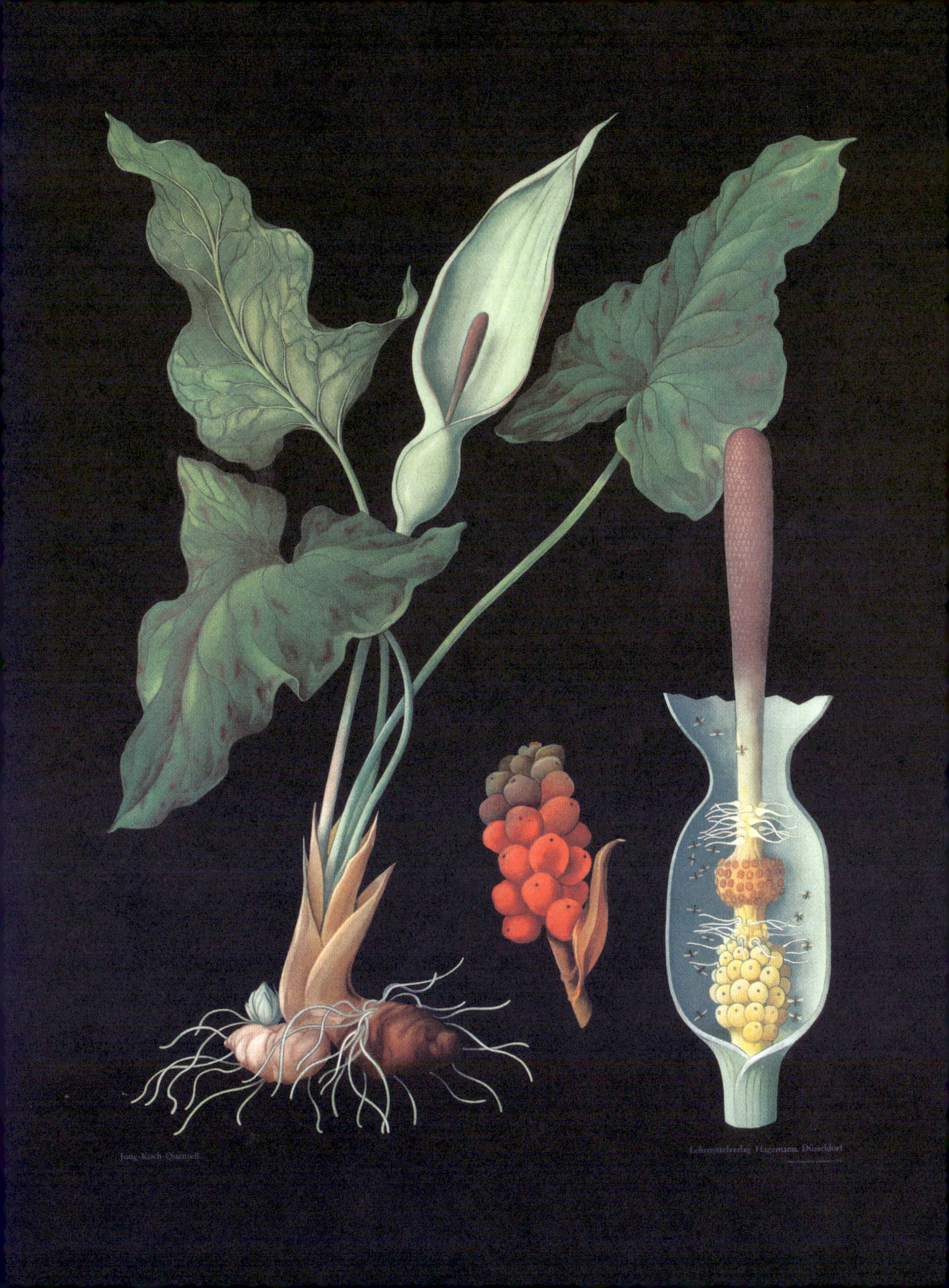
Jung-Koch-Quentell
Lehrmittelverlag Hagemann, Düsseldorf

RIGHT: While today the *Araceae* family may not be especially popular – its species generally lack in both economic value and ubiquity – the arums were somewhat prominent in the latter part of the nineteenth century, when their classification was scrutinized by botanists and their exoticism was elevated in the public eye. The family received its first formal classification in 1858, soon revised in 1876 by German botanist Adolf Engler, an eminent plant taxonomist. The arums were also fairly exotic to contemporary audiences, and for educators they provided an opportunity to teach adaptive traits such as olfactory carrion mimicry and pollinator captivation.

Rather than illustrate the life cycle and morphology of a single species, Henriette Schilthuis chose to portray several species in the family, all of which she characterized by the signature spadix and spathe structures.

From left to right: *Amorphophallus titanum* (titan arum) has a spathe and spadix that are typical of arums, but superlative within the family – it produces one of the largest inflorescences of any plant, often 10 feet (3m) tall. At the base of the cream-coloured spadix, within the chamber formed by the deep purple spathe (far left), is a band of small male flowers above a ring of the larger pink female flowers (left). When the flowers are ready for pollination, the spadix temperature rises and emits a nauseating smell. The odor classifies the arum as a carrion flower, or 'corpse plant,' and this smell is designed to attract pollinating insects.

A second species (centre) illustrates the morphological variance among the *Araceae* family. A female *Arisaema saxatile* without its spathe reveals only female flowers – a dioecious species. The long pale green spadix is a common feature of many arums. Unlike others, *A. saxatile* emits a pleasant lemon fragrance.

Arisaema griffithii (right) is a conspicuous species with its dark purple spathe with a webbed tissue of green veins.

Arisaema intermedium (far right) is monoecious, and also has a very long spadix appendage.

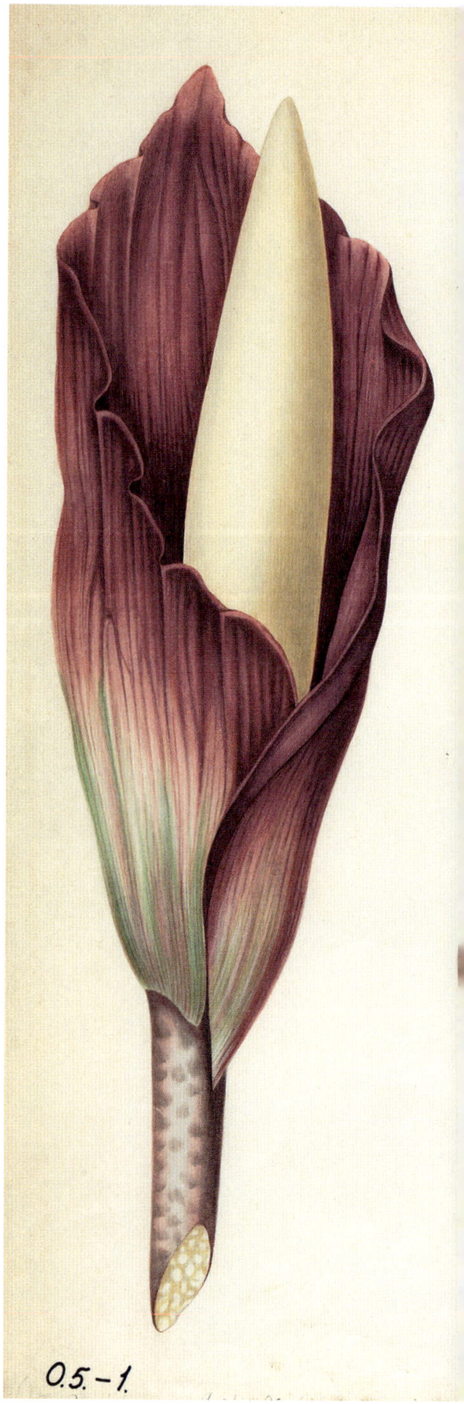

RIGHT:
Title: *Bloeikolven van Aroïdeeën (Blooms and Spathes of the Aroid (Araceae) Family)*
Author: Henriette Schilthuis
Language: N/A
Country: The Netherlands
Series/Book: N/A
Plate: N/A
Publisher: Industrial School for Female Youth (Amsterdam, The Netherlands)
Year: c.1880

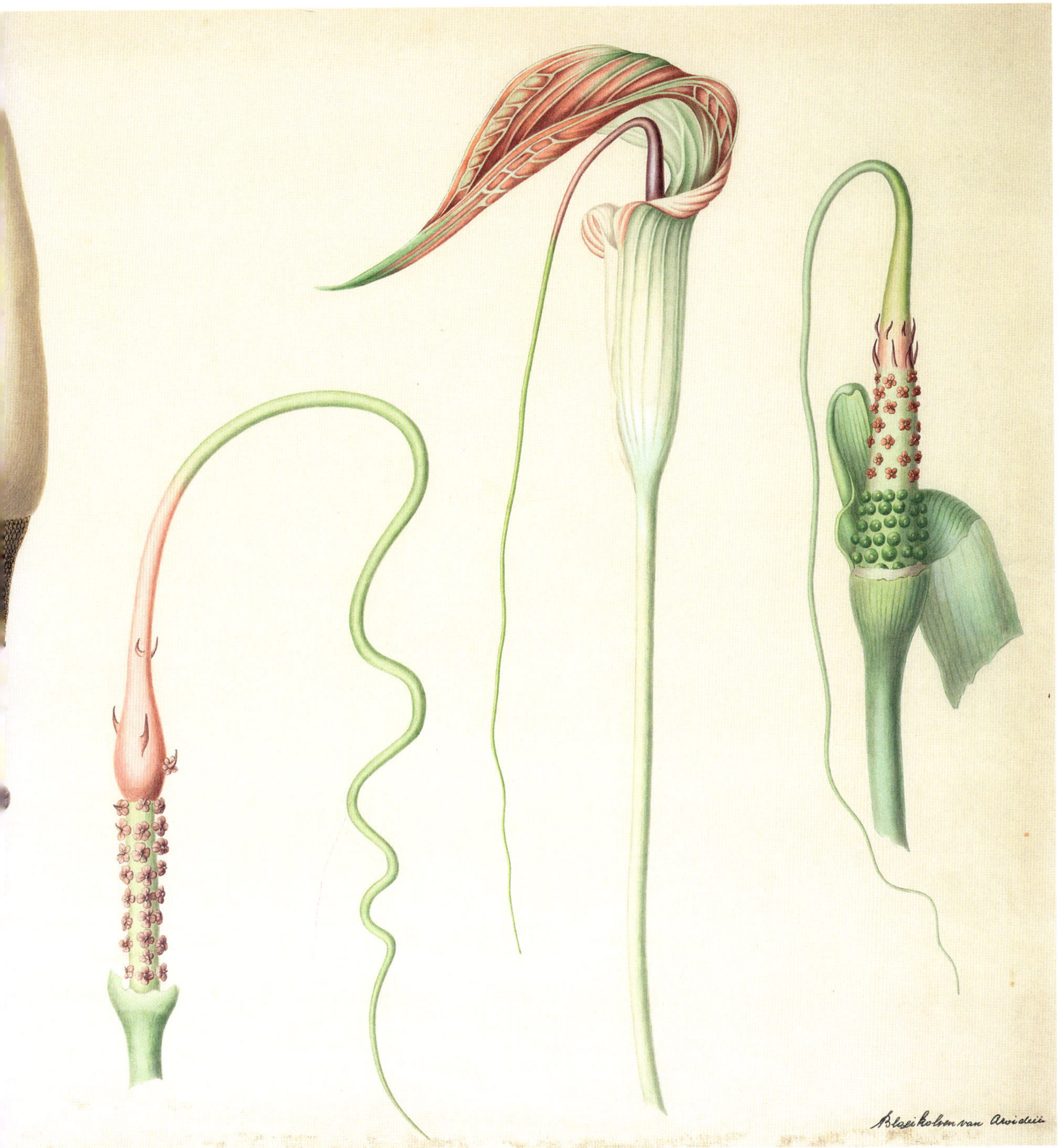
Bloeikolven van Aroideën

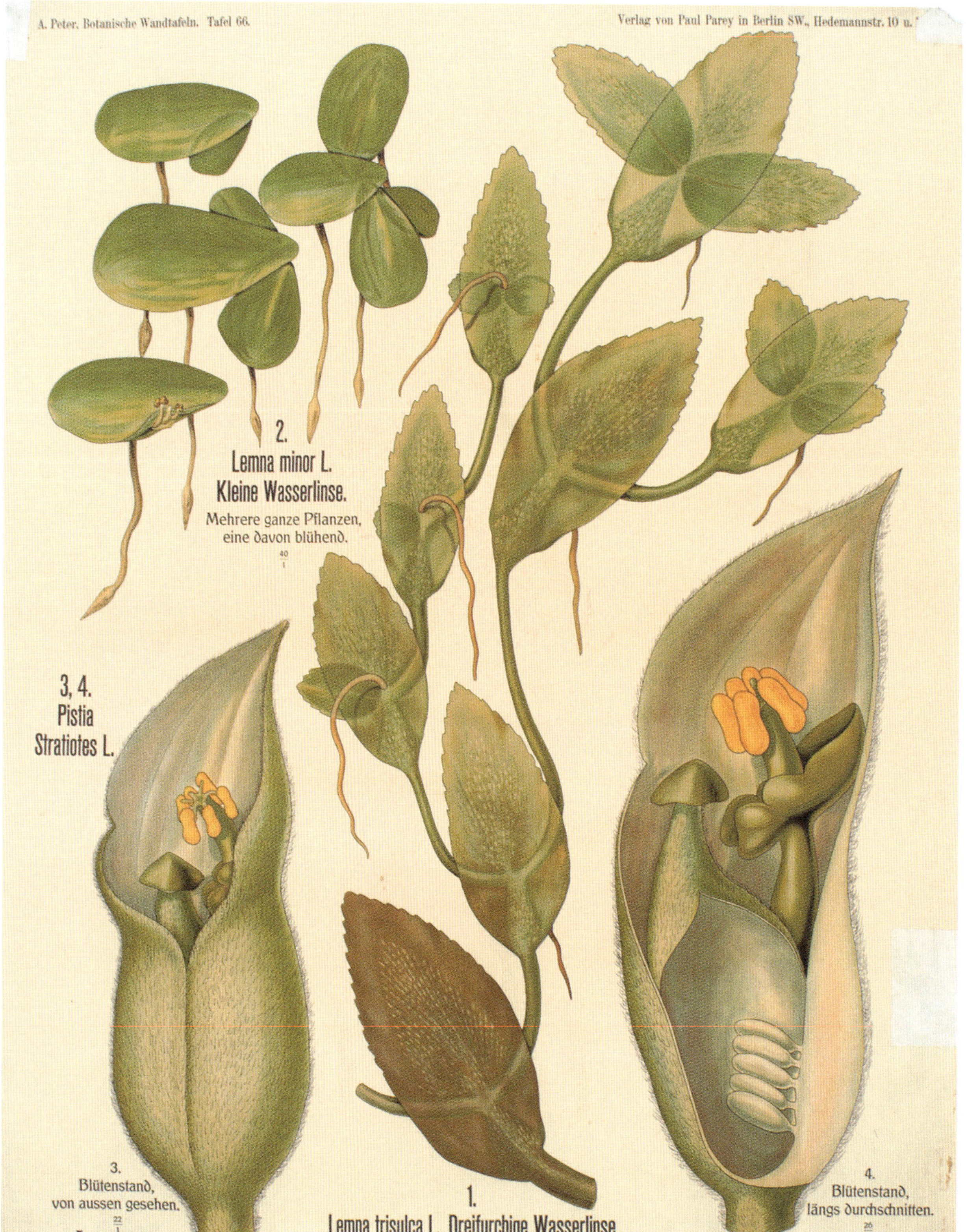
A. Peter, Botanische Wandtafeln. Tafel 66.
Verlag von Paul Parey in Berlin SW., Hedemannstr. 10 u.
2.
Lemna minor L.
Kleine Wasserlinse.
Mehrere ganze Pflanzen, eine davon blühend.
40/1
3, 4.
Pistia Stratiotes L.
3.
Blütenstand, von aussen gesehen.
22/1
Lemnaceae.
1.
Lemna trisulca L., Dreifurchige Wasserlinse.
Hälfte einer Pflanze. 1/26
4.
Blütenstand, längs durchschnitten.
26/1
Pistiaceae.

LEFT:
Title: *Lemnaceae / Pistiaceae*
Author: A. Peter
Language: German
Country: Germany
Series/Book: *Botanische Wandtafeln*
Plate: 66
Publisher: Paul Parey (Berlin, Germany)
Year: 1901

OPPOSITE: Taxonomy has long beguiled botanists and confused laymen. The *Araceae*, a supremely mixed family, provides a good explanation as to why it should be so challenging. By all morphological evidence, duckweeds and water lettuces couldn't be more dissimilar to the arums. Arums are formidable species with extreme morphology and complex reproduction. Duckweeds and water lettuces, however, are aquatic freshwater plants with the simplest, smallest flowers in the world (0.04–0.8 inches, or 1–20mm). When first identified, they were placed in their own families (*Lemnaceae* and *Pistiaceae*, respectively) organized by similar appearances. Not until the late twentieth century, with the advent of molecular phylogenetics, were they reclassified within the *Araceae* family. These boards are thus historic documents charting the evolution of plant families.

A. Peter represents both *Lemna minor* (common duckweed) and *L. trisulca* (star duckweed) as clonal bodies floating gracefully across the chart. Neither leaves nor stems are considered to be differentiated; instead the thallus-like bodies are flattened, usually with a root that trails below. As the bodies grow larger, they separate into new individuals. Unlike *L. minor* (fig. 2), bodies of *L. trisulca* (fig. 1) are serrated and submerged in the water, where the branched chains form tangled masses.

Finally, with *Pistia stratiotes* (water lettuce), we see the feature all three species share with the arums: a spathe and spadix, albeit on a much smaller scale. Echoing the composition's arc, the spathes shield a pistil and stamens, arranged as a spathe. The exterior is covered with tiny hairs; the interior is glabrous. Peter opted not to illustrate the basal rosette (resembling, yes, a head of lettuce), most likely for spatial concerns, as, relative to the size of the flowers, the leaves would span three chapters of this book! The common name will suffice, then, the eponymous feature sacrificed for a magnified flower.

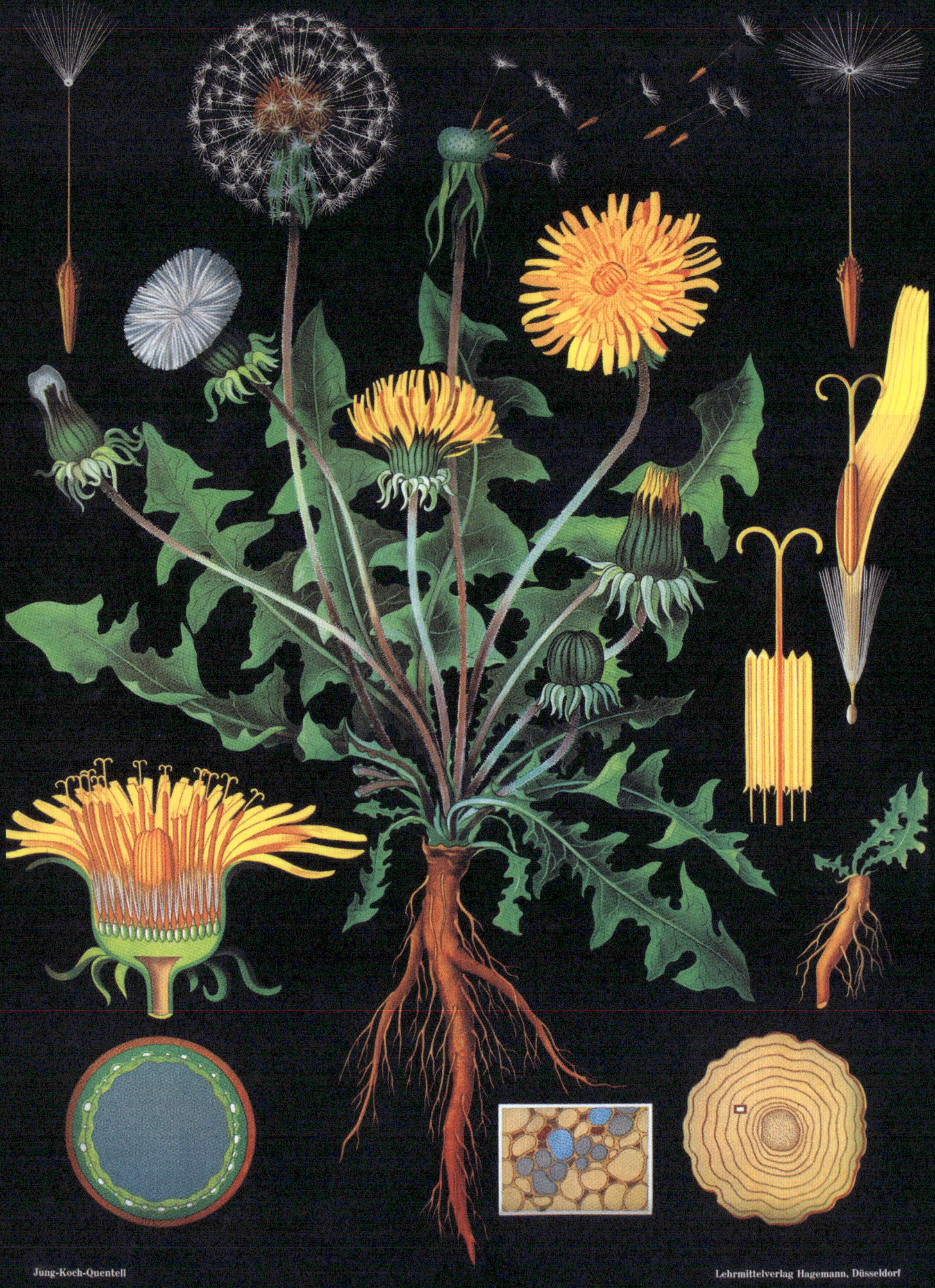

Jung-Koch-Quentell

Lehrmittelverlag Hagemann, Düsseldorf

IV ASTERACEAE / DAISY FAMILY

Also known as the *Compositae* or daisy family, this huge group of over 1,700 genera and 25,000 species is probably the largest family of angiosperms. They are globally widespread in many habitats, extremely variable, and often aromatic. Principally herbaceous annuals, biennials, and perennials, they also encompass some shrubs and small trees.

There are far too many stalwarts of the ornamental garden to mention in this family, but they range from cottage garden favourites such as *Aster*, *Brachyscome*, *Calendula*, *Centaurea*, *Cosmos*, *Dahlia*, *Echinops*, *Tagetes* and *Zinnia* to prairie-style plants like *Echinacea*, *Helenium*, *Heliopsis*, *Solidago* and *Vernoniastrum*. Other ornamentals include *Achillea*, *Anaphalis*, *Argyranthemum*, *Bidens*, *Cineraria*, *Coreopsis*, *Doronicum*, *Erigeron*, *Gazania*, *Inula*, *Liatris*, *Ligularia* and *Senecio*. Many are valued by florists, for example *Chrysanthemum*, *Gerbera*, *Helichrysum* and *Osteospermum*. Shrubby *Asteraceae* include *Artemisia*, *Olearia* and *Pachystegia*.

Among the weeds in this family are *Bellis perennis* (common daisy) and *Taraxacum officinale* (dandelion). In the vegetable garden, you might grow artichokes, endive, lettuce, or scorzonera. The seeds of *Helianthus* provide sunflower oil and *Stevia rebaudiana* yields a natural sweetener.

Characteristics: leaves basal or alternate; sap often milky; inflorescence often mimics a single flower, with various arrangements of ray florets and/or disc florets in a capitulum subtended by involucre of bracts, sometimes with a central disc; sepals generally modified to pappus of scales, bristles, hairs, or awns to aid wind-distribution of seed; fruit a one-seeded cypsela. It can be difficult to differentiate the numerous *Asteraceae* species accurately without examining minute details of anthers, stigma, pappus and fruits.

OPPOSITE:
Title: *Taraxacum officinale*
Authors: Heinrich Jung, Dr. Friedrich Quentell
Illustrator: Dr. Gottlieb von Koch
Language: N/A
Country: Germany
Series/Book: *Neue Botanische Wandtafeln*
Plate: 30
Publisher: Fromann & Morian (Darmstadt, Germany); Hagemann (Düsseldorf, Germany)
Year: 1928; 1951–63

OPENING PAGE: By the 1900s, Jung, Koch, Quentell were beginning to break the rules they had established early in their partnership. Gone were the grids and negative space, the flat dimensionality and structure. While they didn't illustrate an environment, they managed to create an energy that was both surreal and precise.

Here, the common dandelion is represented as a carnival of bright yellow blossoms, parachuting seeds and an electric rosette of leaves. *Taraxacum officinale* is an extremely hardy and ubiquitous species, quickly colonizing through copious seed dispersal and vigorous taproots. The authors have produced a world where a breezy sky is filled with achenes, and robust roots split the earth. As any gardener will attest, the dandelion's roots are stubborn and brittle, and persist despite the most tenacious spade. Rather than yield to uprooting, the dandelion's roots will break into fragments, each producing a new plant.

To the right of a central root system, the authors have included a root segment and young offshoot. The supporting cast includes a flower and stem in cross-section (lower left); a stout disc floret and a slender ray floret (right); and two enlargements of the formidable roots.

OPPOSITE: In their *Pflanzenphysiologische Wandtafeln*, or *Plant Physiology Boards*, Albert Bernhard Frank and Alexander Tschirch include a portrait of *Helianthus annuus* (the common sunflower), offering a view that students would certainly never have seen before. Microscopes were becoming more powerful and accessible for scientists, but the cost would remain prohibitive for schools for many years. So, botanists like Frank and Tschirch would dissect, magnify and illustrate the cellular structure of common plants. This was a significant contribution that wall charts gave to education; while specimens could be collected and studied for basic anatomy, any glimpse beyond a low-power loupe was limited to a research laboratory. Here, Frank and Tschirch offer the magnified cross-section of a mature sunflower stem, comprised of the primary phloem and xylem, medullary rays, cortex and cambium. A boxed annotation at the bottom of the chart notes the magnification ('In nature, 11mm [0.4 inches] diameter') – a consistent feature of *Plant Physiology Boards*.

OPPOSITE:

Title: *Erwachsener Stengel von Helianthus annuus Querchnitte* (Cross-Section of an Adult *Helianthus annuus* Stem)
Authors: Albert Bernhard Frank, Alexander Tschirch
Language: German
Country: Germany
Series/Book: *Pflanzenphysiologische Wandtafeln von Frank und Tschirch (Plant Physiology Boards of Frank and Tschirch)*
Plate: 19
Publisher: Paul Parey (Berlin, Germany)
Year: 1889

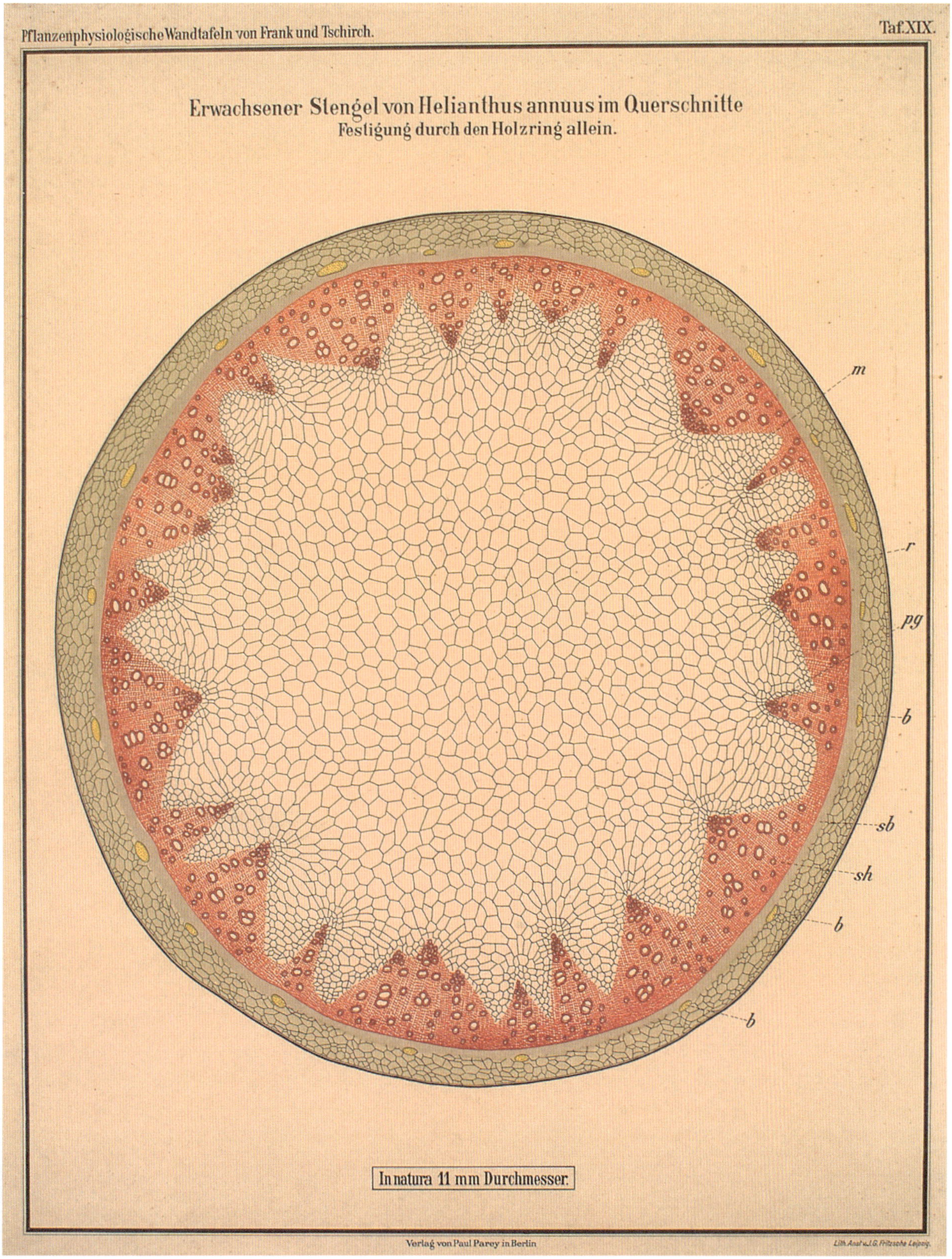
Pflanzenphysiologische Wandtafeln von Frank und Tschirch.
Taf. XIX.
Erwachsener Stengel von Helianthus annuus im Querschnitte
Festigung durch den Holzring allein.
m
r
pg
b
sb
sh
b
b
In natura 11 mm Durchmesser.
Verlag von Paul Parey in Berlin
Lith. Anst. v. J. G. Fritzsche Leipzig.

RIGHT: Buoyant buttons on leafy pedestals, the chamomiles are miniature daisies with a sweet fragrance and efficacious properties. Von Engleder portrays the genus by dissecting a *Matricaria chamomilla* specimen, including a full plant with feathery limbs and petite flower heads. Accompanying illustrations include a cross-section of the capitulum, exposing an aggregation of bowed white ray florets, yellow tubular disc florets and the arch of tiny ovaries below the glowing halo of compound anthers. Engleder further magnifies both types of florets, a row of stamens, the leafy involucre and a seed.

OPPOSITE: A wall chart from the Czech Republic is a contrast to Engleder's organized white space, tidy specimens and systematic dissection. Species include *Matricaria chamomilla*; *M. discoidea* (wild chamomile); and *M. indora*, now known as *Tripleurospermum indorum* (scentless chamomile). Most illustrators hoped to achieve a verisimilitude with nature; here, Czech botanist Otakar Zejbrlík has achieved an even higher level of realism. At first glance, the chart could be an herbarium specimen. The chart is notable for the expertly rendered details and depth, as well as the realism with which the plants sprawl across the board. Zejbrlík also refused to scale down the adult plants, so students could easily compare the measurements of flowers, leaves and stems. This explains the author's unusual decision to bend the tallest stem in two places, rather than crop its roots, scale it down, or sever the stem and portray the roots alongside. The purpose of the chart, however, is to differentiate between the three species – all of these species grow wild and can easily be mistaken for each other. Keeping the same degree of magnification is, therefore, critical, for it allows a student to directly compare flower heads, leaf structure and density, and height of the plant. The three species also produce differently shaped seeds, which the illustrator has generously magnified to show various topographies of miniature mountains and valleys.

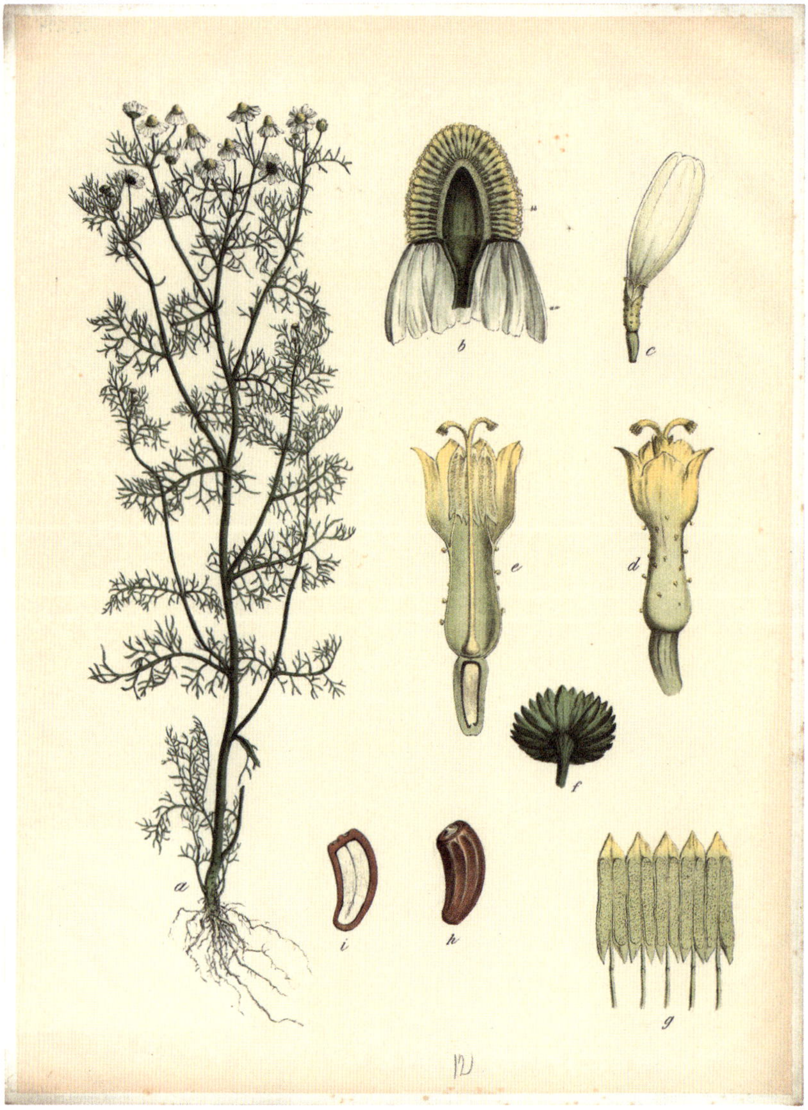

ABOVE:
Title: *Matricaria chamomilla*
Author: Von Engleder; Illustrator: C. Dietrich
Language: German
Country: Germany
Series/Book: *Engleders Wandtafeln für den naturkundlichen Unterricht Pflanzenkunde (Engleder's Wall Charts for Natural History Lessons: Botany)*
Plate: 12
Publisher: J. F. Schreiber (Esslingen, Germany)
Year: 1897

RIGHT:
Title: *Hermánek Pkavy (Matricaria chamomilla); Hermánek Tervovity (Matricaria discoidea); Hermánek nevonný (Matricaria inodora)*
Author: Otakar Zejbrlík
Language: Czech
Country: Czech Republic
Series/Book: Unknown
Plate: Unknown
Publisher: Unknown
Year: 1953

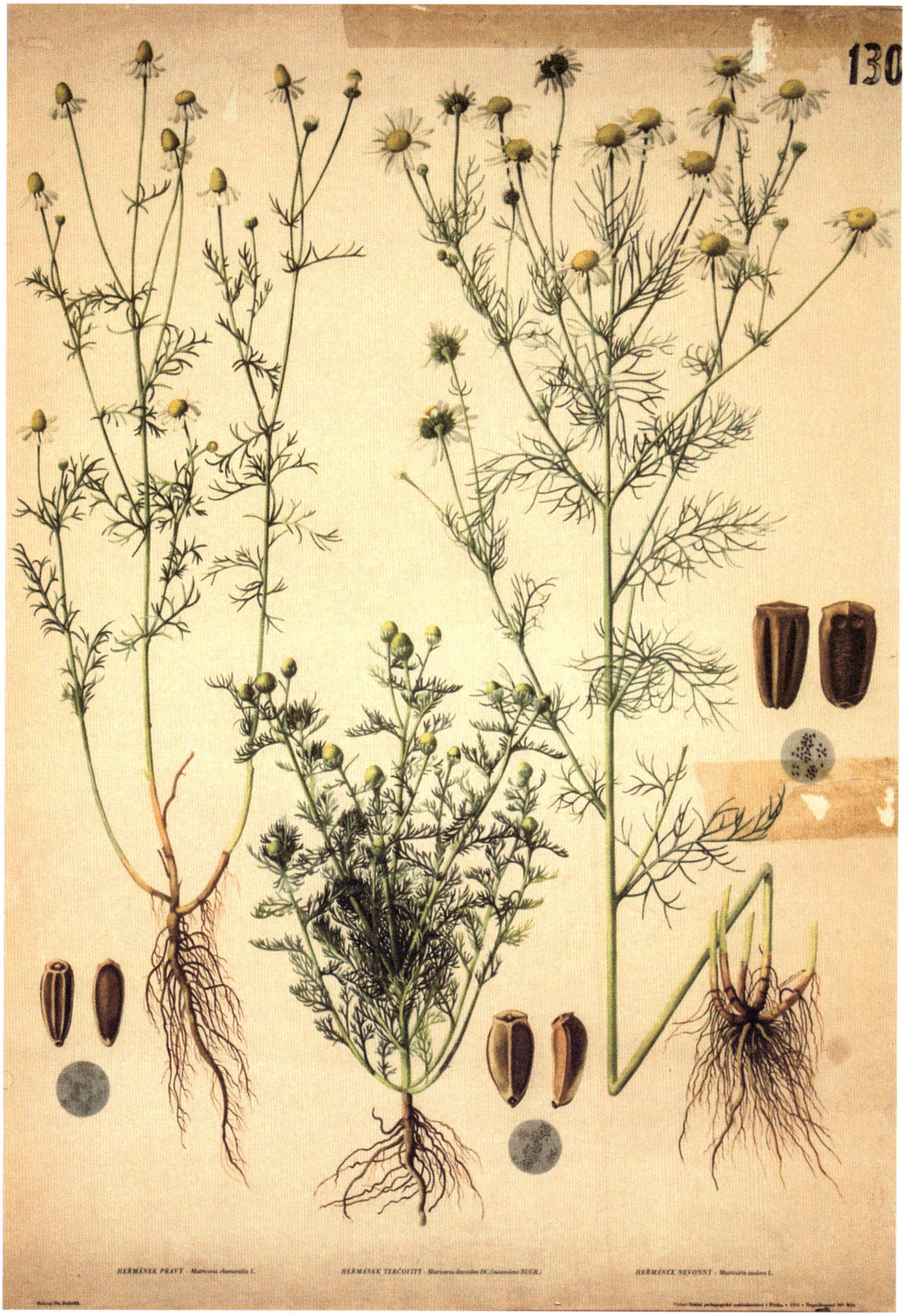
130
HEŘMÁNEK PRAVÝ · Matricaria chamomilla L.
HEŘMÁNEK TERČOVITÝ · Matricaria discoidea DC.
HEŘMÁNEK NEVONNÝ · Matricaria inodora L.

RIGHT: Zippel and Bollmann illustrate the *Köpfchenblüter*, or *Asteraceae* family, with dissected diagrams of *Arnica montana* (mountain tobacco) and *Matricaria chamomilla* (chamomile), with adjacent numerals and a corresponding legend. Notable are the varying relationships between ovaries and sepals (fig. 1, *Arnica montana;* fig. 2, *Matricaria chamomilla)* and the magnified seeds displacing their florets (figs. 3, 4, *Arnica montana).* In the upper right corner, Zippel and Bollman have also included a seed head of *Taraxacum officinale* (dandelion) and a flower head of *Onopordum acanthium* (Scotch thistle), distinguishing morphological stages that assist in identification of the two species.

Focusing on *Arnica montana* and *Matricaria chamomilla*, the authors have portrayed two primary types of *Asteraceae* distinguished by their dispersal strategies. Dandelions, thistles and mountain tobacco all equip their seeds with a pappus, a tufty appendage that aids dispersal by wind. Chamomiles, sunflowers and chrysanthemums, however, are bereft of pappus. Seeds are still produced within the aggregate flower head, but their target audience is different. The seeds of a dandelion will be favoured by a passing breeze, while a sunflower's nutritious seeds wait to be carried away by a hungry visitor.

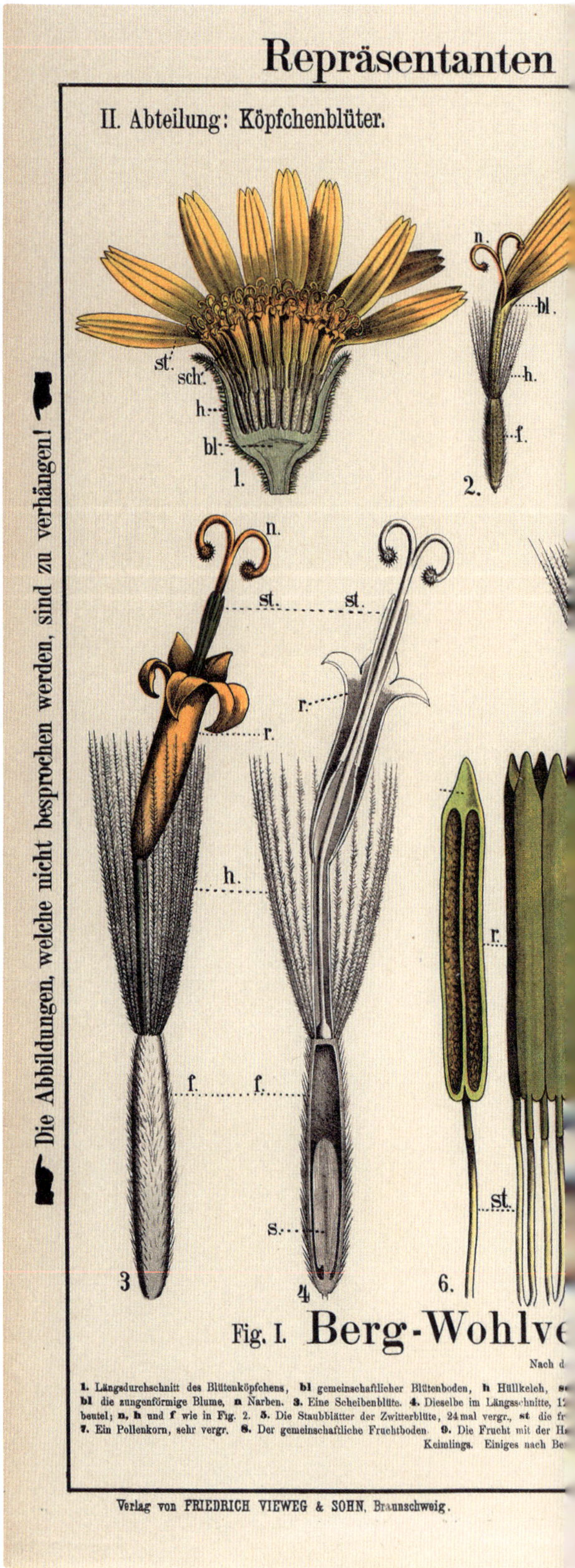

RIGHT:
Title: *Köpfchenblüter (Asteraceae)*
Author: Hermann Zippel; Illustrator: Carl Bollmann
Language: German
Country: Germany
Series/Book: *Repräsentanten einheimischer Pflanzenfamilien (Representatives of Indigenous Plant Families)*
Plate: II Abteilung; 31
Publisher: Friedrich Vieweg & Sohn (Braunschweig, Germany)
Year: 1879

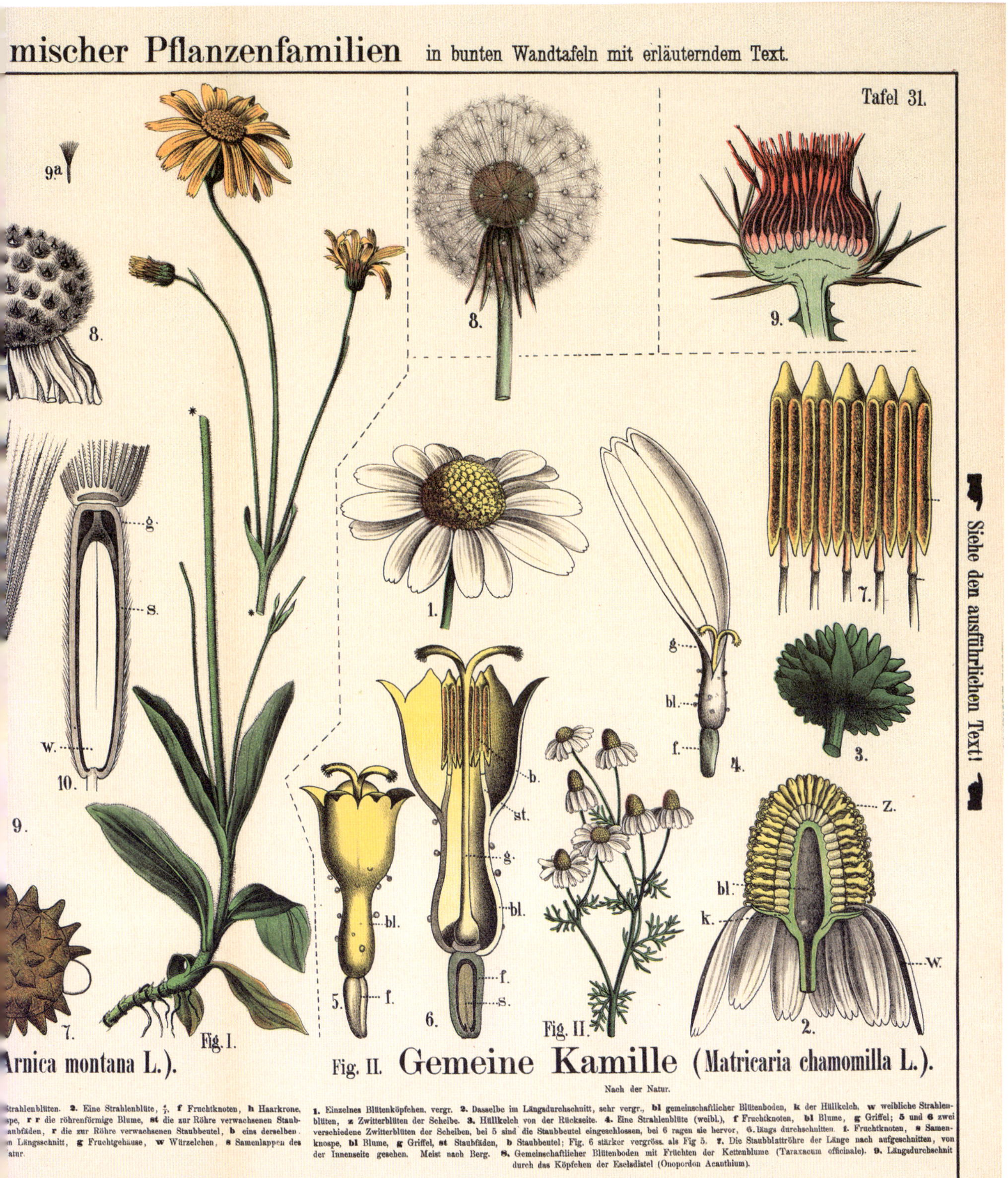
mischer Pflanzenfamilien in bunten Wandtafeln mit erläuterndem Text.
Tafel 31.
9a
8.
8.
9.
7.
1.
g.
bl.
f.
4.
3.
g.
s.
w.
10.
9.
z.
bl.
k.
w.
b.
st.
g.
bl.
bl.
5.
f.
f.
s.
6.
7.
Fig. I.
Fig. II.
2.
Siehe den ausführlichen Text!
Arnica montana L.).
Fig. II. Gemeine Kamille (Matricaria chamomilla L.).
Nach der Natur.
Strahlenblüten. 2. Eine Strahlenblüte, 4/1. f Fruchtknoten, h Haarkrone,
spe, r r die röhrenförmige Blume, st die zur Röhre verwachsenen Staub-
aubfäden, r die zur Röhre verwachsenen Staubbeutel, b eins derselben.
n Längsschnitt, g Fruchtgehäuse, w Würzelchen, s Samenlappen des
atur.
1. Einzelnes Blütenköpfchen, vergr. 2. Dasselbe im Längsdurchschnitt, sehr vergr., bl gemeinschaftlicher Blütenboden, k der Hüllkelch, w weibliche Strahlenblüten, z Zwitterblüten der Scheibe. 3. Hüllkelch von der Rückseite. 4. Eine Strahlenblüte (weibl.), f Fruchtknoten, bl Blume, g Griffel; 5 und 6 zwei verschiedene Zwitterblüten der Scheiben, bei 5 sind die Staubbeutel eingeschlossen, bei 6 ragen sie hervor, 6. längs durchschnitten. f. Fruchtknoten, s Samenknospe, bl Blume, g Griffel, st Staubfäden, b Staubbeutel; Fig. 6 stärker vergröss. als Fig 5. 7. Die Staubblattröhre der Länge nach aufgeschnitten, von der Innenseite gesehen. Meist nach Berg. 8. Gemeinschaftlicher Blütenboden mit Früchten der Kettenblume (Taraxacum officinale). 9. Längsdurchschnit durch das Köpfchen der Eselsdistel (Onopordon Acanthium).
Herausgegeben von HERMANN ZIPPEL und CARL BOLLMANN.
Zeichnung, Lithogr. und Druck des lithogr.-artist. Instituts von Carl Bollmann, Gera

OPPOSITE: In the companion text to their atlas, Arnold and Carolina Dodel-Port offer the highest praise for *Asteraceaes*. As always, they articulate the purpose of their chart, and explain their species selection in wonderfully purple prose: 'Celebrating here ... the triumphs of natural selection' and the reasons for 'her great advantage in the competition for existence.' The *Asteraceae* are experts at attracting pollinators, leveraging those brief visits, and dispersing their seeds. As the Dodel-Ports reflect, these adaptations account for the enormity of the family – boasting some 30,000 species, it is estimated that the *Asteraceae* comprise a tenth of all flowering plants. So the *Asteraceae* are well-adapted organisms indeed; *Centaurea cyanus* (cornflower) is no different and even boasts the enviable ability to reinvent itself – the cornflower's hermaphroditic organs allows it to change its sex at opportune moments.

Let's look under the petals. Before the flower opens, fertile male disc florets have developed dehiscent anthers that form a tube and release pollen. A style patiently waits at the base of the tube, pushing upward, and the stiff hairs below the stigma sweep pollen to the top of the tube. The flower opens and white pollen masses protrude from the tube. The touch of a visiting insect causes the motion-sensitive anthers to contract and deposit pollen onto its body. After the pollen is dispersed, *C. cyanus* transitions to its female form. The style resumes its growth until it breaches the end of the tube, exposing the receptive stigma to pollen-laden insects.

OPPOSITE:
Title: *Centaura Cyanus*
Authors: Arnold and Carolina Dodel-Port
Language: German
Country: Switzerland
Series/Book: *Anatomisch physiologische Atlas der Botanik* (*The Anatomical & Physiological Atlas of Botany*)
Plate: 41
Publisher: J. F. Schreiber (Esslingen, Germany)
Year: 1878–93

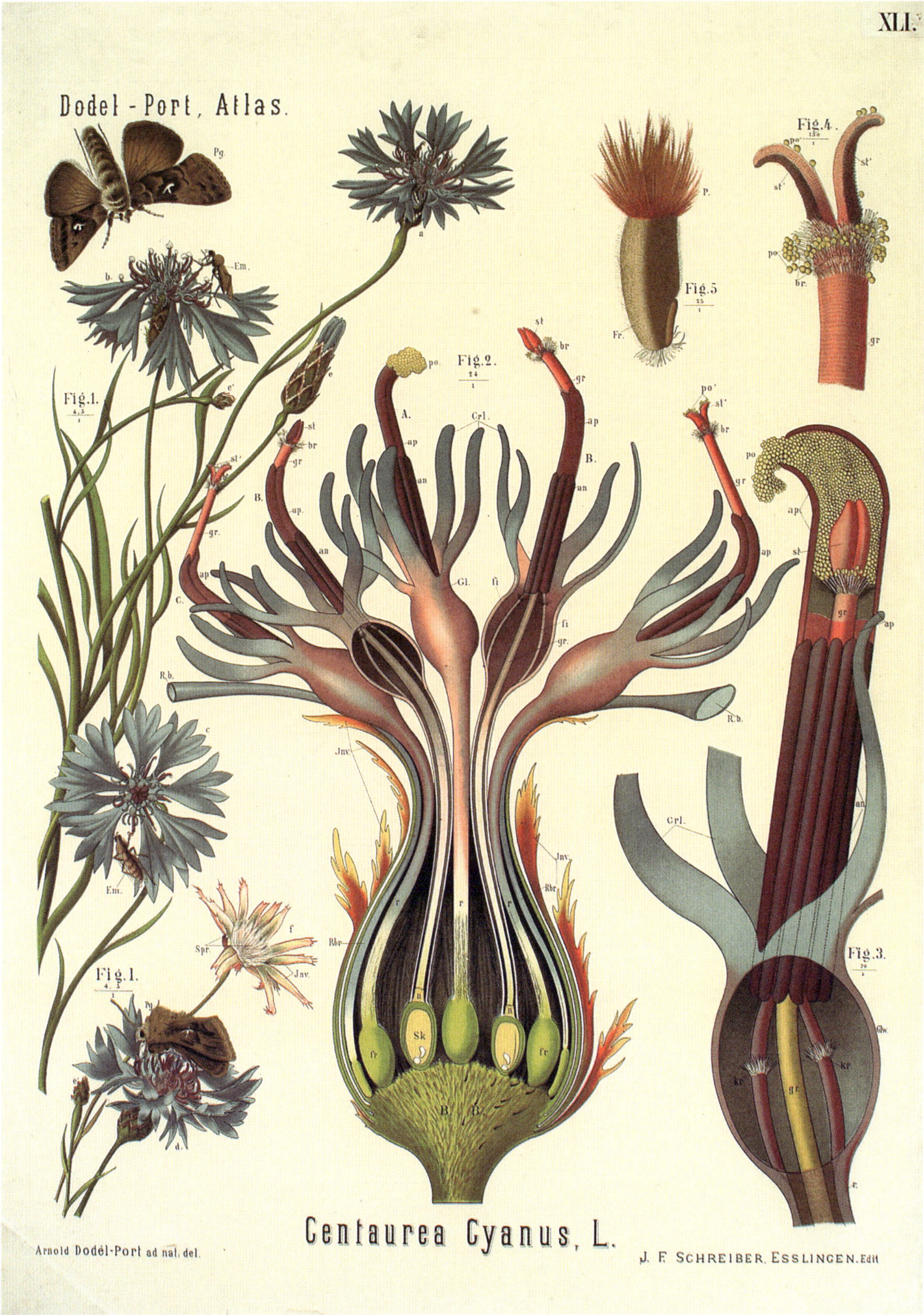
XLI.
Dodel-Port, Atlas.
Fig.1.
Fig.1.
Fig.2.
Fig.3.
Fig.4.
Fig.5
Centaurea Cyanus, L.
Arnold Dodel-Port ad nat. del.
J. F. SCHREIBER. ESSLINGEN. Edit

RIGHT: A Russian chart offers an unusual *Asteraceae* lineup: a succulent, a low-growing ground cover, and a globe thistle.

Tussilago farfara (coltsfoot, fig. I) is a curious species; stems have red-tipped scales and flowers emerge long before the leaves – unusual, because leaves allow plants to photosynthesize and receive nutrients from the soil. But *T. farfara*'s bright yellow dandelion-esque flowers have already cycled through pollination and dispersal by the time each stem grows its single heart-shaped leaf. The upper surface is sea green; the undersides are a blanket of woolly white fibres and veins, like a white glove, palmate and furry. While the illustration includes two tufted seeds on a desolate head, *T. farfara* doesn't usually grow from seed, rather from the overwintering root buds pictured in the chart.

Antennaria dioica (pussytoes, fig. II) can be identified by its ground cover of soft, grey foliage. The common name derives not from the leaves, but from the tight flower clusters that grow up from the low bed at the end of spring, which are said to resemble the pads of a cat's paw.

Echinops sphaerocephalus (pale globe-thistle, fig. III) blooms with a lollipop inflorescence of white or blue-grey disc florets. The stems are slightly wrinkled, grey and hairy, and support an array of large, sharp-toothed leaves with sticky, hairy upper surfaces and white woolly underbellies. Like many *Asteraceae*, seeds are hairy achenes dispersed by wind.

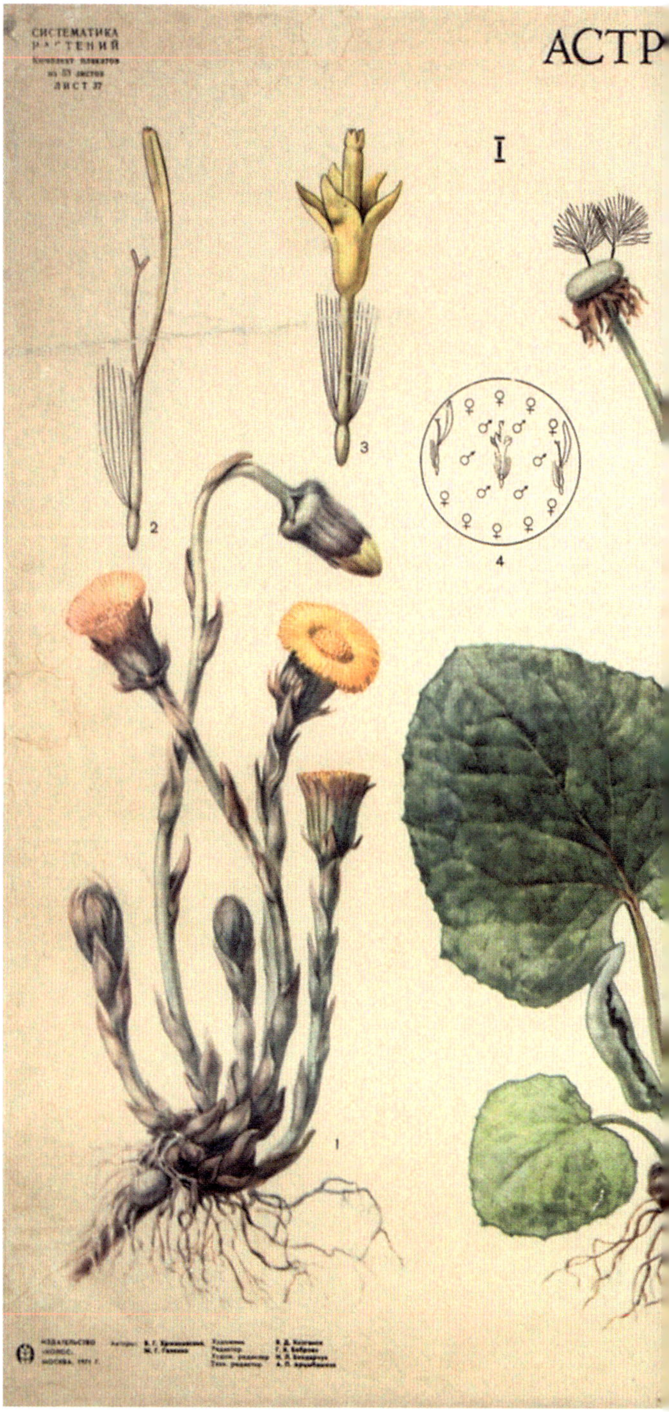

RIGHT:
Title: *Asteraceae*
Author: V. G. Chrzhanovskii
Language: Russian
Country: Russia
Series/Book: *Sistematika Rasteniy Komplekt Plakatov iz 53 Listov (Systematics of Plants – Posters from 53 Sheets)*
Plate: 37
Publisher: Kolos Publishing
Year: 1971

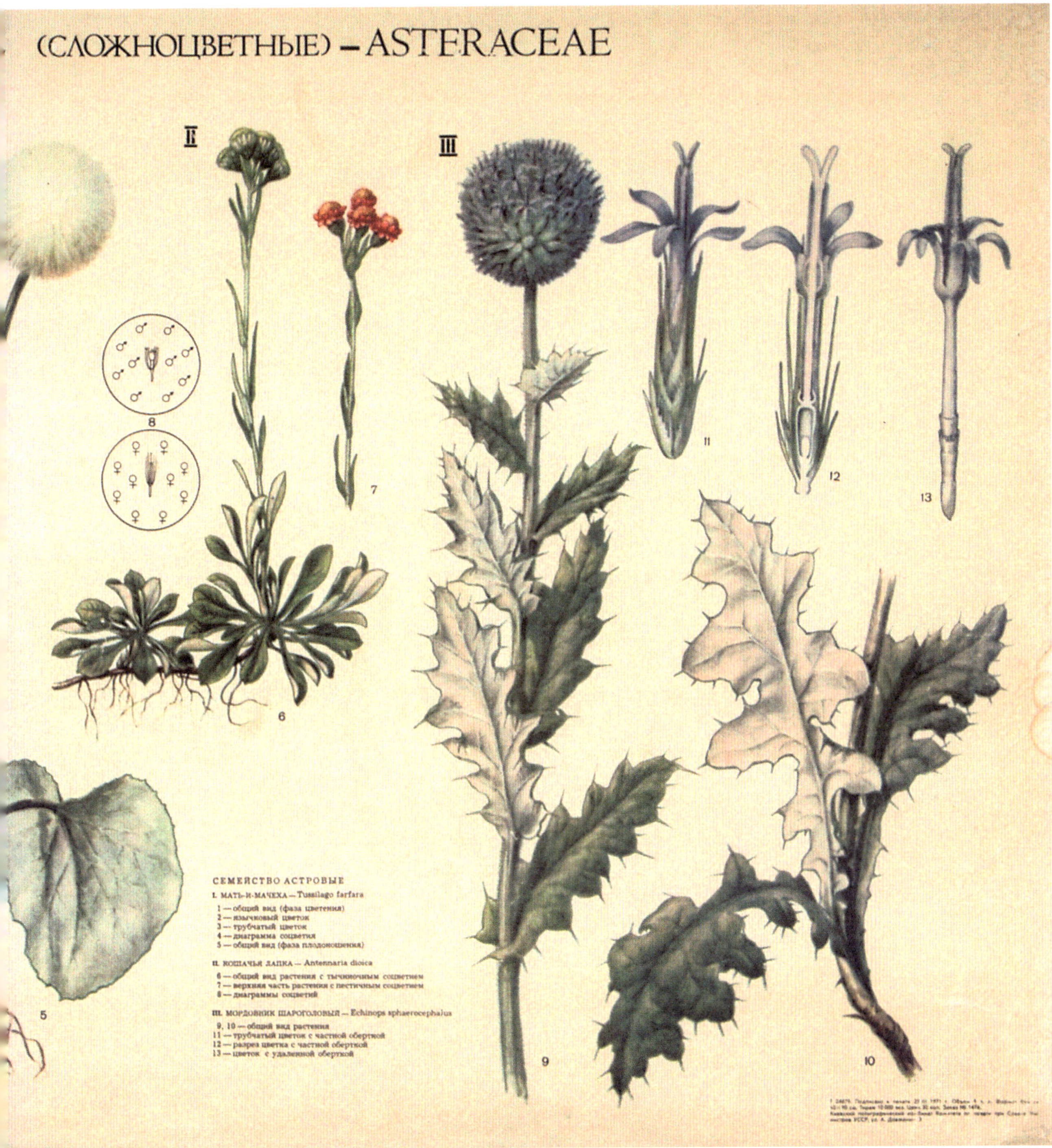
(СЛОЖНОЦВЕТНЫЕ) – ASTERACEAE
II
III
8
7
6
11
12
13
5
9
10
СЕМЕЙСТВО АСТРОВЫЕ
I. МАТЬ-И-МАЧЕХА – Tussilago farfara
1 — общий вид (фаза цветения)
2 — язычковый цветок
3 — трубчатый цветок
4 — диаграмма соцветия
5 — общий вид (фаза плодоношения)
II. КОШАЧЬЯ ЛАПКА – Antennaria dioica
6 — общий вид растения с тычиночным соцветием
7 — верхняя часть растения с пестичным соцветием
8 — диаграммы соцветий
III. МОРДОВНИК ШАРОГОЛОВЫЙ – Echinops sphaerocephalus
9, 10 — общий вид растения
11 — трубчатый цветок с частной оберткой
12 — разрез цветка с частной оберткой
13 — цветок с удаленной оберткой

V BRASSICACEAE / CABBAGE FAMILY

The cabbage family is large, with about 350 genera and 4,000 species found across the world. They are mostly herbaceous, ranging from annuals to perennials, with typically pungent and sometimes peppery foliage; a few are shrubby. Brassicas have been cultivated widely for centuries, including many staple vegetables and fodder crops such as broccoli, Brussels sprouts, cabbage, cauliflower, kale, kohlrabi, radish, swede and turnip. This family also provides many salad vegetables, for example cress, komatsuna, mizuna, purslane, rocket and watercress, as well as spices, including horseradish, mustard and wasabi.

In the ornamental garden common *Brassicaceae* include *Crambe*, *Erysimum*, *Hesperis*, *Lobularia*, *Lunaria*, *Matthiola* and *Nasturtium*, as well as rock-garden favourites *Alyssum*, *Arabis* and *Iberis*. *Cardamine hirsuta* (hairy bittercress) and *Capsella bursa-pastoris* (shepherd's purse) number among the many annual weeds.

Although now considered a noxious weed in parts of North America, *Isatis tinctoria* (woad) was highly prized as a source of indigo-blue dye by Europeans as far back as the ancient Britons.

Characteristics: leaves simple or divided; flowers very uniform, spike or raceme, corymbose, usually without bracts, often white or yellow; four free sepals alternating with four free petals, usually in form of a cross, hence the alternative name '*Cruciferae*'; stamens – usually six; fruit a dry capsule, either a siliqua (three times longer than it is broad) or silicula (less than three times longer than it is broad).

OPPOSITE:
Title: *Cardamine pratensis*
Authors: Heinrich Jung, Dr. Friedrich Quentell
Illustrator: Dr. Gottlieb von Koch
Language: N/A
Country: Germany
Series/Book: *Neuen Wandtafeln*
Plate: 5
Publisher: Fromann & Morian (Darmstadt, Germany)
Year: 1902–3

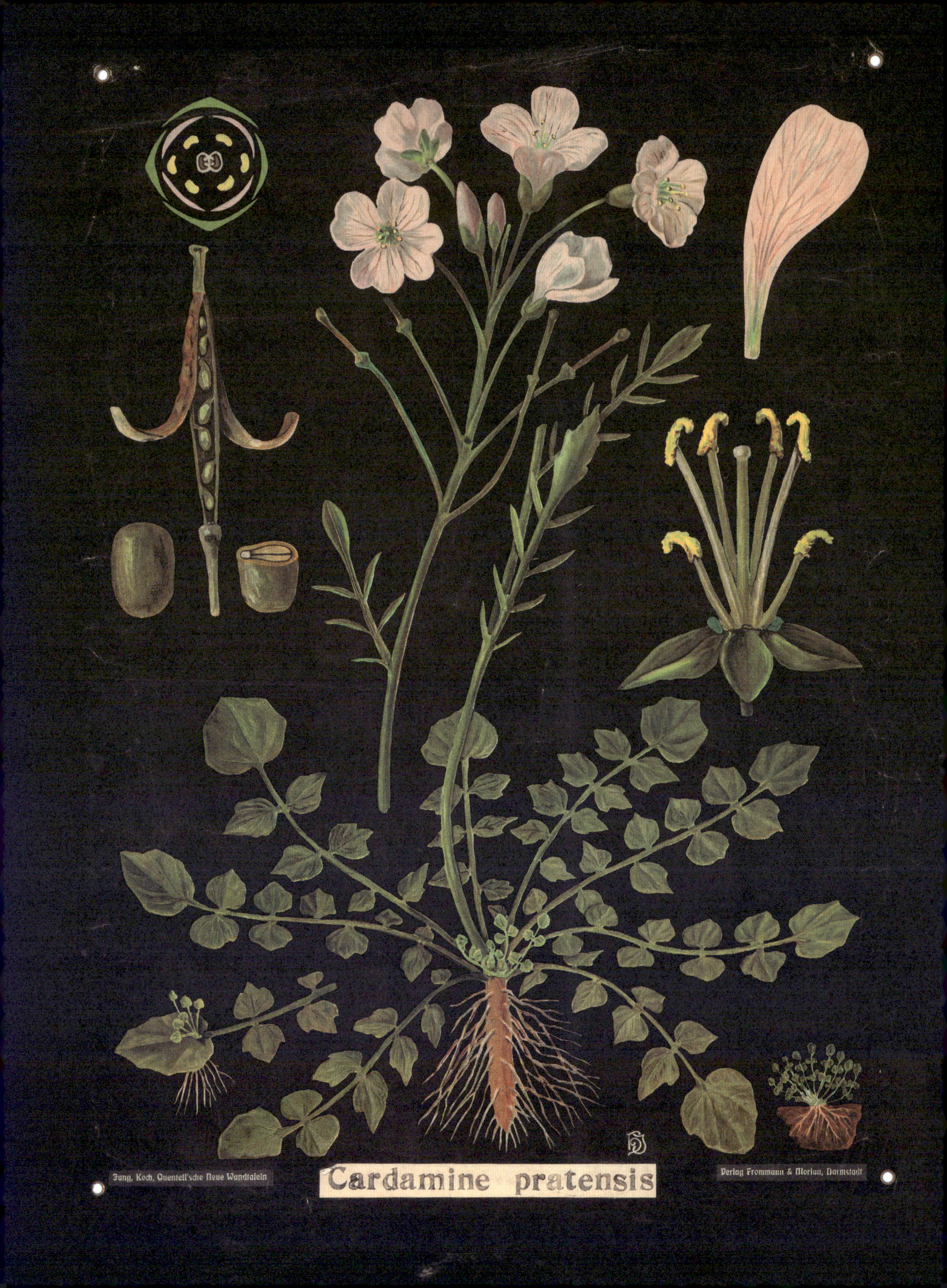
Jung, Koch, Quentell'sche Neue Wandtafeln
Cardamine pratensis
Verlag Frommann & Morian, Darmstadt

OPENING PAGE: Perhaps the most distinguishing features of *Cardamine pratensis* (cuckooflower or lady's smock) are a cluster of delicate lilac flowers atop a long stem, a wet grassland habitat and (perhaps apocryphally) a coincidence with the arrival of spring's first cuckoo bird. But the focus of Jung, Koch, Quentell's wall chart is the glorious basal rosette of paired leaves, compared to which all other elements would recede when viewed across a classroom. Gently arcing with the upper stems above, the rosette sprawls across a large portion of the chart, with hypnotic movement.

Other elements include a coquettish fruit, peeling away her siliqua to reveal a neat row of seeds; a whole and halved mature seed; and a nascent bulbil and emergent roots in the lower-left corner – although *Brassicaceaes* are known for their long slender seed pods, some species rely primarily on bulbils that fall to the ground and establish a new plant.

OPPOSITE: Another curious aspect of taxonomy is the distinction between cultivar and species. A species is distinguished by common genetics, and a cultivar is a variety that has been domesticated from a wild progenitor. Usually, a cultivar's appearance doesn't dramatically deviate from its parent species. Not so with *Brassicaceae*. Despite fantastically different morphologies (at least from the blossom down), ten of the most common cruciferous vegetables belong to a single species, *Brassica oleracea*. Taxonomically these different vegetables are distinguished only as cultivars, meaning that they've all been domesticated from their wild progenitor, *B. oleracea*. Here, a Russian chart illustrates a few of their varieties.

From the left: a *B. oleracea* Capitata Group (cabbage) leads the lineup with a broad, toothed leaf and tall flower spike, which only emerges in the second year of growth, and the cross-section of a head of tight leaves, produced in the first year. Next, *B. oleracea* var. *garioides*, now *B. oleracea* Gongylodes Group (kohlrabi). Presuming the illustrator has imagined about the same soil level through the chart, a hint of roots at the base indicates that the misshapen purple globe is not a taproot, but an enlarged lateral meristem (thus agricultural novices would have no trouble distinguishing from a beetroot). Adjacent is *B. oleracea* Botrytis Group (cauliflower). Unlike its neighbours, this brassica has been cultivated not for its leaves or stalk but for its flowers; the textured head of a cauliflower is actually an undifferentiated mass of modified inflorescences, called an 'inflorescence meristem' or 'curd.' To its right, a frond of leaves atop a stalk of tiny cabbages – *B. oleracea* Gemmifera Group is Brussels sprout, and the little globes are axillary buds that have been cultivated to whorl against the elongated stem. Finally, the only vegetable on this chart that is not a variant of *B. oleracea* – *Raphanus sativus* (radish). The swollen taproot is edible and identifiable when a little white pate, peering above the soil line, deepens to red.

As important as the meristems and taproots are the inflorescences at both ends of the chart. These blossoms remind viewers that, despite different appearances, this cast of characters are rooted in the family formerly known as *Cruciferae*, for the four-petalled flowers perpendicularly oriented in the form of a cross.

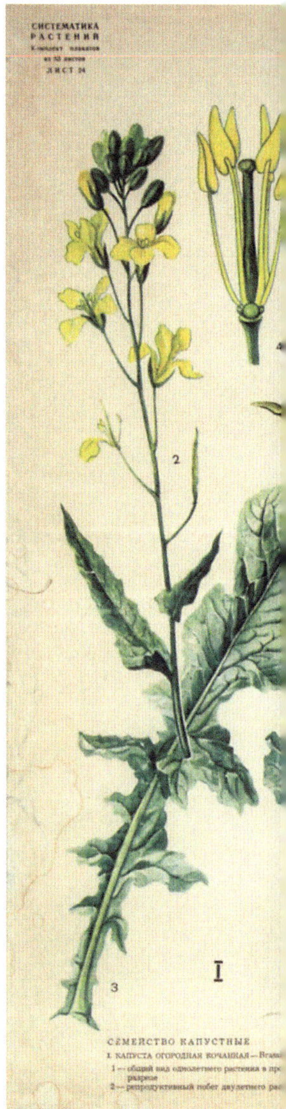

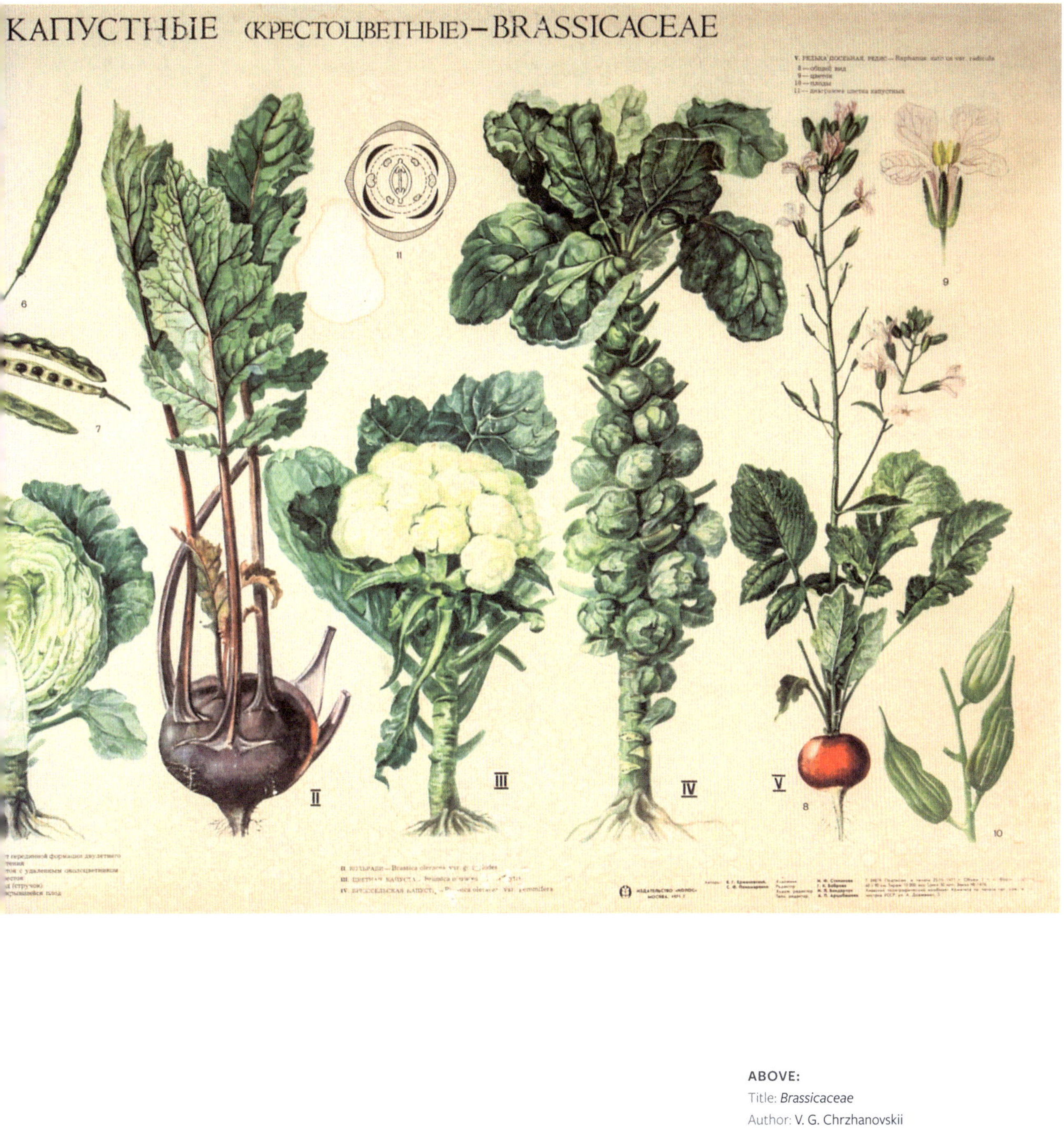

ABOVE:

Title: *Brassicaceae*

Author: V. G. Chrzhanovskii

Language: Russian

Country: Russia

Series/Book: *Sistematika Rasteniy Komplekt Plakatov iz 53 Listov* (*Systematics of Plants – Posters from 53 Sheets*)

Plate: 24

Publisher: Kolos Publishing

Year: 1971

OPPOSITE: When he wasn't developing a new process to reproduce botanical pressings, Alois Pokorny illustrated educational charts that were as beautiful as they were clean and precise. Unlike some other illustrators, Pokorny's designs managed a breathtaking degree of verisimilitude, with the hints of an artist's hand. His use of subtle shadows, compositional asymmetries, torques, and his exquisite sensitivity to the power of negative space, are all hard to quantify individually, but the gestalt is an aesthetic distinctly Pokorny's.

Here, the nineteenth-century naturalist (and one-time professor to Sigmund Freud) analyzes the sexuality and progeny of *Brassica napus* (oilseed rape). At the top of the chart is a bright yellow blossom, newly fertilized, from front and side, showing a slender green young fruit peering from among the stamens. Below, a cross-section reveals developing ovules nestled tightly in their ovary. The illustrations at the bottom show a faceted siliqua, both whole – attached and removed from the parent plant – and halved – horizontally and vertically, respectively. The seeds, however, are not yet mature.

An oilseed rape clipping at the base of the chart displays a mature seed pod; a dehiscent siliqua splits to reveal an interior membrane, releasing the plant's progeny – the hard brown seed beside it on the right. The dispersed seed produces new life in descending order on the right-hand side of the chart, from a new seedling to a young rooted plant. At left, a flowering adult plant spans the height of its own illustrated life cycle.

OPPOSITE:
Title: *Brassica napus*
Author: Alois Pokorny
Language: N/A
Country: Germany
Series/Book: *Botanische Wandtafeln*
Plate: N/A
Publisher: Smichow (Neubert, Germany)
Year: 1894

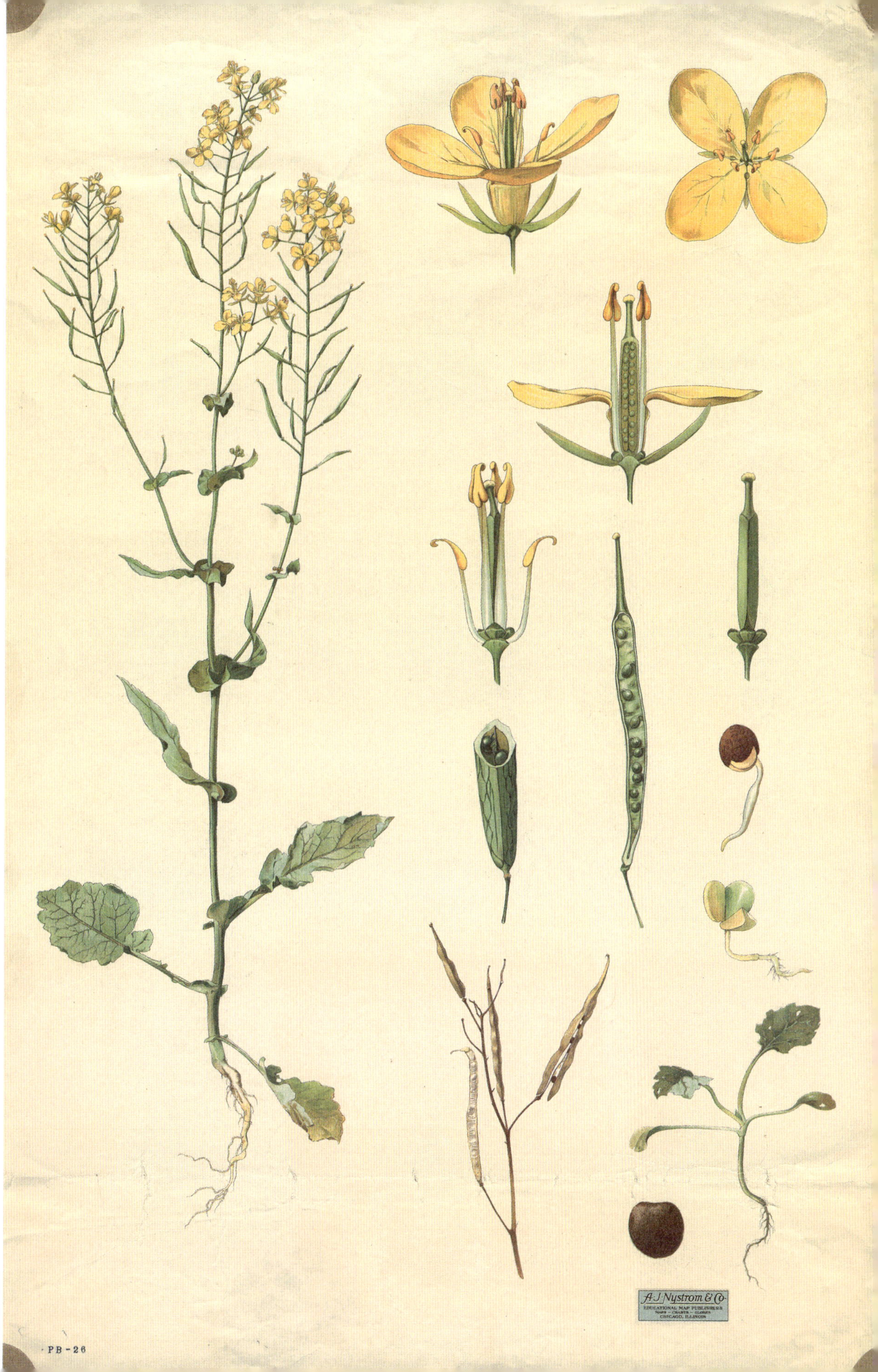
PB-26
A·J·Nystrom & Co·
EDUCATIONAL MAP PUBLISHERS
MAPS – CHARTS – GLOBES
CHICAGO, ILLINOIS

RIGHT: As previously mentioned, *Brassicaceae* are often identified by a classic siliqua seed pod. Many illustrators understandably included this distinctive fruit, but one can't help but delight at A. Peter's decision to design an entire chart of *Brassicaceae* with unique dehiscent fruits, most of which bear little resemblance to the family's classic pod. In fact, the only species in this chart that bears the classic siliqua (*Hesperis matronalis*, or dame's rocket) is instead portrayed by its flower. A. Peter further dispenses with tradition by slicing through a second characteristic trait (the four-petalled flower), thus dissembling the cruciform for which the family was named '*Cruciferae*'. His species all exhibit other qualities typical of the family, but Peter has instead opted to illustrate the exotic territory of unconventional fruits (doubtlessly aware that cruciforms and siliquas would have already been addressed in a student's botany lessons).

In this chart, therefore, are the lesser-known family fruits: *Sinapis alba* (white mustard) with a saber-like taper, yields just a few seeds – up to half a dozen. Rather than peel away tip-to-tip, only the lower segments of the pod fall away like two bristled lids of a box.

Raphanistrum lampsana, now called *R. raphanistrum* subsp. (wild radish), at the far right, produces fleshy seedpods with baubles of decreasing size, concluding in a torqued tip, dehiscing into a corky pod that, unlike most *Brassicaceaes*, does not split when the seeds are mature.

Draba aizoides (yellow whitlow-grass) produces an ovate, flattened fruit perhaps most similar to a typical elongated siliqua. Both sides peel away to allow seed dispersal, while the translucent inner membranes – equal to the length of its petiole – remain on the stem like so many tiny, handled mirrors.

Thlaspi arvense (field pennycress) was bestowed its common name for its fruits. Flattened and copper-coloured, the tarnished coins split vertically to disperse seeds. The plant retains the papery fruits until all seeds have fallen. *Neslea paniculata*, now *Neslia* (ball mustard), was also named for its fruits.

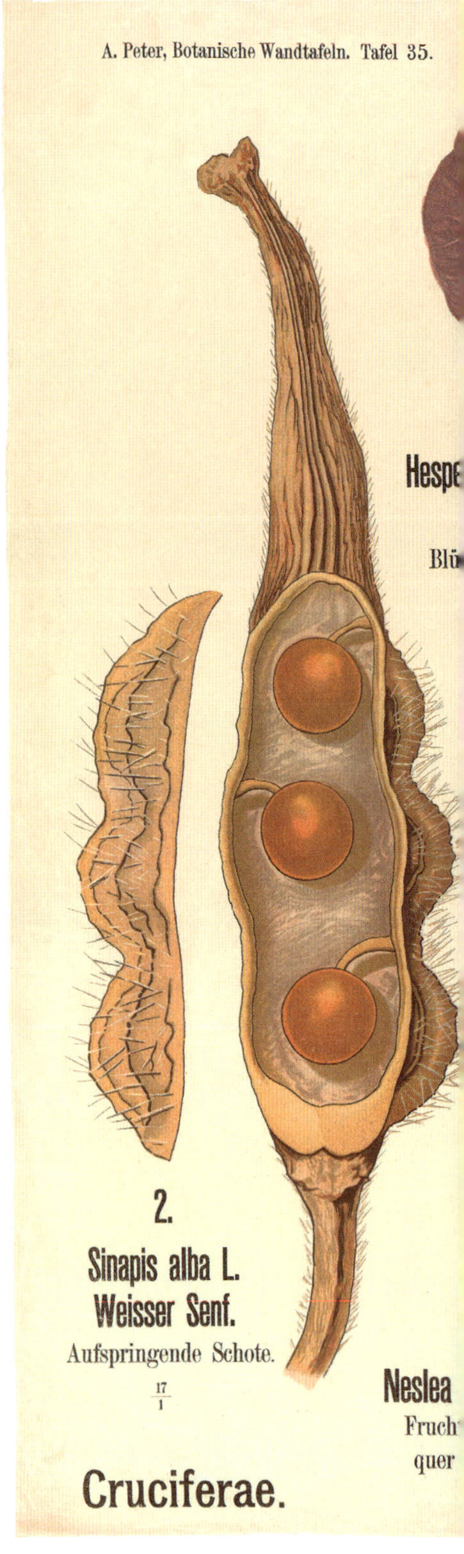

RIGHT:
Title: *Cruciferae*
Author: A. Peter
Language: German
Country: Germany
Series/Book: *Botanische Wandtafeln*
Plate: 35
Publisher: Paul Parey (Berlin, Germany)
Year: c.1900

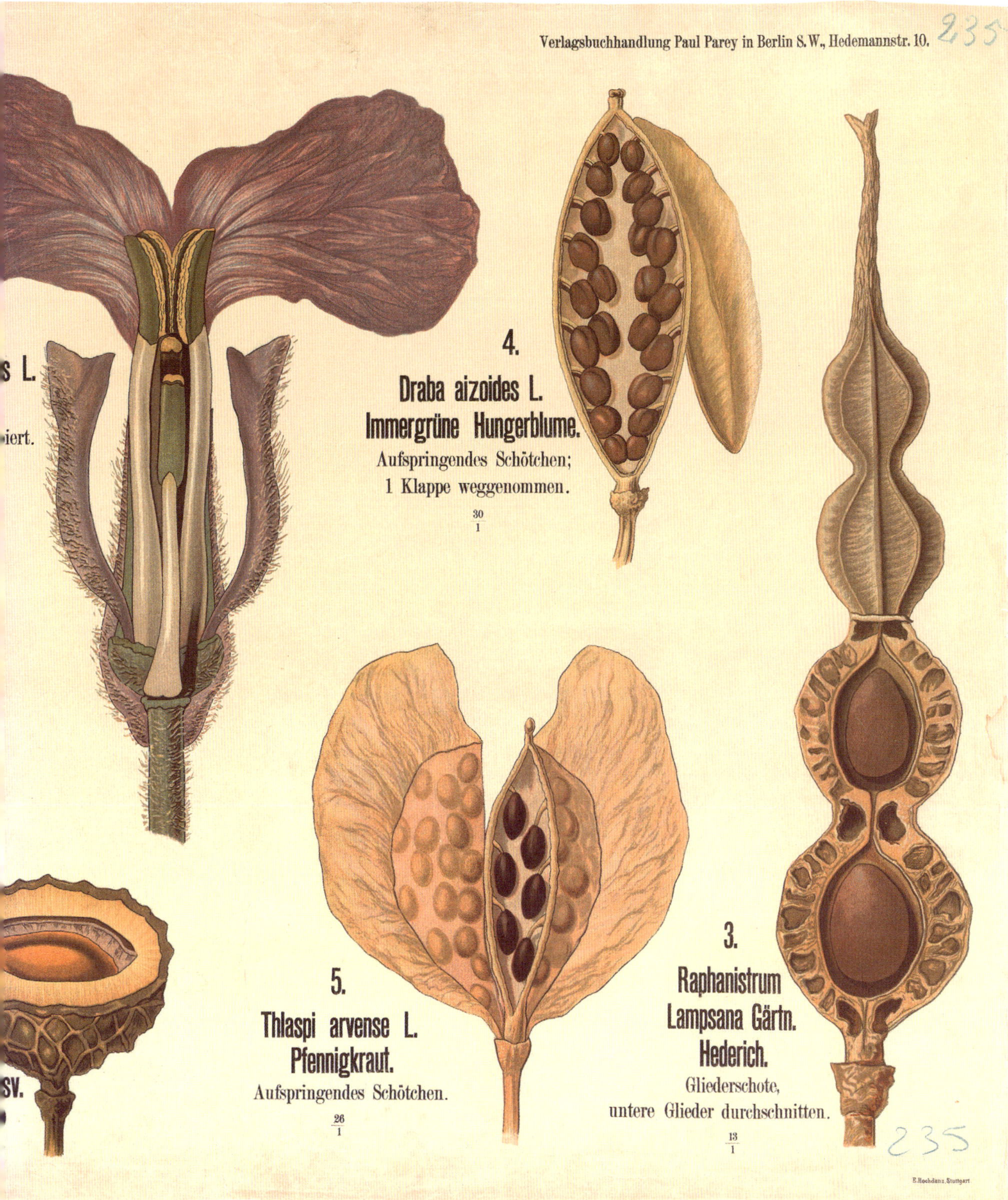

Verlagsbuchhandlung Paul Parey in Berlin S.W., Hedemannstr. 10.
s L.
iert.
4.
Draba aizoides L.
Immergrüne Hungerblume.
Aufspringendes Schötchen;
1 Klappe weggenommen.
30/1
5.
Thlaspi arvense L.
Pfennigkraut.
Aufspringendes Schötchen.
26/1
3.
Raphanistrum
Lampsana Gärtn.
Hederich.
Gliederschote,
untere Glieder durchschnitten.
13/1
sv.
E. Hochdanz, Stuttgart

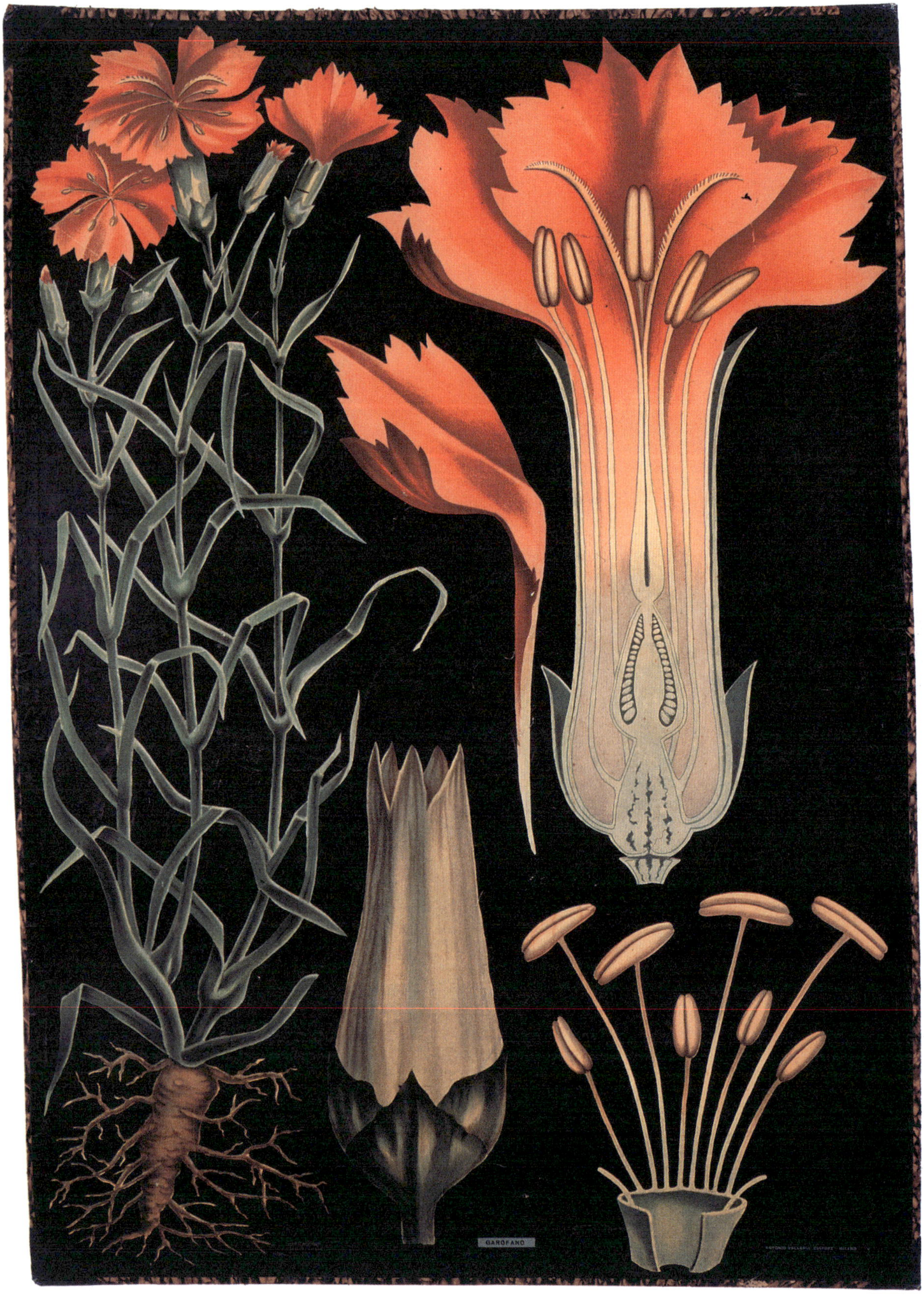
GAROFANO

VI CARYOPHYLLACEAE / PINK FAMILY

This family with approximately 2,500 species in more than 100 genera occurs globally, especially in the northern temperate regions of the northern hemisphere. The family comprises mainly herbaceous annuals or perennials, of which plants from the genera *Agrostemma*, *Gypsophila*, *Saponaria*, and *Silene*, among others, are grown in the garden. Several genera belong to the chickweeds: *Drymaria*, *Holosteum*, *Moenchia*, *Paronychia* and *Stellaria*.

The best-known genus, which gives the 'pink family' its common name, is *Dianthus*, including border and perpetual flowering carnations, garden and alpine pinks. Border pinks, then called gillyflowers, featured in secluded late-medieval gardens, and also in the famous Renaissance painting *Madonna of the Pinks* by Raphael as symbols of the tears of the Virgin Mary. Some alpine pinks, such as *D. callizonus* and *D. webbianus* are now endangered in their wild habitats. Also under threat is *Silene regia* (royal catchfly), so-called because it traps insects with sticky hairs on its calyx; the North American prairies in which it grows are gradually disappearing.

Characteristics: leaves usually opposite, simple, exstipulate; stems generally weak or brittle, with prominent nodes; four or five sepals, free or fused into long tube; four or five petals; usually eight to ten stamens; inflorescence solitary or cyme, typically dichasial (two-branched); fruit an achene, berry, or nutlet.

OPPOSITE:

Title: *Garofano* (*Dianthus caryophyllus*)
Author: Paola Manfredi
Language: Italian
Country: Italy
Series/Book: *Botanica Spicciola: Boschi, stagni, praterie* (*Simple Botany: Woods, ponds, prairies*)
Plate: Unknown
Publisher: Antonio Vallardi (Milan, Italy)
Year: 1923

OPENING PAGE: Besides its prosaic association with *boutonnières* and Mother's Day, *Dianthus caryophyllus* (carnation) is known as a far more powerful symbol with international resonance. Since the late nineteenth century, the carnation has been held aloft to represent the labor movement, left-wing political parties, the revolutionary and the proletariat. In 1974, the flower appeared on the streets of Lisbon. Celebrating a successful military coup and civil resistance movement that overthrew the authoritarian government, jubilant soldiers and citizens filled the streets, carnations adorning the muzzles of rifles and the hands of children. The near-bloodless victory was called the Carnation Revolution.

In Italy, the carnation was historically the symbol of May Day, a celebration of the working class, spring, youth and growth. Here, Paola Manfredi, an Italian botanist, has depicted *D. caryophyllus* as a tall stalk of exultant inflorescences with undulating leaves and a healthy root system. To its right, the cross-section of a flower is depicted as a near-symmetric body with stigma, stamen, and ovary. Below, a splayed androecium with eight stamens and a dehiscent capsule.

Of course, before Mother's Day, May Day, and the Carnation Revolution, there was Shakespeare. 'The fairest flowers of the season, Are our carnations...' (*The Winter's Tale*, Act iv. scene. 3)

OPPOSITE: Zippel and Bollmann illustrate the *Nelkenartige* or *Caryophyllales*, an order of plants including 33 families, of which the eponymous *Caryophyllaceae* is only the fourth largest (behind *Aizoaceae*, *Amaranthaceae* and *Cactaceae*). This isn't as surprising as it seems; remember, taxonomy is fluid, so families and orders can be repositioned as easily as species are rechristened. It's also unsurprising, therefore, that three of the four species represented here are within the *Caryophyllaceae* family, and one is of a different family within the order. After all, the chart is titled *Caryophyllales*, referring to the entire order; the authors meant no deceit.

A common perennial, *Saponaria officinalis* (wild sweet William or soapwort) can be found in disturbed areas, especially under hedgerows and along roadsides. An 1847 guide to agricultural plants describes it as 'a conspicuous weed, spreading by root [in] large bunches by buildings and giving a slovenly appearance to the [American] farm' (*Agricultural Botany*, William Darlington, M. D.). However, in its native Europe, soapwort has long been used as a skin salve ('*officinalis*' refers to the medical qualities of a plant) and a gentle soap ('*Saponaria*' derives from the Latin '*sapo*' for soap).

Stellaria holostea (now called *Rabelera holostea*), a native of western and central Europe and the British Isles, received its common name, stitchwort, for its supposed efficacy in relieving the pain of a stitch. As recorded in 1863, 'They are woont to drinke it in wine with the powder of acornes, against the pain in the side' (*On the Popular Names of British Plants*, R. C. Alexander Prior).

Of the German knotweed, a 1796 text confirms 'Nothing can be more common on a sandy soil than *Scleranthus annuus* ... especially in fallow fields' (*Coloured Figures of British Plants, with their Essential Characters, Synonyms, and Places of Growth*, James Edward Smith and James Sowerby). The species is, as the Latin name hints, an annual, though the author assures that each spring one can can look for a new crop of the hardy species.

The odd one out here is the succulent *Portulaca oleracea* (common purslane), of the *Portulacaceae* (purslane family).

RIGHT:
Title: *Nelkenartige*
Author: Hermann Zippel; Illustrator: Carl Bollmann
Language: German
Country: Germany
Series/Book: *Repräsentanten einheimischer Pflanzenfamilien (Representatives of Indigenous Plant Families)*
Plate: II Abteilung, 41
Publisher: Friedrich Vieweg & Sohn (Braunschweig, Germany)
Year: 1879

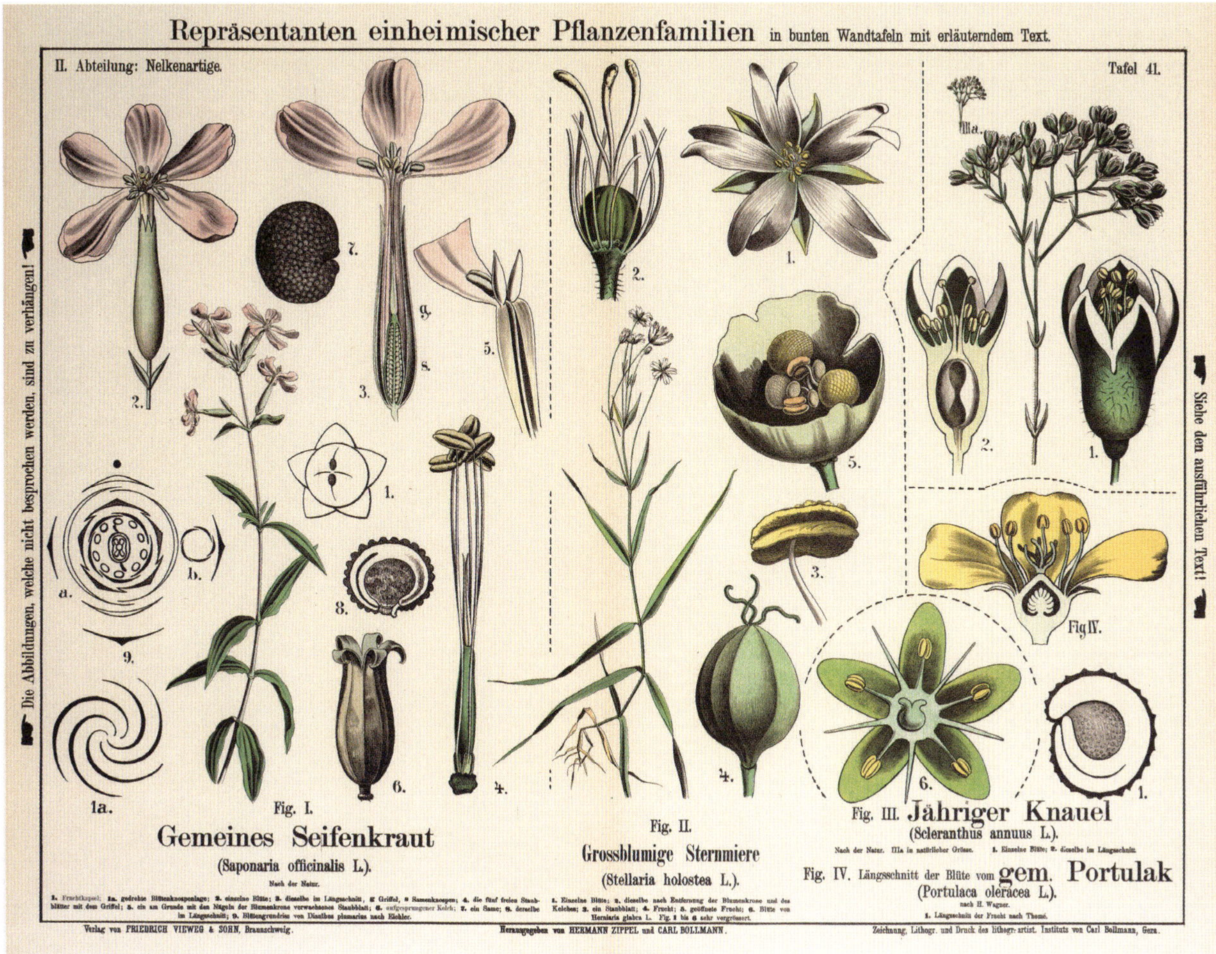
Repräsentanten einheimischer Pflanzenfamilien in bunten Wandtafeln mit erläuterndem Text.
II. Abteilung: Nelkenartige.
Tafel 41.
Die Abbildungen, welche nicht besprochen werden, sind zu verhängen!
Siehe den ausführlichen Text!
Fig. I.
Gemeines Seifenkraut
(Saponaria officinalis L.).
Nach der Natur.
Fig. II.
Grossblumige Sternmiere
(Stellaria holostea L.).
Fig. III. Jähriger Knauel
(Scleranthus annuus L.).
Fig. IV. Längsschnitt der Blüte vom gem. Portulak
(Portulaca oleracea L.).
Fig IV.
Verlag von FRIEDRICH VIEWEG & SOHN, Braunschweig.
Herausgegeben von HERMANN ZIPPEL und CARL BOLLMANN.

THIS SPREAD: While bladder campion, catchfly and ragged-robin are neither the flowers of revolution nor portrayed as such, a closer look at Emil Korsmo's three *Caryophyllaceae* wildflowers (or, weeds, per the series title) reveals morphological and compositional similarities to Paola Manfredi's chart on this chapter's opening page – both morphologically and compositionally.

Most notably, in cross-section and in full, all three of the flowers represented in Korsmo's charts show an almost symmetric arrangement of petals, stamens and pistils, and an androecium of exceptional height. On the chart on this page, the central *Silene venosa* (now called *S. vulgaris* subsp., bladder campion) blossom is portrayed upright, despite the bowed inflorescences. An inflated calyx also distinguishes the species, though Korsmo's treatment echoes the tubular *Dianthus caryophyllus*.

On the second chart, *Viscaria vulgaris* (sticky catchfly), native to sandy meadows and dry hillsides, is illustrated with soaring grass-like leaves and tall flower stalks. A viscous secretion below the flower panicle is quite accomplished at trapping herbivorous insects and nectar thieves, hence the flower's common name.

To its right, *Lychnis flos-cuculi* (now called *Silene flos-cuculi*, ragged-robin) is the perfect companion to elucidate distinguishing features of the two species portrayed on this chart. Korsmo has dissected and arranged both species identically, emphasizing the deeply articulated star-shaped flowers, black seeds, and lance-shaped leaves that will identify *S. flos-cuculi* in wild meadows and fields.

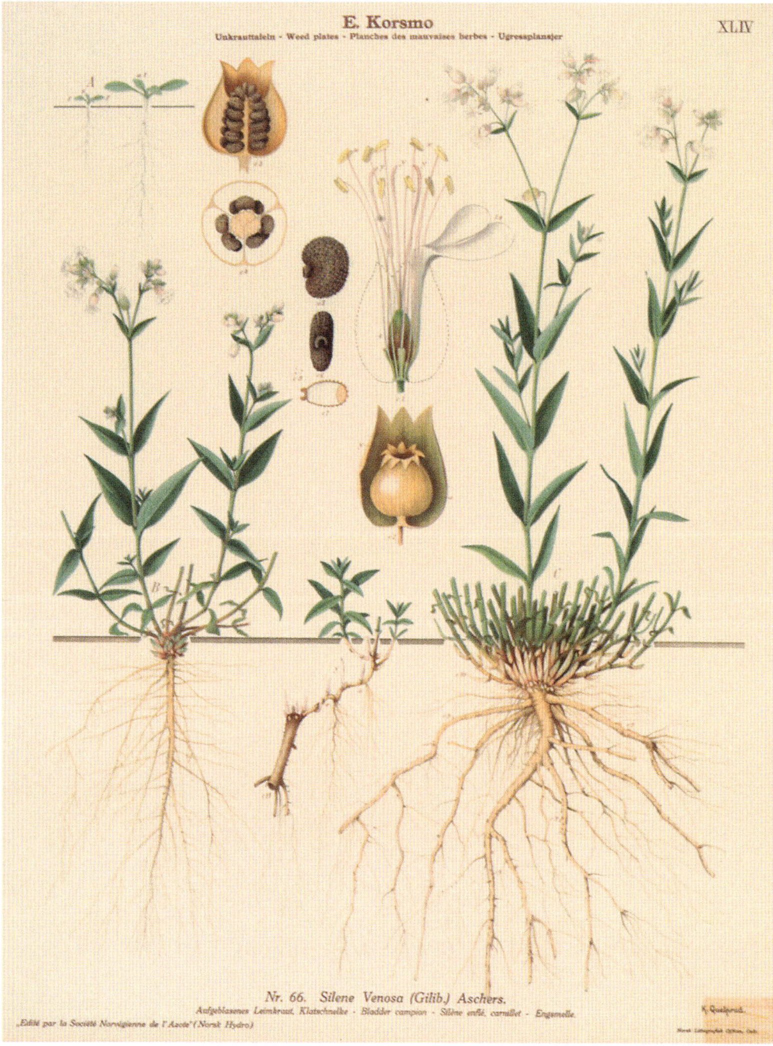

ABOVE & OPPOSITE:
Title, above: *Nr. 66 Silene Venosa*; Opposite: *Nr. 127 Viscaria vulgaris*; *Nr. 128 Lychnis flos cuculi*
Author: Emil Korsmo; Illustrator: Knut Quelprud
Language: German, English, French, Norwegian
Country: Norway
Series/Book: *Unkrauttafeln* (*Weed Plates*)
Plates: 5; 81
Publisher: Norsk Hydro (Oslo, Norway)
Year: 1934

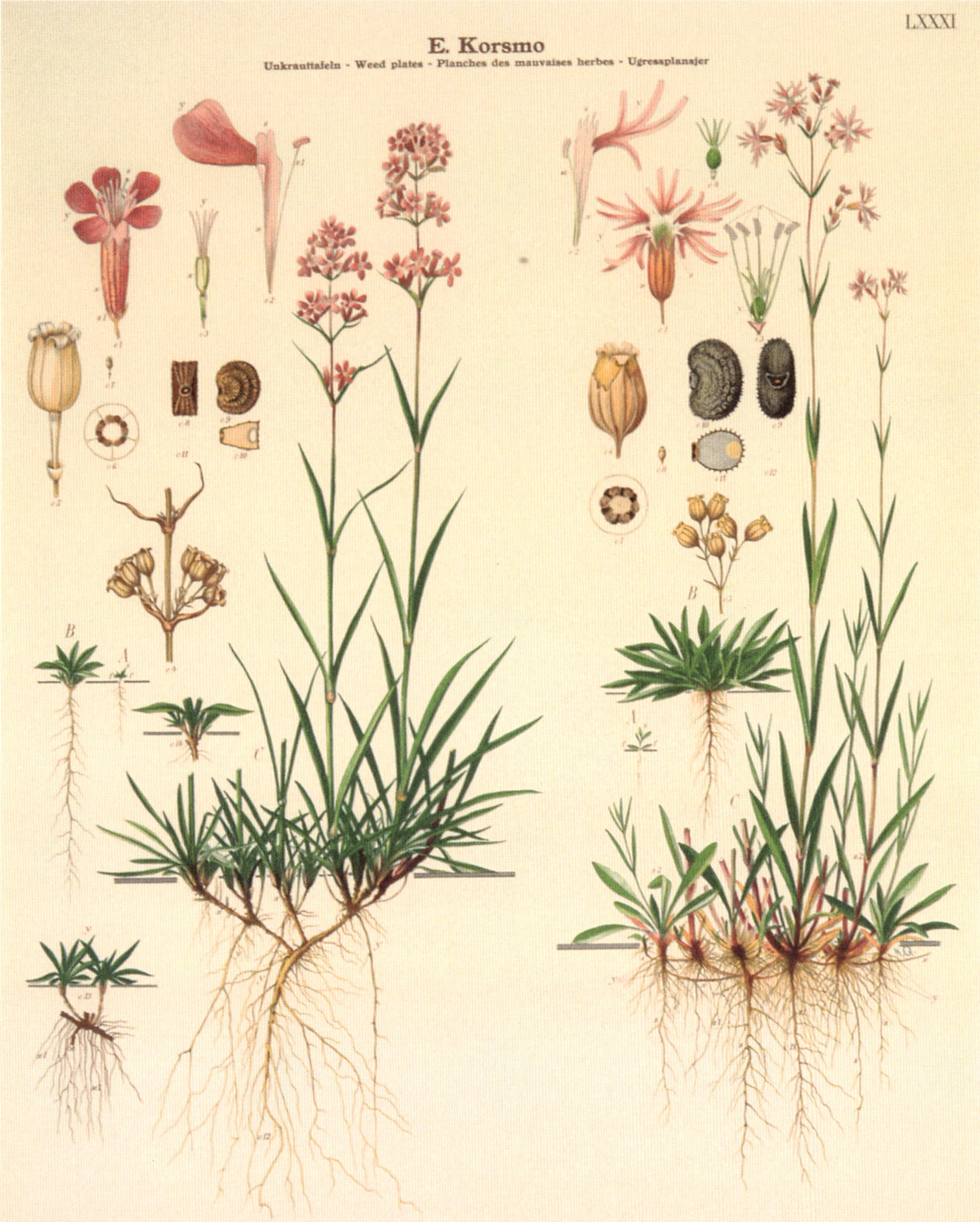
LXXXI
E. Korsmo
Unkrauttafeln - Weed plates - Planches des mauvaises herbes - Ugressplansjer
Nr. 127. Viscaria vulgaris Roehl.
Gemeine Pechnelke - Viscid campion - Oeillet de Janséniste - Engtjæreblom.
Nr. 128. Lychnis flos cuculi L.
Kuckucks-Lichtnelke, Gauchraden - Ragged robin - Fleur de coucou - Hanekam.
„Edité par la Société Norvégienne de l'Azote" (Norsk Hydro)

VII CONVOLVULACEAE / MORNING GLORY FAMILY

A fairly uniform family, with approximately 1,600 species in 60 genera from temperate and tropical regions around the world, of usually annual and perennial, herbaceous or woody climbers. It is also called the morning glory family after the popular garden genus, *Ipomoea*, grown for their striking trumpet-shaped blooms. While species of the *Ipomoea* genus tend to be valued, the closely related *Convolvulus* species are mainly invasive garden and agricultural weeds, for example *C. arvensis* (field bindweed), with a few cultivated garden varieties. In tropical climates, *Ipomoea batatas* (sweet potato) and *I. aquatica* (water spinach) are grown as food crops. *Cuscuta* (dodder) is an unusual genus: the plants are parasitic vines; they lack chlorophyll and therefore cannot photosynthesize. Instead of leaves they have scales, and they twine around host plants in order to tap into their vascular systems to obtain nutrients.

Characteristics: usually simple, alternate leaves; five sepals, free or united; five petals, forming a tubular, campanulate, or infundibuliform corolla; five stamens alternate with corolla lobes; cymose or solitary flower or clustered inflorescence; fruit a capsule or berry. The stem often has milky latex and the plant climbs by twining.

OPPOSITE:
Title: *Cuscuta glomerata, Choisy*
Authors: Arnold and Carolina Dodel-Port
Language: German
Country: Switzerland
Series/Book: *Anatomisch physiologische Atlas der Botanik* (*The Anatomical & Physiological Atlas of Botany*)
Plate: 30
Publisher: J. F. Schreiber (Esslingen, Germany)
Year: 1878–93

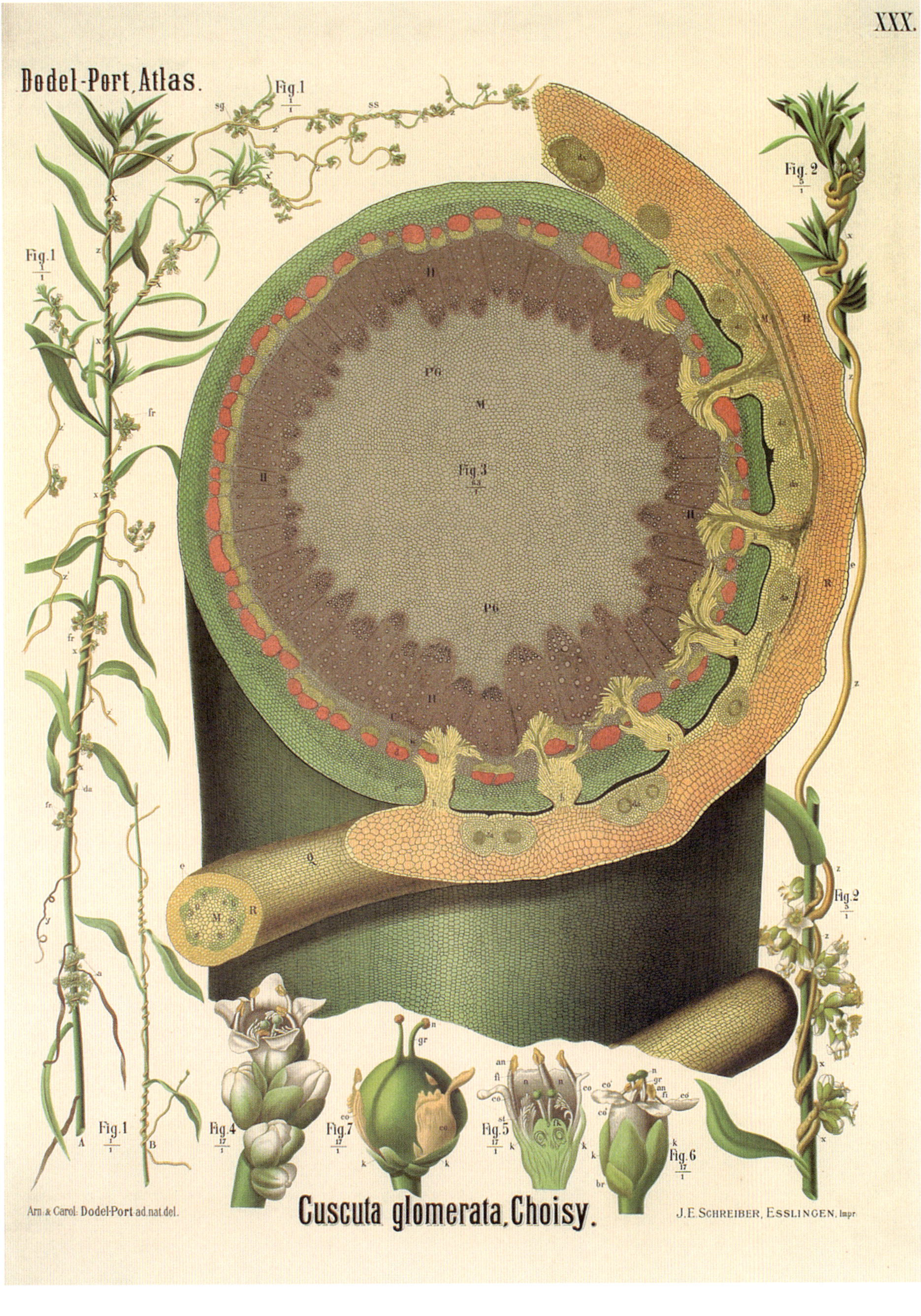
XXX.
Dodel-Port, Atlas.
Fig. 1
Fig. 2
Fig. 3
Fig. 4
Fig. 5
Fig. 6
Fig. 7
A
B
Arn. & Carol: Dodel-Port ad.nat.del.
Cuscuta glomerata, Choisy.
J.E. Schreiber, Esslingen. Impr.

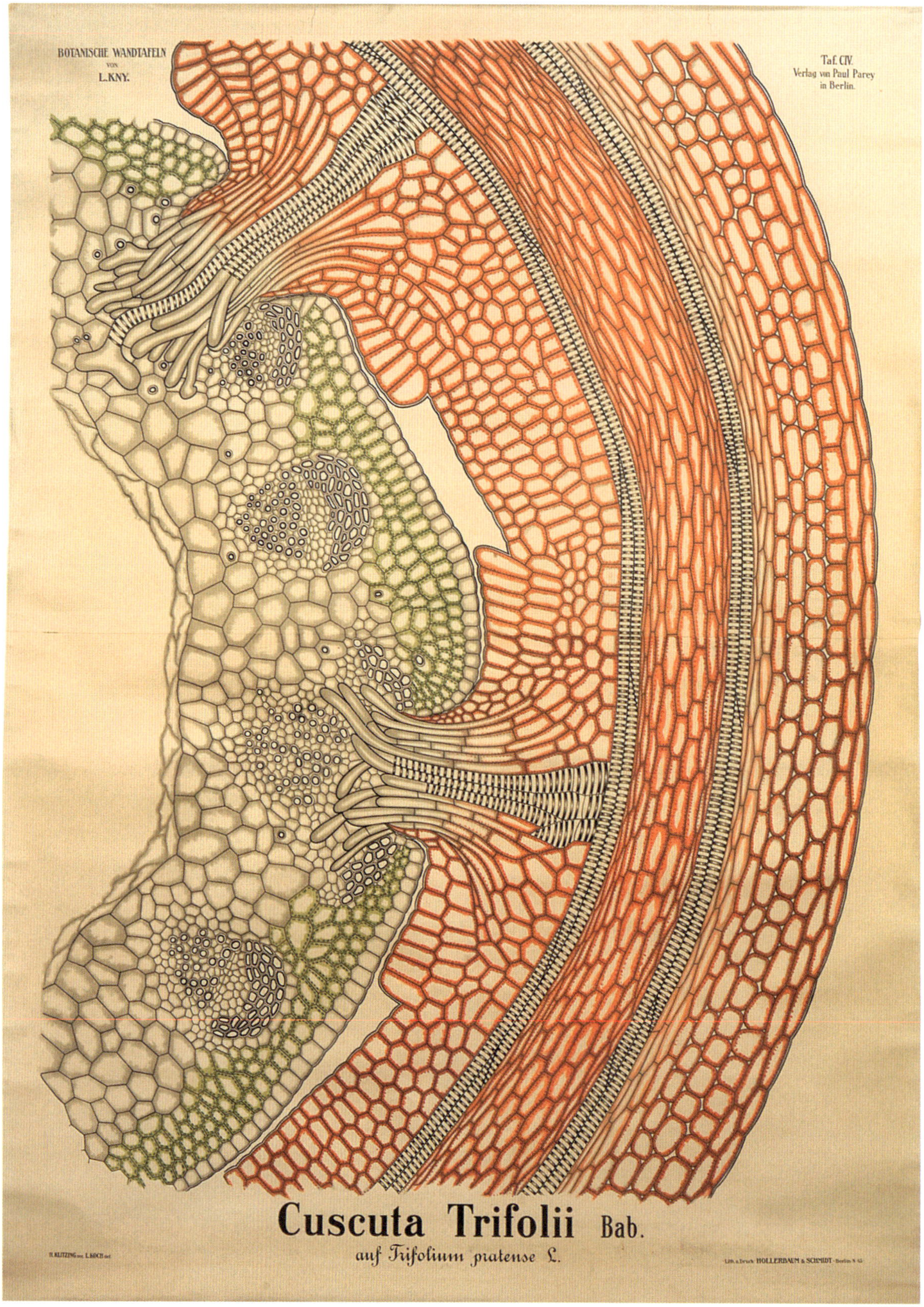
BOTANISCHE WANDTAFELN
VON
L. KNY.
Taf. CIV.
Verlag von Paul Parey
in Berlin.
Cuscuta Trifolii Bab.
auf Trifolium pratense L.
HOLLERBAUM & SCHMIDT

OPENING PAGE, LEFT & OVERLEAF: By the nineteenth century, botanists had defined parasitic plants by their ability to derive nutrients from another living plant. Despite centuries of observation, however, exactly how this occurred remained a mystery. Not until the light microscope was introduced in the 1800s were biologists able to observe this mechanism, called a haustorium. Finally, the anatomy of these peculiar adaptations could be viewed, an opportunity embraced by Carolina and Arnold Dodel-Port, who were determined to understand *Cuscuta glomerata*, the rope dodder. Not only did they succeed, they were hailed for demystifying a species that confounded botanists. With the dissemination of their *Botanik Atlas*, classrooms were now walled with new research.

Dodders have neither roots nor chlorophyll (or very little), so they seek nutrients from other plants. *Cuscuta glomerata* is no different. The Dodel-Ports spent ten years studying its behavioural anatomy, then illustrated their microscopic illustrations in a wall chart. Here, the host stem, the dodder, and the haustorium are writ large with cellular structures revealed. This cross-section shows the seven haustoria through which *C. glomerata* penetrates the host's tissue and accesses nutrients. A haustorium is a fungal appendage that enlarges when it senses an appealing host, releasing enzymes that dissolve the cell wall and allow the haustoria to grow within the host plant, siphoning water, minerals and carbohydrates.

Rigorous scientists and consummate artists Carolina and Arnold Dodel-Port produced a chart as beautiful as it is precise and comprehensive. To the left, a filiform *C. glomerata* twines around its host (fig. 1); below are diagrams of the anatomy of male and female flowers, and a developing fruit. To the right is an enlarged stem segment with flowers forming parallel rows on either side of the vine. This rope-like inflorescence is, of course, *C. glomerata*'s distinguishing characteristic.

On the facing page, Leopold Kny offers a further magnified view of a dodder and its host. Here, *C. trifolii* (now called *C. epithymum* subsp.), the clover dodder, embraces a stem of *Trifolium pratense* (red clover). In his highly detailed illustration, one can distinguish the epidermis, xylem and phloem, measure the width of cellular walls and observe the breached membrane and two invading haustoria.

And overleaf, we have Jung, Koch, Quentell's portrayal of *Cuscuta europaea* (greater dodder or European dodder). As well as a microscopic view of the haustorium at work on a host plant, Jung, Koch, Quentell also step back to give an overview of the dodder and host, depicting its counter-clockwise-winding filiform stem (along with a neat clipping of this) and clustered inflorescences. On the right-hand side of the chart we have three views of the flower; one view from the outside, and two views of its reproductive organs. *Cuscuta* are usually insect-pollinated, and each flower may produce two to three seeds, which may seem a small number until you realize that each plant may put out thousands of tiny flowers. At the bottom of the chart, five little seedlings creep up from the soil. With no other way to acquire nutrients, if the seedlings do not find a host within a few days, then they will die. The seeds, however, are formed with very hard coats, and will not germinate until this is broken or softened by water absorption. Only around five percent of seeds will germinate the year they were produced – a clever back-up plan in case of the absence of a suitable host – and seeds can lie dormant in the soil for more than twenty years.

OPPOSITE:
Title: *Cuscuta trifolii*
Author: L. Kny
Language: German
Country: Germany
Series/Book: *Botanische Wandtafeln*
Plate: 104
Publisher: Paul Parey (Berlin, Germany)
Year: 1874

RIGHT:

Title: *Cuscuta europaea*

Authors: Heinrich Jung, Dr. Friedrich Quentell

Illustrator: Dr. Gottlieb von Koch

Language: N/A

Country: Germany

Series/Book: *Neuen Wandtafeln*

Plate: 31

Publisher: Fromann & Morian (Darmstadt, Germany)

Year: 1902–3

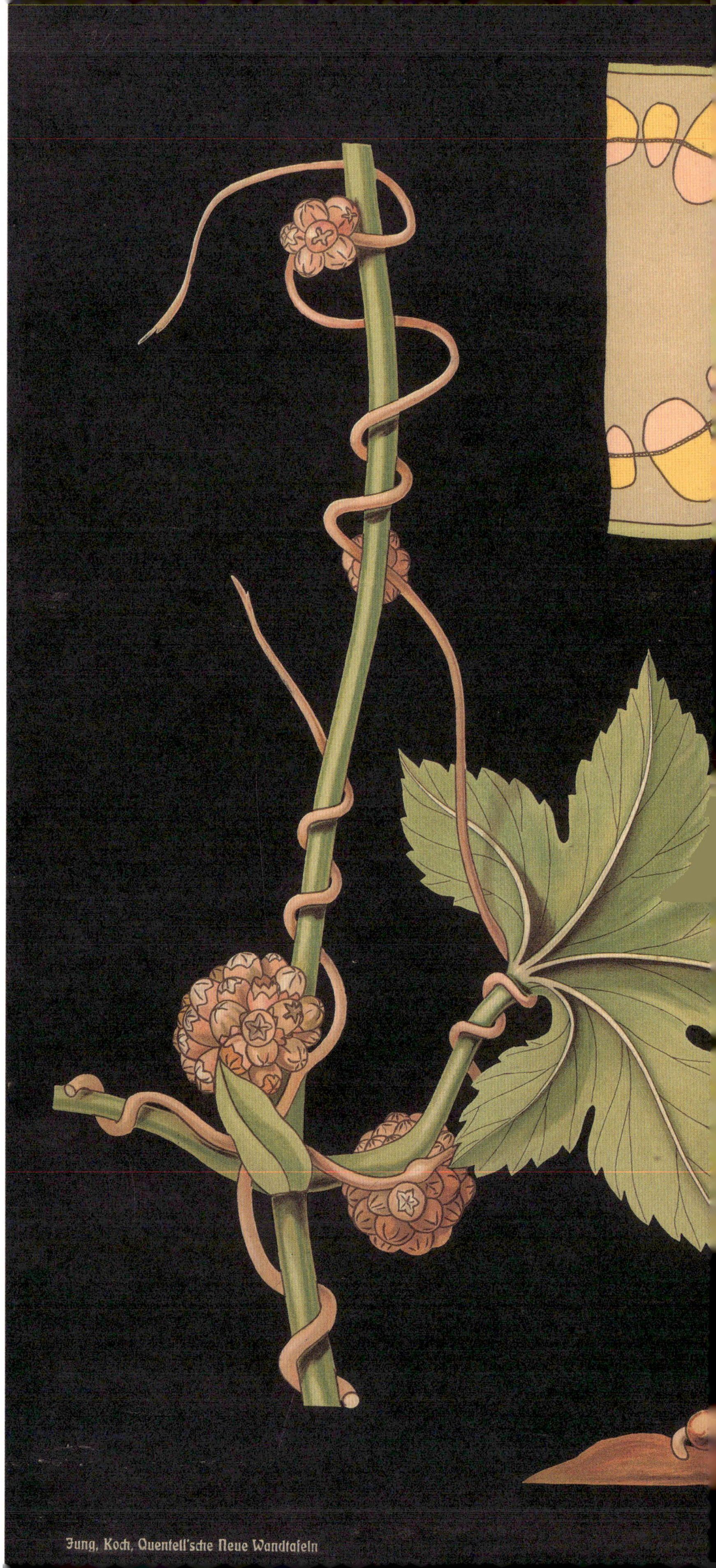

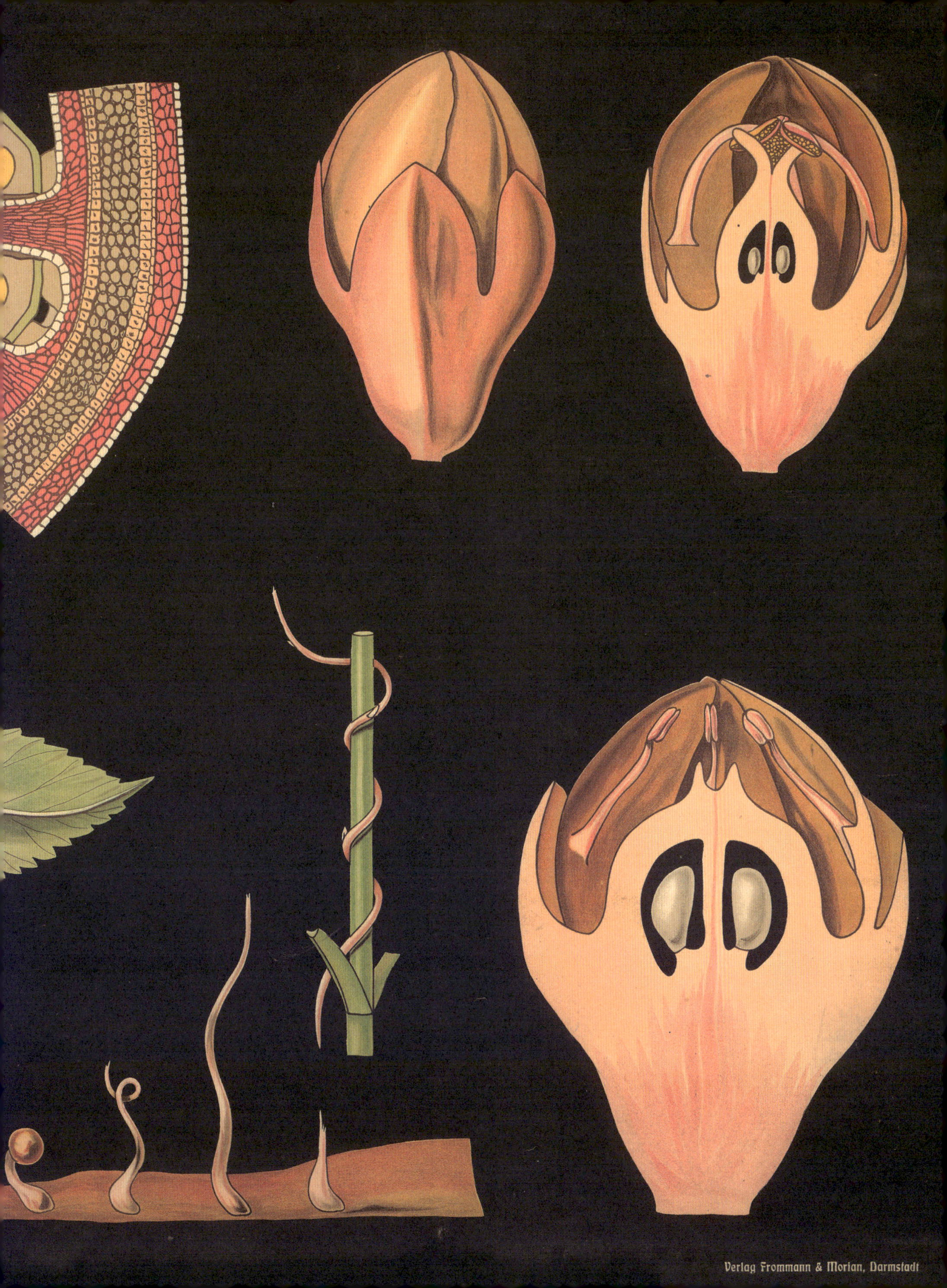
Verlag Frommann & Morian, Darmstadt

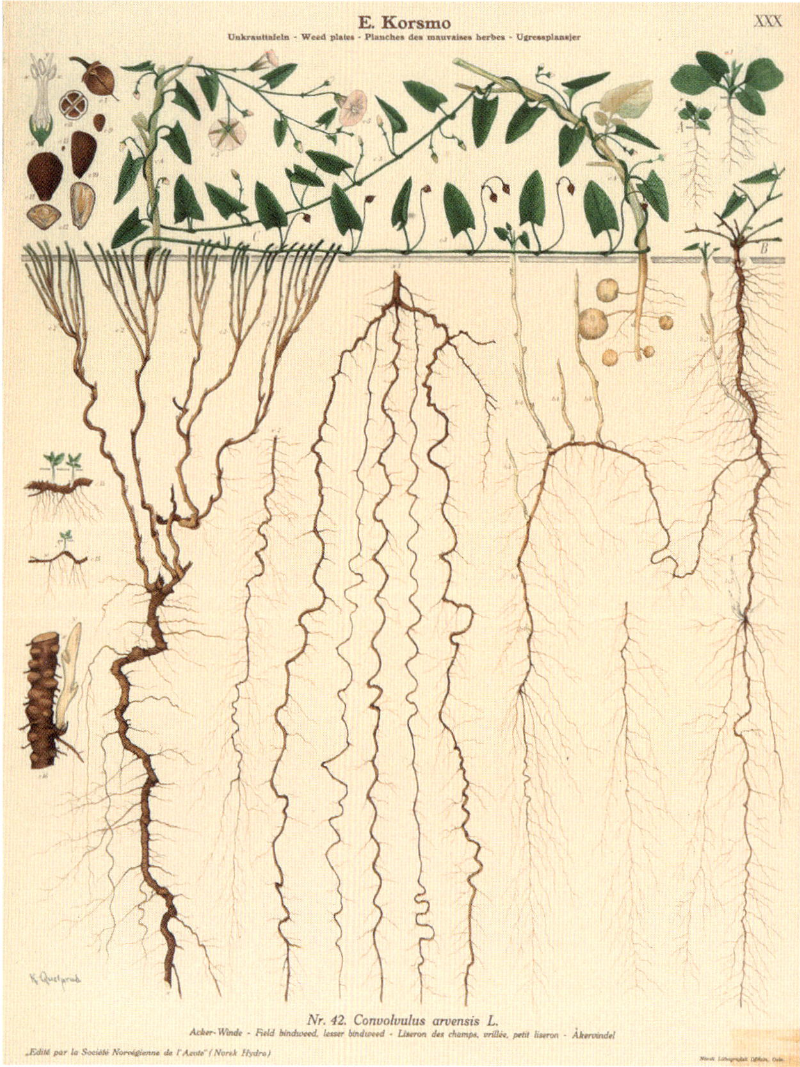

ABOVE & OPPOSITE:

Title, above: *Nr. 42 Convolvulus arvensis*; Opposite: *Nr. 37 Convolvulus sepium*
Author: Emil Korsmo; Illustrator: Knut Quelprud
Language: German, English, French, Norwegian
Country: Norway
Series/Book: *Unkrauttafeln* (*Weed Plates*)
Plate: 25; 30
Publisher: Norsk Hydro (Oslo, Norway)
Year: 1934

THIS SPREAD: In the garden, morning glories (*Ipomoea*) and bindweeds (*Convolvulus*) often arrive with very different receptions: the former, nurtured; the latter, cursed. While morning glories can surpass bindweeds' invasive sprawl, horticulturalists have always needed to distinguish the two genera. However, it's not always easy to tell them apart. There are only a couple of clues to differentiate them, one of which is the leaves – those of *Ipomoea* are heart-shaped, while those of *Convolvulus* are pointed, like arrows.

Emil Korsmo's comprehensive collection of weeds includes charts of two similar *Convolvulus* species. By no means a pedantic decision; Korsmo used the two charts to emphasize common characteristics of the genus, as well as notable differences between the two species. Effectively, they form one chart in two parts: rather than compress a full visual story of the plant's anatomy and life cycle into one (as he tasked himself throughout the series), he's parsed common information between the two.

On this page, *C. arvensis* (field bindweed) shows the capacity of roots to produce shoot growth. On the facing page, *C. sepium*, now called *Calystegia sepium* (hedge bindweed), includes a side view of the genera's signature funnel-shaped flower. Korsmo's extended canvas allows both plants ample space to root, wind, and bind – as bindweeds will do. *Convolvulus arvensis* is a historically notorious species, so an extensive list of aliases is no surprise: devil's guts, creeping Jenny, European bindweed, hedge bells, corn lily, withwind, bellbine, laplove, sheepbine, corn-bind, bearbind, green vine... A native of Eurasia, the temperate species has been unwanted in almost fifty countries for centuries. Korsmo includes seeds, the cross-section of a fruit, and a mature plant. More importantly, he articulates the arrow-shaped leaves, which distinguish the genera from *Ipomoea*, and maps an extensive perennial root system that may penetrate the soil to a depth of 20 feet (6m). The roots store carbohydrates and proteins and allow a new plant to grow from rhizomes and fragments even when the plant seems to have been destroyed.

Calystegia sepium (opposite) looks similar to *Convolvulus arvensis*, but enlarged. Most structures, including flowers, leaves and seeds, are larger. The root system, however, is not as expansive. The seeds disperse from a four-chambered spherical capsule (see the horizontal cross-section of the fruit, top centre) and can remain viable for decades.

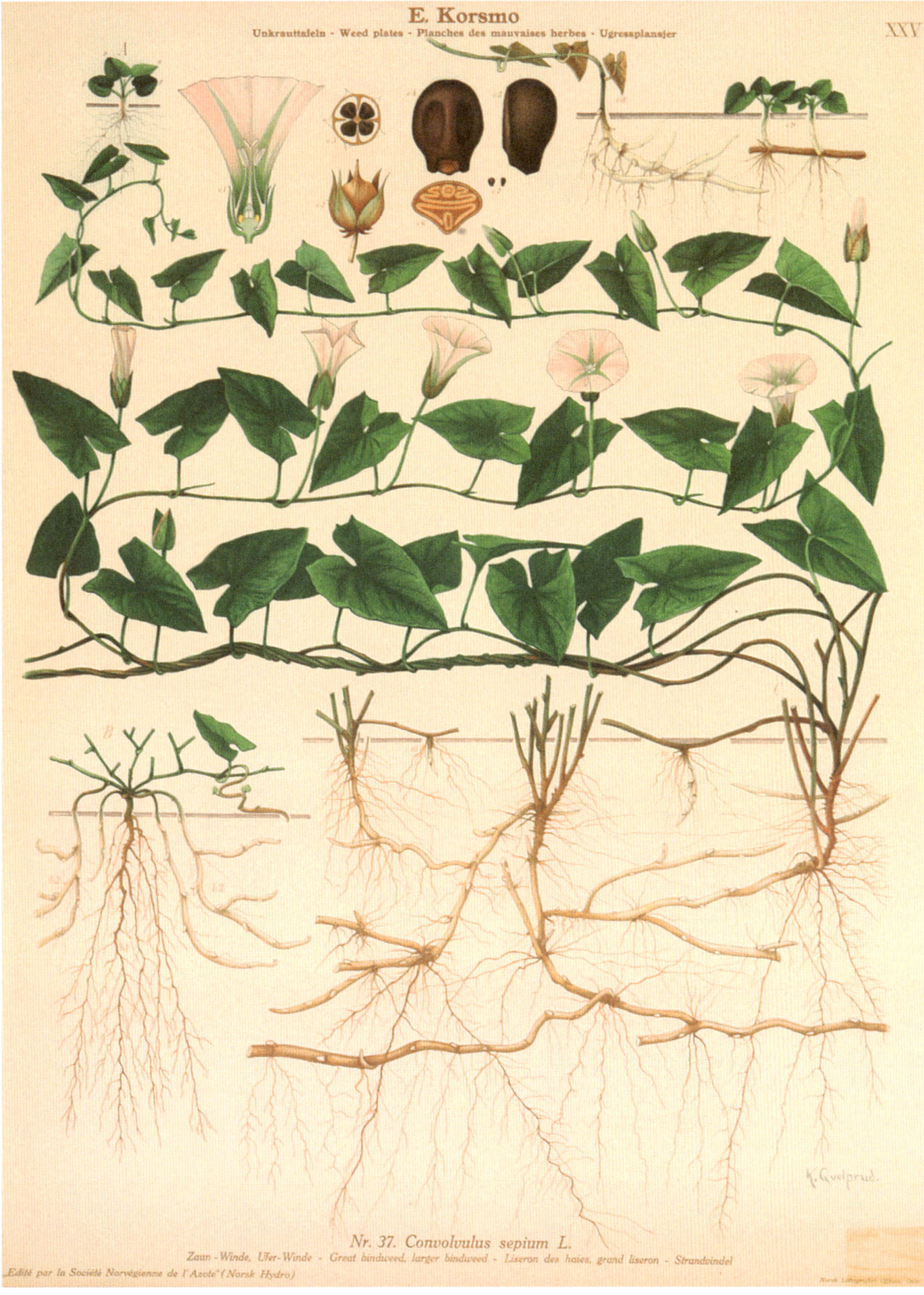
E. Korsmo
Unkrauttafeln - Weed plates - Planches des mauvaises herbes - Ugressplansjer
XXV
K. Quelprud.
Nr. 37. Convolvulus sepium L.
Zaun-Winde, Ufer-Winde - Great bindweed, larger bindweed - Liseron des haies, grand liseron - Strandvindel
„Edité par la Société Norvégienne de l'Azote" (Norsk Hydro)

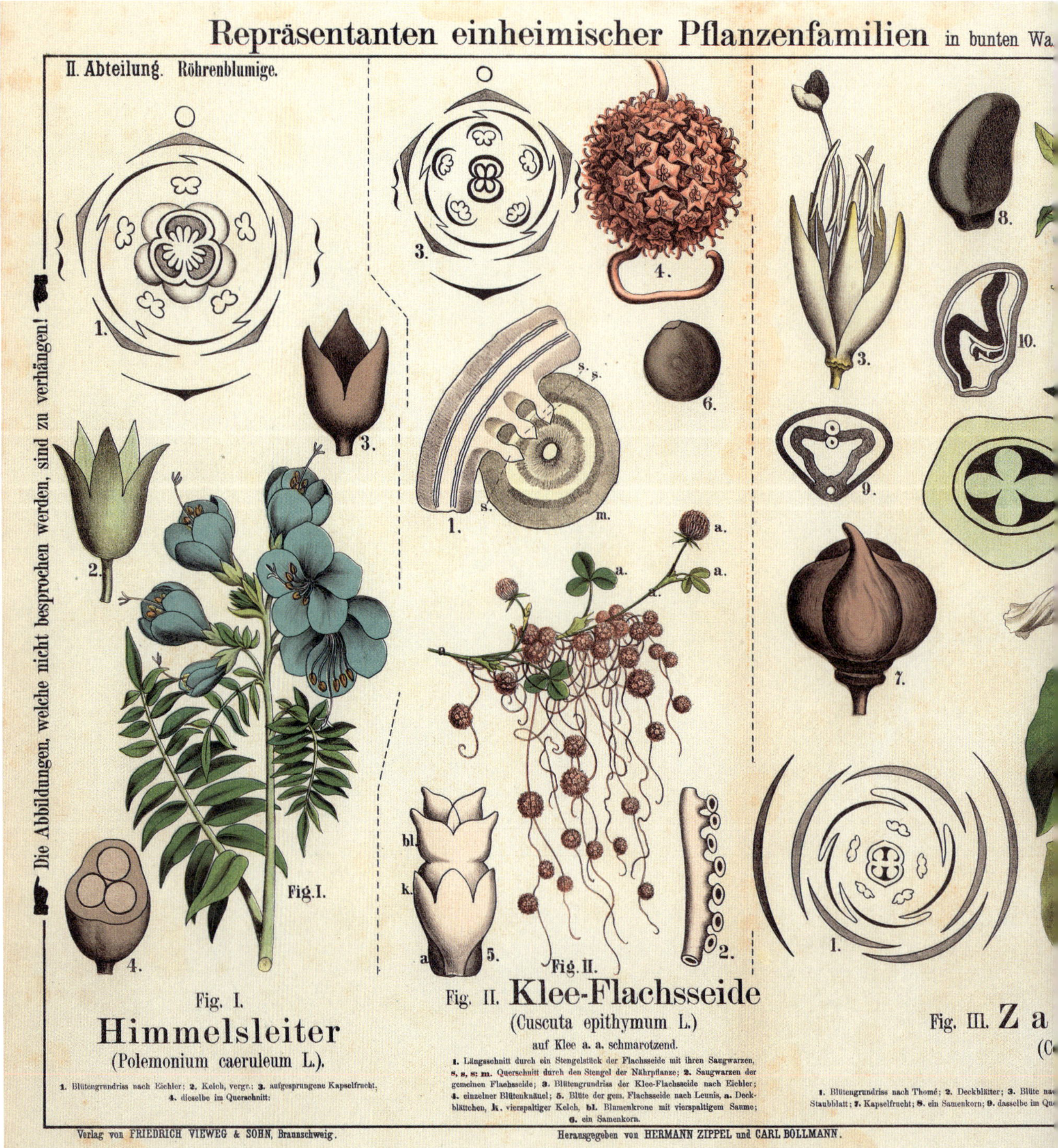
Repräsentanten einheimischer Pflanzenfamilien in bunten Wa
II. Abteilung. Röhrenblumige.
Die Abbildungen, welche nicht besprochen werden, sind zu verhängen!
Fig. I.
Fig. II.
Fig. I.
Himmelsleiter
(Polemonium caeruleum L.).
1. Blütengrundriss nach Eichler; 2. Kelch, vergr.; 3. aufgesprungene Kapselfrucht.
4. dieselbe im Querschnitt:
Fig. II. Klee-Flachsseide
(Cuscuta epithymum L.)
auf Klee a. a. schmarotzend.
1. Längsschnitt durch ein Stengelstück der Flachsseide mit ihren Saugwarzen, s, s, s; m. Querschnitt durch den Stengel der Nährpflanze; 2. Saugwarzen der gemeinen Flachsseide; 3. Blütengrundriss der Klee-Flachsseide nach Eichler; 4. einzelner Blütenknäuel; 5. Blüte der gem. Flachsseide nach Leunis, a. Deckblättchen, k. vierspaltiger Kelch, bl. Blumenkrone mit vierspaltigem Saume; 6. ein Samenkorn.
Fig. III. Z a
1. Blütengrundriss nach Thomé; 2. Deckblätter; 3. Blüte na
Staubblatt; 7. Kapselfrucht; 8. ein Samenkorn; 9. dasselbe im Qu
Verlag von FRIEDRICH VIEWEG & SOHN, Braunschweig.
Herausgegeben von HERMANN ZIPPEL und CARL BOLLMANN.

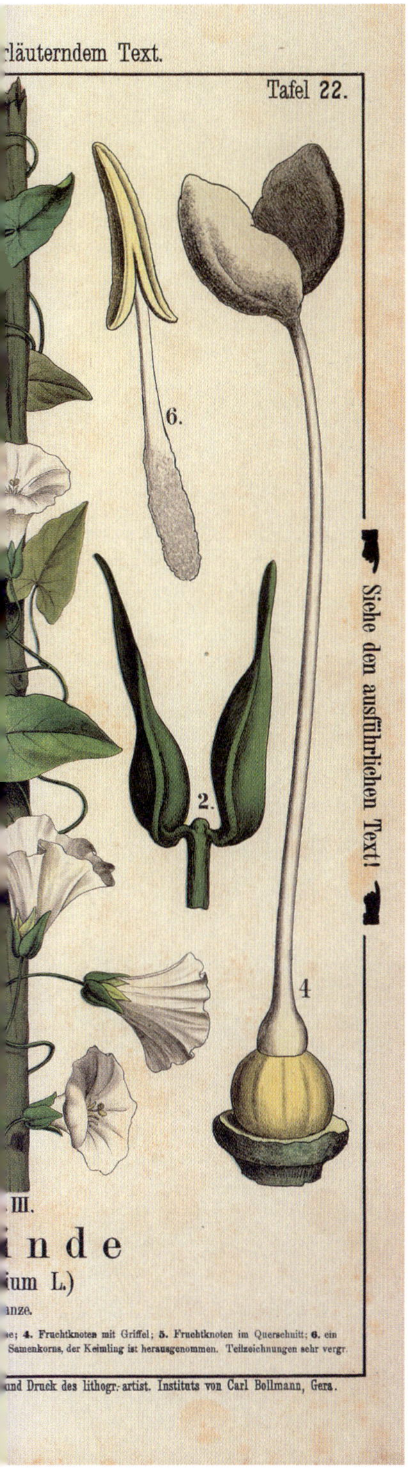

LEFT:
Title: *Röhrenblumige (Funnel-shaped Flowers)*
Author: Hermann Zippel; Illustrator: Carl Bollmann
Language: German
Country: Germany
Series/Book: *Repräsentanten einheimischer Pflanzenfamilien (Representatives of Indigenous Plant Families)*
Plate: II Abteilung, 22
Publisher: Friedrich Vieweg & Sohn (Braunschweig, Germany)
Year: 1879

LEFT: Zippel and Bollmann didn't always follow common taxonomy when populating the charts in their series *Repräsentanten einheimischer Pflanzenfamilien* (*Representatives of Indigenous Plant Families*). Here, they include two species from the *Convolvulaceae* family, with one outlier from the *Polemoniaceae* family. The chart's title offers an explanation: *Abteilung: Röhrenblumige*, or *Department: Funnel-shaped Flowers* – more precisely, an involucre of bracts. And the flowers of *Polemonium caeruleum* (Jacob's ladder) do indeed appear funnel-shaped when nestled in the involucre (fig. 2). The illustrations include a diagram of the flower (at top) and a cross-section of the fruit (fig. 4).

Cuscuta epithymum (clover dodder) is a rootless parasitic plant with tiny pink flowers and involucres. The globed inflorescences drape over a host plant like a string of lights, no less colourful or tangled. More than just a vining plant, the dodder is also able to absorb nutrients through tiny haustoria (fig. 2) that insert themselves into the host's vascular system (fig. 1).

The most interesting aspect of Zippel and Bollmann's treatment of *Convolvulus sepium* (now *Calystegia sepium*) is at the style depicted at the far right of the chart (fig. 4), enlarged to a higher degree than any other element. Of course, this is no accident. Morphologically, *C. sepium* (and other bindweeds) can be distinguished from the more esteemed *Ipomoea* genus (morning glories) by their leaves, as we saw on the previous page, and by their styles. *Ipomoea* has a small style with one to three lobes, while *Convolvulus* has a linear-to-oblong style, always with two lobes. Rarely will an illustrator portray a style with equal height to a mature plant; when it happens, there's invariably a good reason for the reader to deduce.

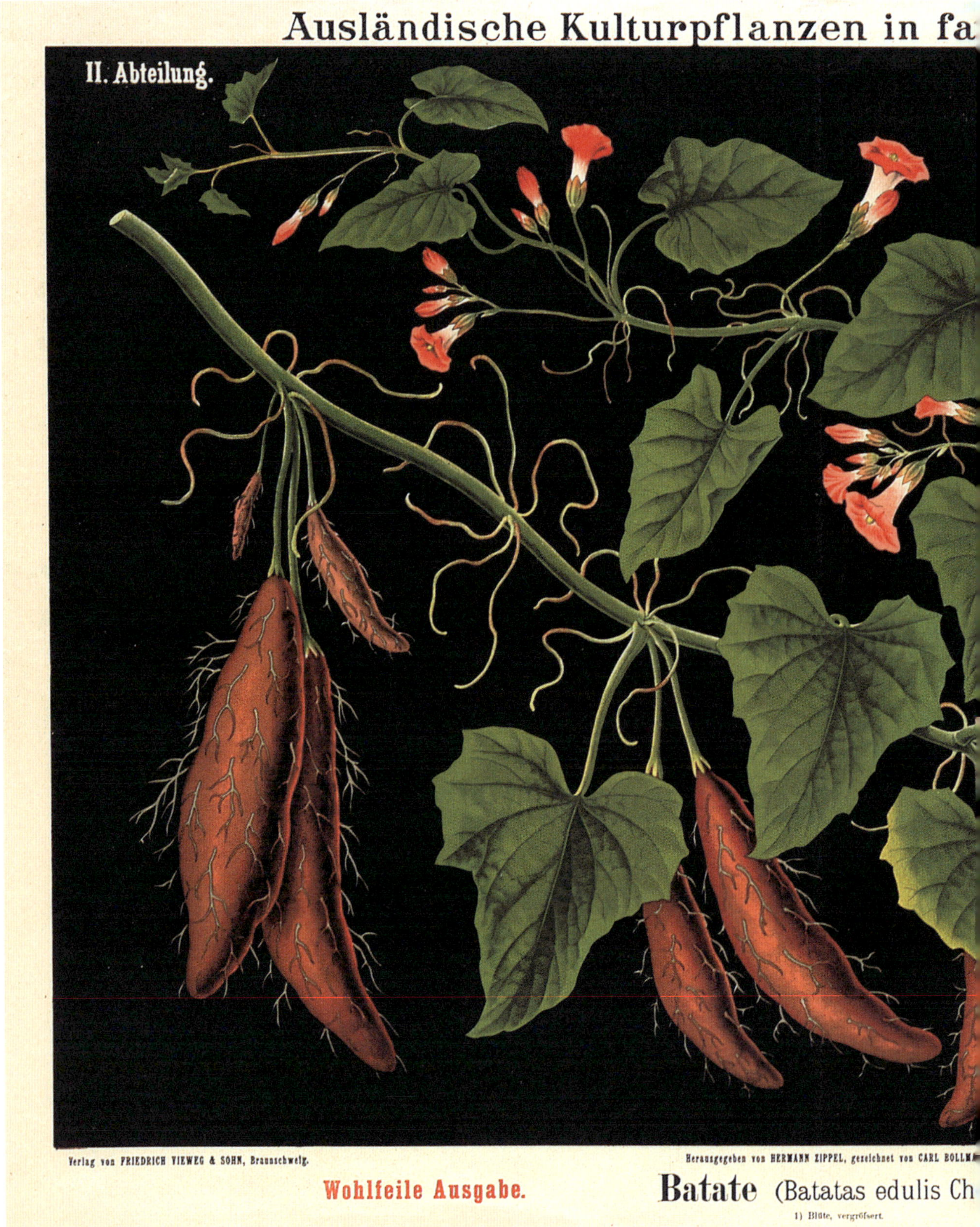
Ausländische Kulturpflanzen in fa
II. Abteilung.
Verlag von FRIEDRICH VIEWEG & SOHN, Braunschweig.
Herausgegeben von HERMANN ZIPPEL, gezeichnet von CARL BOLLM
Wohlfeile Ausgabe.
Batate (Batatas edulis Ch
1) Blüte, vergrößert.

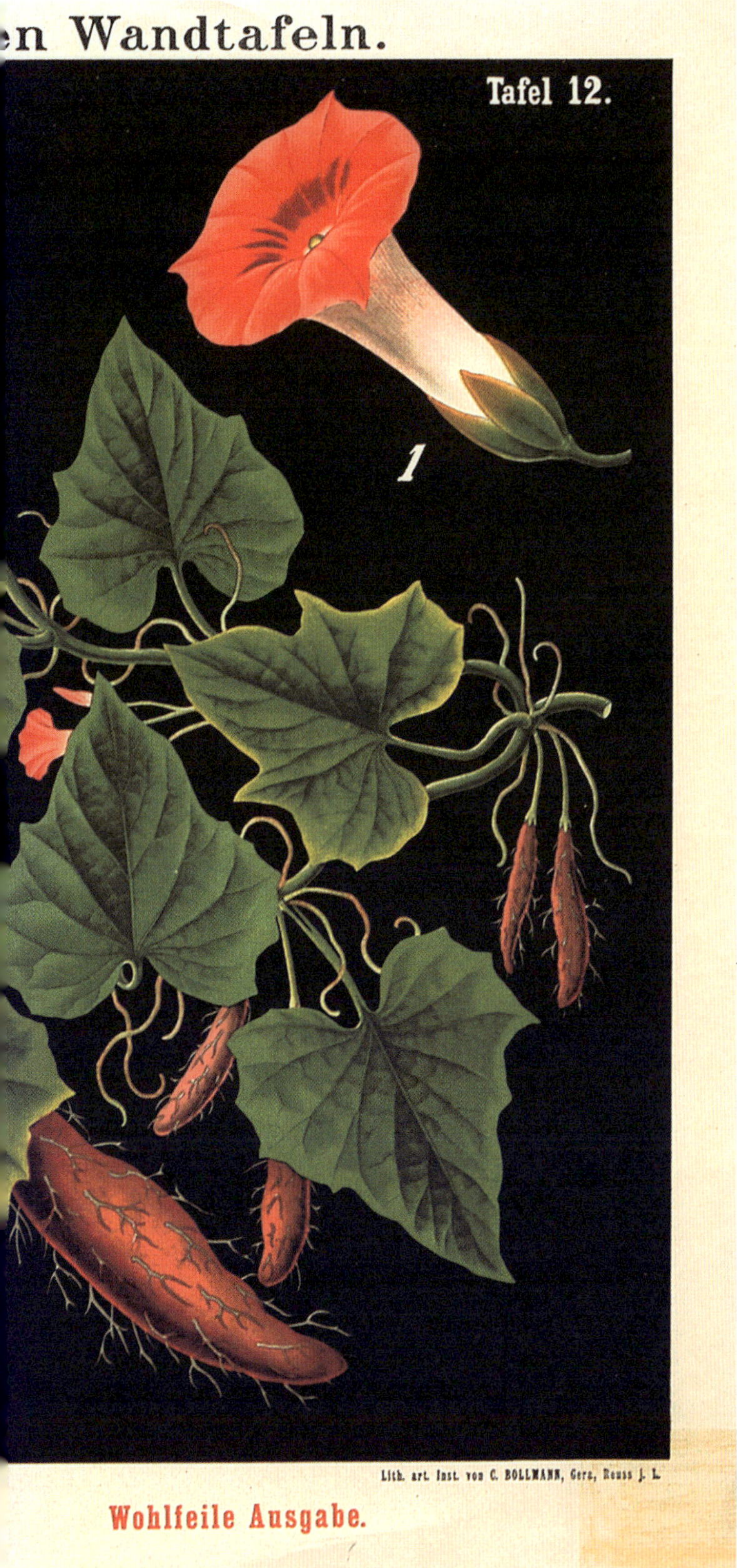

LEFT: A complement to the previous page – here Zippel and Bollmann illustrate a foreign crop, the sweet potato (*Ipomoea batatas*). The purpose of their *Ausländische kulturpflanzen in farbigen wandtafeln* (*Foreign Crops in Coloured Wall Panels*) was not to produce travel brochures of exotic species, but to illustrate the morphology of important non-native economic crops that were not cultivated within Germany's borders. Originally domesticated in Central or South America at least 5,000 years ago, *I. batatas* still hadn't found its way to Europe (and, in fact, production today is still very small on the European continent).

This chart is one of their most straightforward. Since the species rarely reproduces from seed, Zippel and Bollmann dispense with reproductive anatomy and familiarize the student with several above-ground cues – a sprawl of veined leaves and tubular flowers, characteristic of the *Convolvulaceae* family – that belie below-ground clusters of modified stems terminating in large, starchy, sweet tubers.

LEFT:
Title: *Batate (Batatas edulis Chois)*
Author: Hermann Zippel; Illustrator: Carl Bollmann
Language: German
Country: Germany
Series/Book: *Ausländische Kulturpflanzen in farbigen Wandtafeln (Foreign Crops in Coloured Wall Panels)*
Plate: II Abteilung, 12
Publisher: Friedrich Vieweg & Sohn (Braunschweig, Germany)
Year: 1879

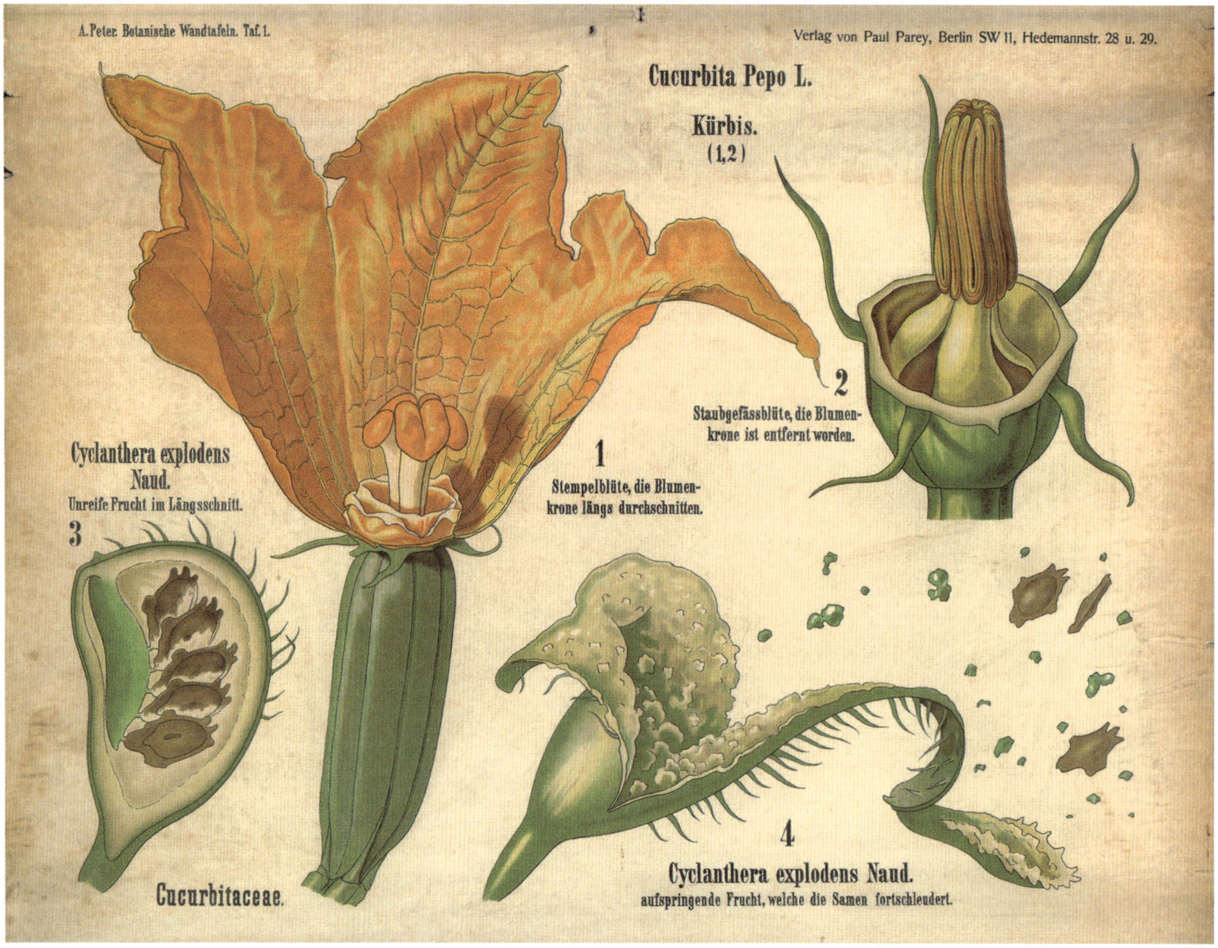
A. Peter. Botanische Wandtafeln. Taf. 1.
Verlag von Paul Parey, Berlin SW 11, Hedemannstr. 28 u. 29.
Cucurbita Pepo L.
Kürbis.
(1,2)
2
Staubgefässblüte, die Blumenkrone ist entfernt worden.
Cyclanthera explodens Naud.
Unreife Frucht im Längsschnitt.
1
Stempelblüte, die Blumenkrone längs durchschnitten.
3
4
Cyclanthera explodens Naud.
aufspringende Frucht, welche die Samen fortschleudert.
Cucurbitaceae.

VIII CUCURBITACEAE / GOURD FAMILY

The gourd family has roughly 100 genera and just under 1,000 species. They are not at all frost-hardy, so are found only in the tropics and warm,, temperate climates. Herbaceous annual vines that climb by means of coiled tendrils, they are cultivated widely for their large fruits. *Citrullus*, *Cucumis*, and *Cucurbita* give us the zucchini, cucumber, pickle, marrow, melon, pumpkin, squash and watermelon. Other gourds, such as *Lagenaria*, have been used as ornamental plants and fashioned into containers, utensils, and musical instruments since ancient times. The fruits of two *Luffa* species – *L. aegyptiaca* and *L. acutangula* – cleaned down to the xylem fibres provide natural sponges.

Characteristics: leaves often palmately lobed or compound; tendrils at 90 degrees to the nodes of leaf petioles, some modified to spines; five free or united sepals and petals; most often three fused stamens; flowers solitary or in axillary cymes, often white or yellow, with deeply lobed calyx and corolla; stems sometimes succulent. Fruit, sometimes very large, is a modified fleshy berry, or pepo, with a thick skin formed from the receptacle; the numerous seeds are often large and flat.

OPPOSITE:
Title: *Cucurbitaceae*
Author: A. Peter
Language: German
Country: Germany
Series/Book: *Botanische Wandtafeln*
Plate: 1
Publisher: Paul Parey (Berlin, Germany)
Year: 1901

Beilage zur Deutschen Landwirtschaftlichen Presse.

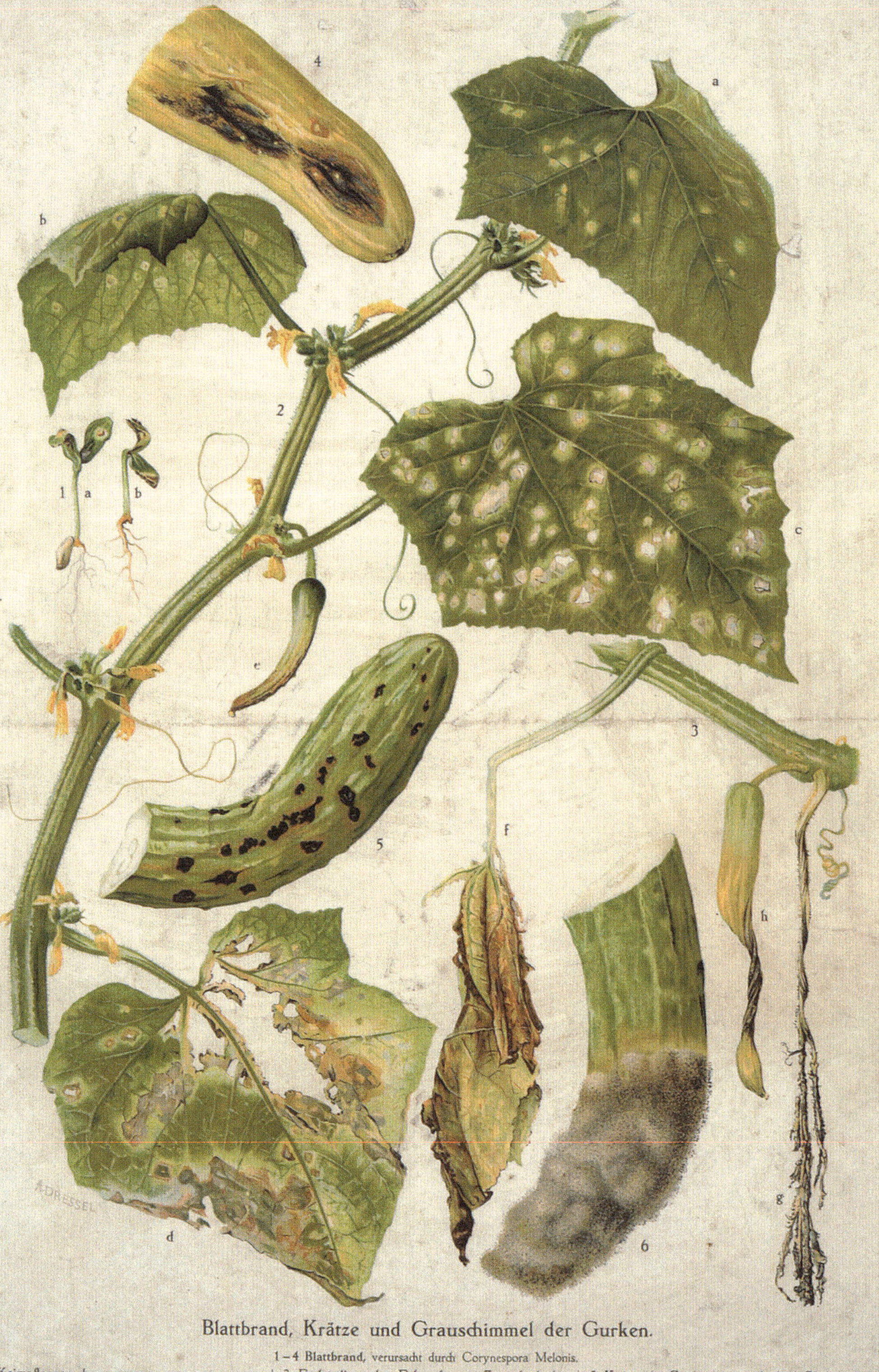

Blattbrand, Krätze und Grauschimmel der Gurken.

1–4 **Blattbrand,** verursacht durch Corynespora Melonis.

1. **Keimpflanzen,** hervorgegangen aus corynespora-kranken Samen,
2. **Gurkentrieb** mit verschiedenen Stadien des Blattbrandes (a–d) und einer corynesporafaulen Gurke (e),
3. **Endstadium der Erkrankung.** Zwei durch den Pilz zerstörte Blätter (f und g) und eine Frucht (h),
4. **Reife Gurke,** an zwei Stellen von Corynespora infiziert,
5. **Krätze der Gurken,** verursacht durch Cladosporium cucumerinum,
6. **Grauschimmel der Gurken,** verursacht durch Botrytis cinerea.

Königl. Universitätsdruckerei H. Stürtz A.G., Würzburg

Verlag von Paul Parey in Berlin SW., Hedemannstr. 10–11.

LEFT:
Title: *Blattbrand Krätze und Grauschimmel der Gurken (Leaf Burn, Scab and Grey Mould of Cucumber)*
Author: Otto Appel; Illustrator: August Dressel
Language: German
Country: Germany
Series/Book: *Atlas der Krankheiten der landwirtschaftlichen Kulturpflanzen (Atlas of Diseases of Agricultural Crops)*
Plate: N/A
Publisher: Paul Parey (Berlin, Germany)
Year: 1924

OPENING PAGE: Rather than represent the *Cucurbitaceae* family with a ubiquitous weed, an economically important crop, or a notably toxic vine, A. Peter gives us a sideshow wonder: *Cyclanthera explodens* (now *C. brachystachya*), a vigorous vining species that grows and disperses with equal enthusiasm. Aptly named the 'exploding cucumber,' this Mediterranean native produces orange-yellow flowers, followed by small green fruits that have evolved to disperse their seeds ballistically. When the seed pod (covered with prickles to deter foragers from consuming its fleshy, edible fruit and seeds) swells to 2 inches (5cm) long, tension has accumulated and the fruit becomes turgid, finally exploding with remarkable force – an adapted strategy to counter a plant's fundamental dilemma: immobility.

OPPOSITE: The beauty of wall charts is that they are as scientific as they are exquisite. It might come as a surprise to see a diseased cucumber portrayed with the same artistry as a thriving daffodil (for instance), but wall charts were produced to instruct, not to entertain. All subjects received equal craftsmanship, whether a weed or an orchid, a parasite or a cultivar, and this series of illustrations titled *Atlas der Krankheiten der Landwirtschaftlichen Kulturpflanzen*, or *Atlas of Diseases of Agricultural Crops*, is a beautiful example of this principle. German botanist and agriculturalist Otto Appel, an expert in potato diseases, extended his knowledge to other crops and produced this series when appointed director of the recently minted Biological Reich Institute for Agriculture and Forestry. German artist August Dressel drew the boards. Unlike many chart illustrators, Dressel was not a scientist, rather a landscape painter, which influenced his compositions and detail. Instead of following a gridded layout, Dressel's cucumbers sprawl across the chart like a kinetic sculpture, engaging the eye in a zigzag across the fruits and leaves with their various afflictions.

Corynespora, depicted opposite, is a fungal disease that afflicts the *Cucurbitaceae* family, especially cucumbers and cantaloupes, and manifests on stems, roots, flowers and fruits. Diseased seeds will produce diseased seedlings (fig. 1); a stricken plant (fig. 2) will break at the stems (fig. 3), develop sunken necrotic lesions, shrivelled young fruits and withered leaves (fig. 3: g, h, i). The poor plants here are also afflicted with various stages of leaf burn (fig. 2: a, b, c); *cladosporium*, a parasitic fungus that leaches nutrients from the fruit's exterior (fig. 5); and the dreaded grey mould, or *Botrytis cinerea* (fig. 6), a necrotrophic fungus that will envelop almost all parts of the plant in a powdery grey decay, sparing only the roots.

RIGHT: Though the first written record of the plant dates to around 1528, it is believed that the cucumber was widespread in Russia from the middle of the seventh century. In 1629 the English herbalist John Parkinson reported that 'in many countries they use to eate coccumbers as wee doe apples or Peares,' and they are similarly eaten and relished in contemporary Russia and Japan. Pickled and marinated cucumbers have been an essential part of traditional Russian cuisine for many centuries.

Starting from the left-hand side, this chart illustrates a cucumber vine (*Cucumis sativus*) in varying stages of growth, beginning with new flowers at the base, to pollinated flowers and emerging fruits at the tip. Surrounding the full plant are unisexual flowers, both male (figs. 2, 3) and female (figs. 4, 5), a mature fruit (fig. 6) and its cross-section (fig. 7).

Next, a herbaceous, perennial vine, the English mandrake (*Bryonia alba*) is native to central and eastern Europe, the Balkans, Turkey and Iran, and sprawls into central Asia. Perplexingly situated between the benign and edible cucumber and watermelon, this species is both aggressive and extremely toxic, smothering other plants and poisoning anyone who unwittingly plucks its onyx berry.

Finally the *Citrullus edulis* (now *C. lanatus*)– a watermelon. To this day watermelons are prized in Russia for their detoxifying properties, and their ubiquity may be compared to that of apples in the West: the Russian word for watermelon – 'arbuz' – corresponds to the first letter of the Russian alphabet.

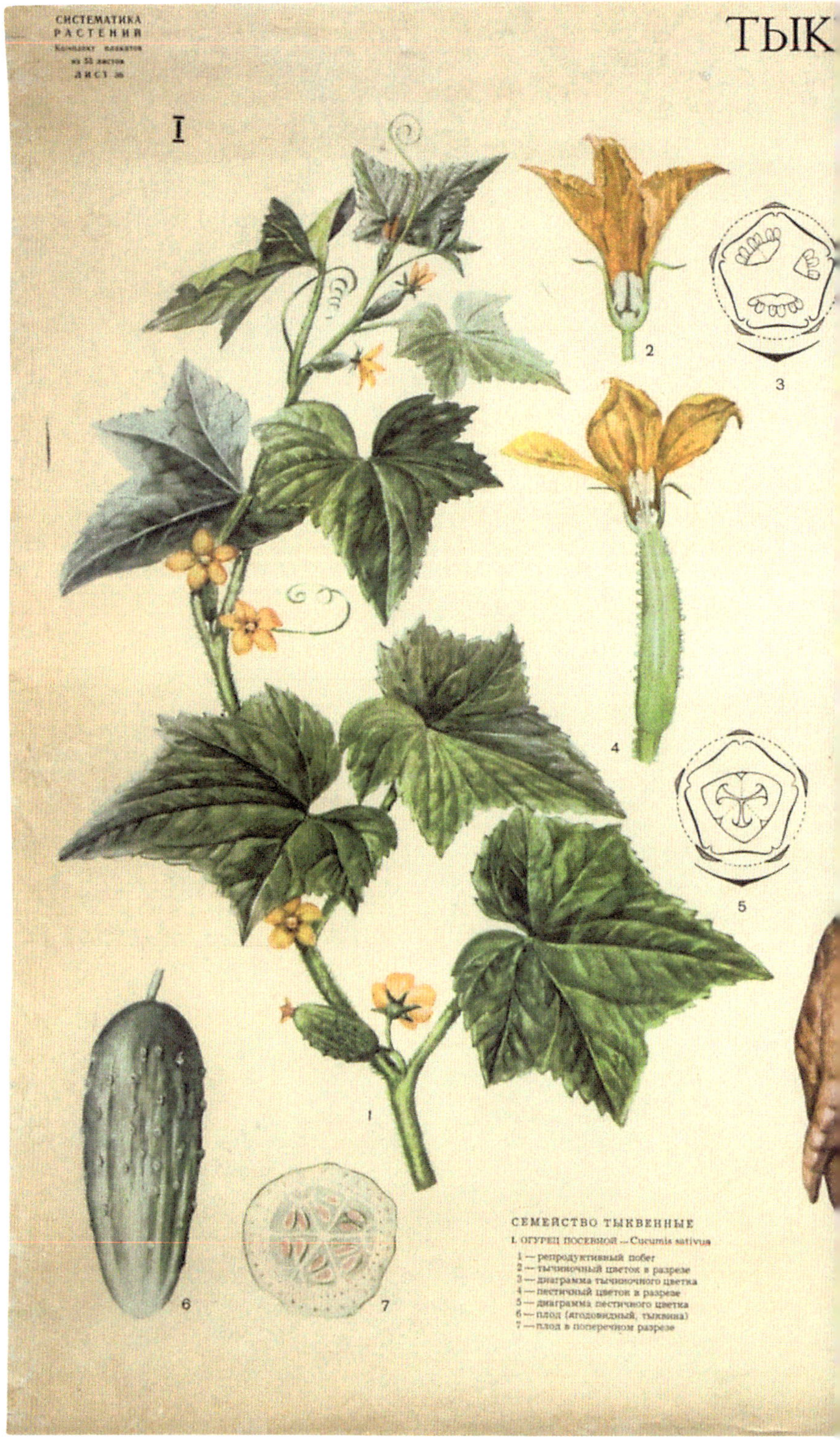

RIGHT:
Title: *Cucurbitaceae*
Author: V. G. Chrzhanovskii
Language: Russian
Country: Russia
Series/Book: *Sistematika Rasteniy Komplekt Plakatov iz 53 Listov (Systematics of Plants – Posters from 53 Sheets)*
Plate: 36
Publisher: Kolos Publishing
Year: 1971

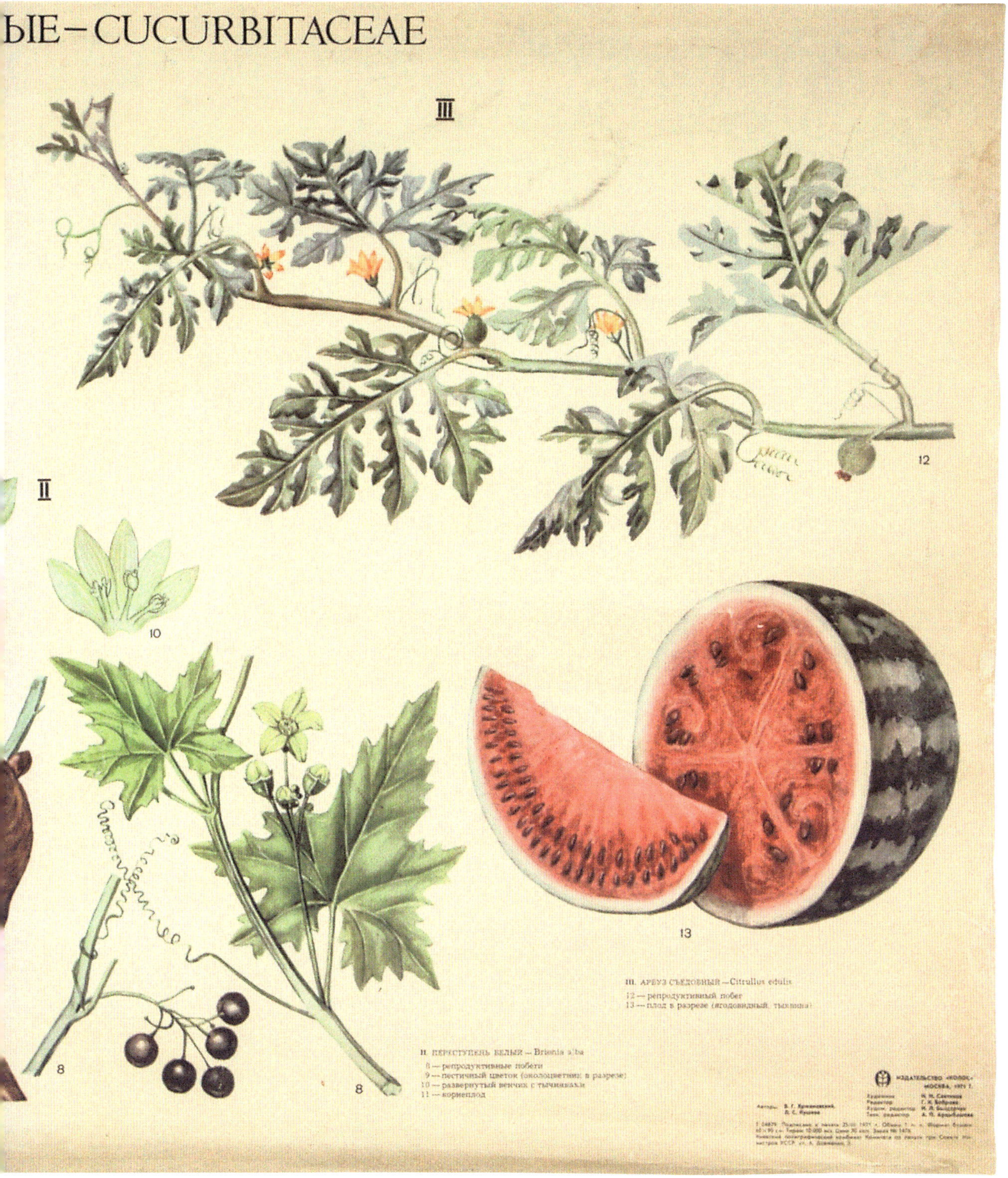
ЫЕ – CUCURBITACEAE
III
12
II
10
13
8
8
III. АРБУЗ СЪЕДОБНЫЙ – Citrullus edulis
12 — репродуктивный побег
13 — плод в разрезе (ягодовидный, тыквина)
II. ПЕРЕСТУПЕНЬ БЕЛЫЙ – Brionia alba
8 — репродуктивные побеги
9 — пестичный цветок (околоцветник в разрезе)
10 — развернутый венчик с тычинками
11 — корнеплод

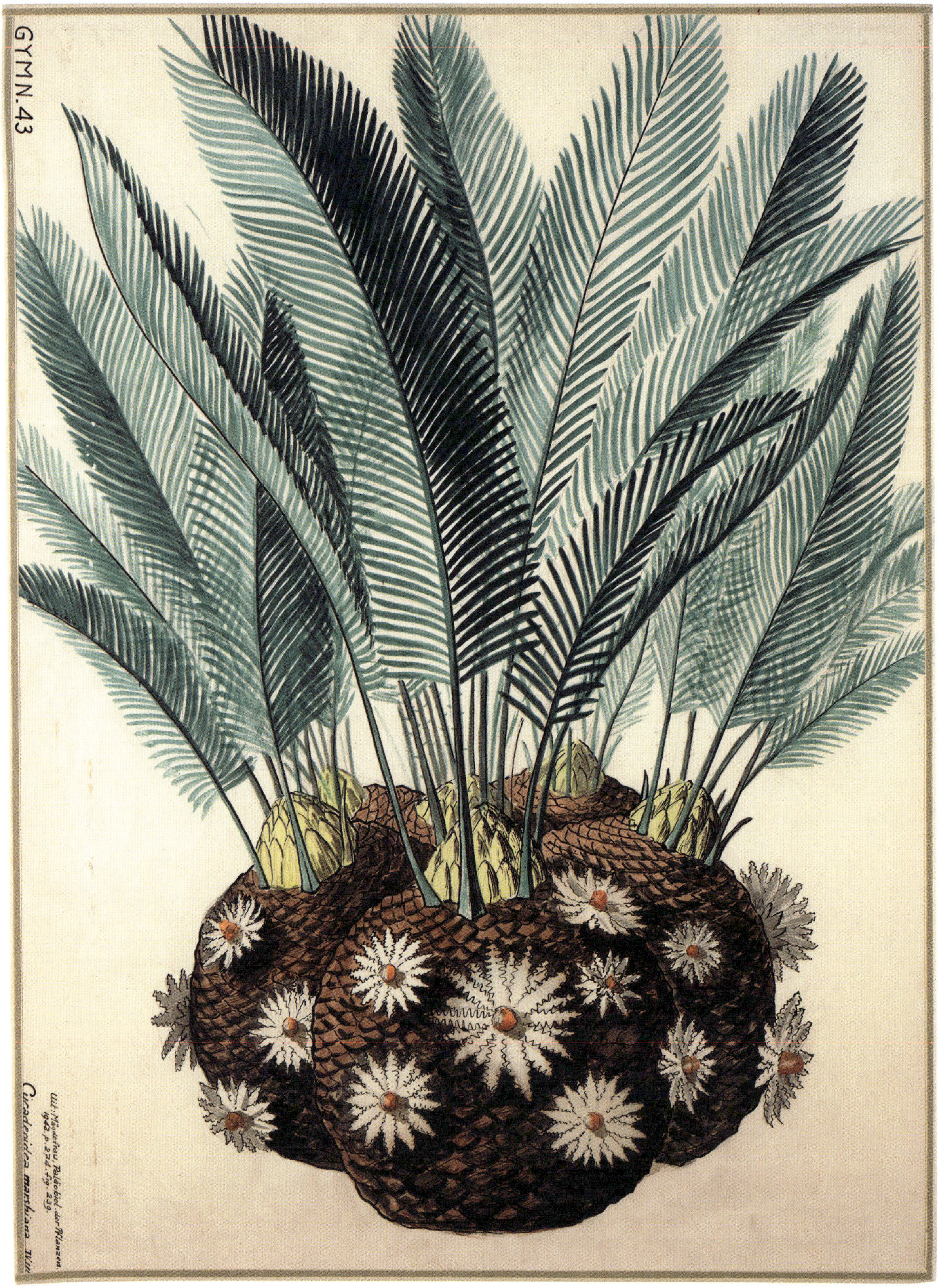
GYMN. 43
Lit: Mägdefrau, Paläobiol. der Pflanzen.
1942. p. 274. fig. 239.
Cicadeoidea marshiana Wil

IX CYCADACEAE / CYCAD FAMILY

There has been much debate about the taxonomy of the *Cycadales* order and until fairly recently, all living examples of cycads were regarded as part of the *Cycadaceae* family. In recent years, however, the *Cycadaceae* has become a monotypic family – that is, it consists of only one genus, *Cycas*, of which there are just under 120 species. All other cycads now have their own family: *Zamiaceae*. Both families are gymnosperms and among the most ancient of plants, yet are all now endangered in their native tropical and subtropical habitats. Fortunately, they are valued in horticulture for their attractive, palm-like appearance.

Cycads all contain toxins, although the seeds and pithy stems have been used for food, fodder, alcohol and medicine, as with *Cycas revoluta* (king sago), which is grown for starch in the Ryukyu Islands of Japan. Indeed, there have been cases of illness and death as a result of careless processing of cycad material.

Characteristics: Cycad species have a straight trunk, either smooth or clad with cataphylls (scale-like, reduced leaves), crowned with a rosette of circinate leaves; leaves are flat, pinnate, each on a rachis; leaves and cataphylls are spirally arranged. Male cones consist of tightly overlapping sporophylls (modified leaves), female cones of loose, pendent sporophylls with up to seven pairs of seeds on the margins; large seeds may have sarcotesta (fleshy coat) to aid buoyancy and dispersal by water.

OPPOSITE:
Title: *Cycadeoidea marshiana (fossielen)*
Author: J. Vuijk
Language: German
Country: Germany
Series/Book: *Paläobiol. der Pflanzen* (Karl Mägdefrau, *Paleobiology of Plants*), 1942 (reprint)
Plate: 43
Publisher: Hagemann (Düsseldorf, Germany)
Year: 1905

OPENING PAGE: *Bennettitales*, or *cycadeoids*, is an extinct order of seed plants that flourished during the early Cretaceous period. Fossil evidence has allowed paleobotanists to triangulate a hazy morphology of the species; however, the task of pencilling between these data points is always left to the botanic illustrator. Here, J. Vuijk has drafted a wonderful profile of *Cycadeoidea marshiana*, a *Bennettitales* that may or may not have been carpeted with daisies (very few renderings of *C. marshiana* exist; at least one is a replication of Vuijk's). Were the ancient plumed cone not extraordinary enough as an individual, Vuijk has illustrated a group of six.

OPPOSITE: Throughout the nineteenth century, the Carboniferous Period was often referred to as the 'Age of Ferns' for the great number of ferns believed to have flourished then. When a descendant of *Calymmatotecha heoninghausii* (now known as *Lyginopteris oldhamia*) was discovered, however, botanists realized that they had misread the fossil record. Extinct species previously accepted as ferns (which reproduce via spores) were actually seed-bearing plants with naked ovules or seeds – or 'seed ferns' or 'fern-cycads.' And so was born a new group of extinct gymnosperms known as *Pteridospermae* or *Cycadofilicales*, launching a zealous period of paleobotanical research and a spate of new genera.

Here, a profile of *Calymmatotecha heoninghausii* (*Lyginopteris oldhamia*, top right) includes a feathery stalk of ovules and leaves; a section of microsporophylls, where tiny epaulets produced pollen; an ovule; and a cross-section of the ovule.

An extinct order of seed ferns, *Medullosales* were characterized by large ovules (0.4–4 inches, or 1–10cm, in diameter), complex pollen-organs and frond leaves; their nearest extant relatives are the cycads. The left-hand section of the chart represents the group with three species; the main image represents *Medullosa noei* as a plucky tree bejewelled with radish-like pollen capsules and six fronds. The other two species, *M. stellata* and *M. solmsu*, are both portrayed by a cross-section of the stem, where black and cross-hatch represent primary and secondary wood, respectively, and vascular bundles wander.

It's possible that angiosperms evolved from seed ferns (pteridosperms), either *Lyginopteridales* (the earliest-classified order of *Pteridosperm*) or *Caytoniales* – a group that breezes through the fossil record from the late Triassic to Cretaceous Period. When first discovered, the *Caytoniales* were seen as a likely ancestral candidate due to their remarkable reproductive structure: a seed-bearing rachis of curled cupules, within which several ovules nestled. This seemed analogous to an angiosperm's seed-protecting pistil, until paleobotanists discovered that they had incorrectly identified a pollen tube within the cupule, and a fleshy quality within the curls. Perhaps the caytonia were pollinated ovaries that had matured into fruits (carpels) with many seeds. Perhaps. For now, however, the case remains open.

In the third section of the chart, a *Caytonia nathorstii* is represented by a fossil imprint of a palmately compound leaf; a stalk of tight curled cupules; an enlarged cupule with four teardrop ovules; two bits of pollen; a stalk of pollen-producing microsporangia.

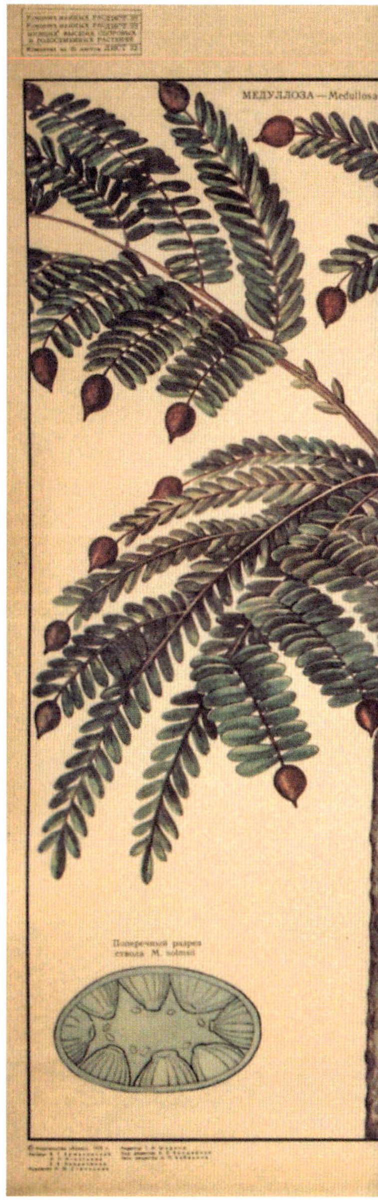

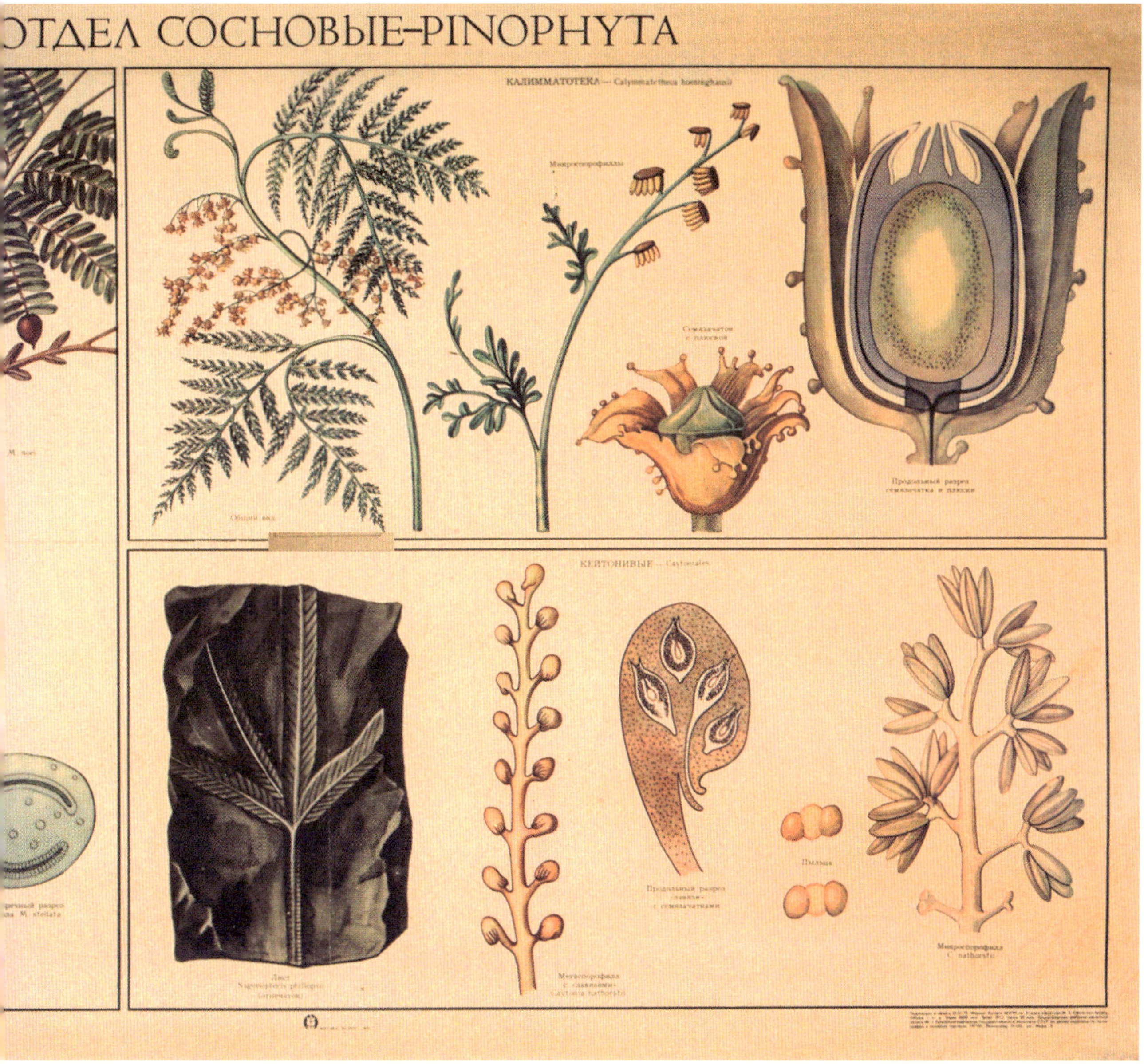

ABOVE:

Title: *Pinophyta*

Author: V. G. Chrzhanovskii

Language: Russian

Country: Russia

Series/Book: *Morfologia i sistematika nizšich i vyssich sporovych i golosemenych rastenij* (*Morphology and Taxonomy of Gymnosperms and Angiosperms*)

Plate: 22

Publisher: Kolos Publishing

Year: 1979

OPPOSITE: Zippel and Bollmann illustrate queen sago, endemic to India and a decidedly foreign species to late nineteenth-century Germany. Like other cycads, *Cycas circinalis* resembles a fern atop a stout palm-tree trunk. Cycads are dioecious; the female yields seeds and the male produces cones of pollen. The chart features a breezy female plant (fig. 1); a young male flower (fig. 2); a single stamen (fig. 3); a group of closed pollen sacs (fig. 4); a group of opened pollen sacs (fig. 5); a female flower (fig. 6); one carpel with six ovules (fig. 7); one carpel with eight seeds (fig. 8); a cross-section of a fruit, revealing the outer flesh and woody seed (fig. 9); and a cross-section of a seed, revealing endosperm and seedling. A slow-growing plant, the seed can take between six and eighteen months to germinate.

This cycad is named for a starchy pith, called sago. The chart's companion text provides an anthropological account of a sago harvest on Seram island, Indonesia. A mature tree is cut, its leaves removed, and a lateral cut is made in the bark. The exposed pith is then softened with a club, starch is separated from fibre, and reddish cakes of sago are shaped. However, the edibility of *C. circinalis* has since been discouraged; many parts of the plant (including the pith) contain toxins that sometimes remain even after proper washing or cooking.

OPPOSITE:

Title: *Eingerollte Farnpalme (Cycas circinalis Linné)*
Author: Hermann Zippel; Illustrator: Carl Bollmann
Language: German
Country: Germany
Series/Book: *Ausländische Kulturpflanzen in farbigen Wandtafeln (Foreign Crops in Coloured Wall Panels)*
Plate: II Abteilung, 1
Publisher: Friedrich Vieweg & Sohn (Braunschweig, Germany)
Year: 1897

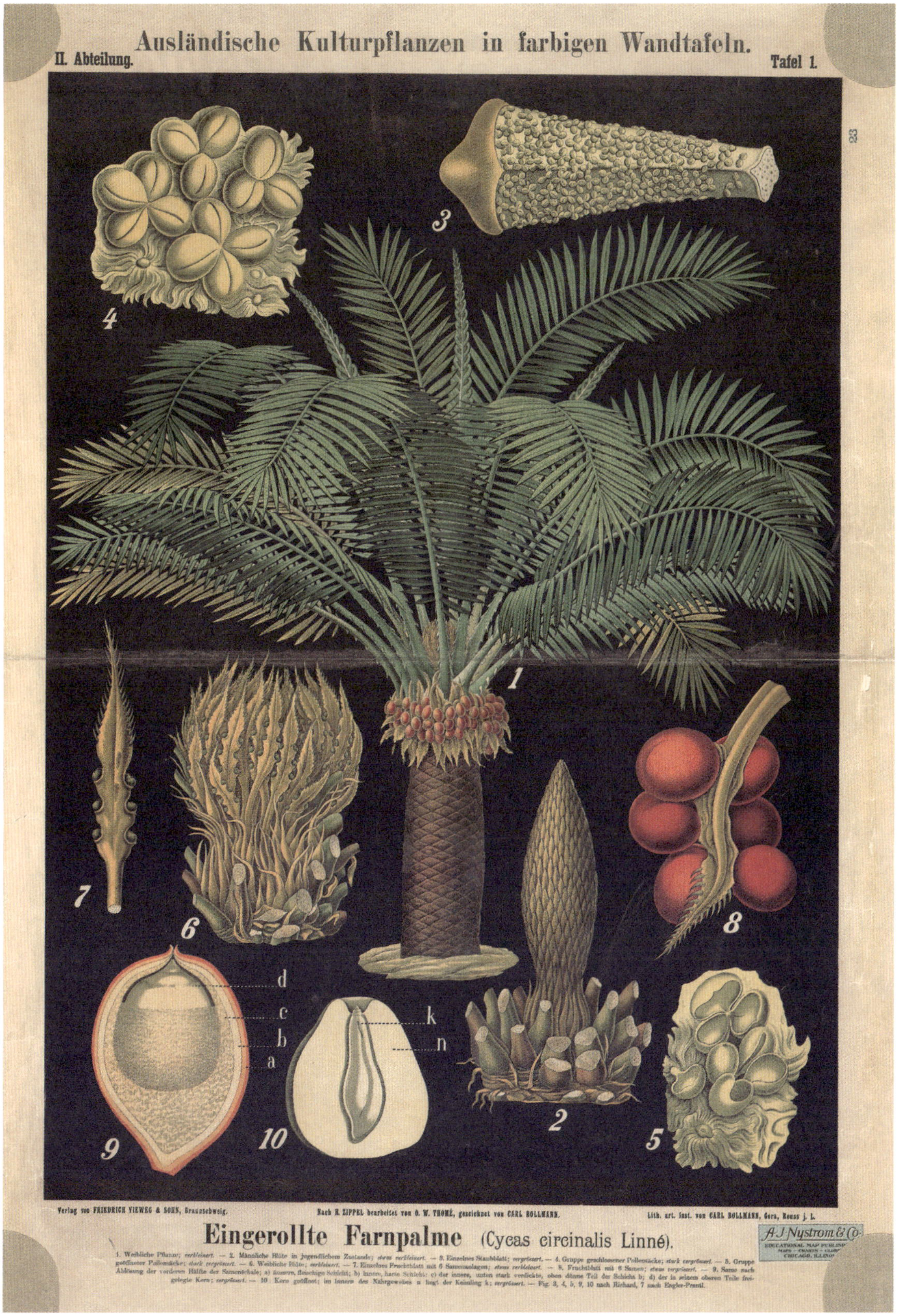
Ausländische Kulturpflanzen in farbigen Wandtafeln.
II. Abteilung.
Tafel 1.
23
1
2
3
4
5
6
7
8
9
10
d
c
b
a
k
n
Verlag von FRIEDRICH VIEWEG & SOHN, Braunschweig.
Nach R. ZIPPEL bearbeitet von O. W. THOMÉ, gezeichnet von CARL BOLLMANN.
Lith. art. Inst. von CARL BOLLMANN, Gera, Reuss j. L.
Eingerollte Farnpalme (Cycas circinalis Linné).
1. Weibliche Pflanze; verkleinert. — 2. Männliche Blüte in jugendlichem Zustande; etwas verkleinert. — 3. Einzelnes Staubblatt; vergrössert. — 4. Gruppe geschlossener Pollensäcke; stark vergrössert. — 5. Gruppe geöffneter Pollensäcke; stark vergrössert. — 6. Weibliche Blüte; verkleinert. — 7. Einzelnes Fruchtblatt mit 6 Samenanlagen; etwas verkleinert. — 8. Fruchtblatt mit 6 Samen; etwas vergrössert. — 9. Same nach Ablösung der vorderen Hälfte der Samenschale; a) äussere, fleischige Schicht; b) innere, harte Schicht; c) der innere, unten stark verdickte, oben dünne Teil der Schicht b; d) der in seinem oberen Teile freigelegte Kern; vergrössert. — 10. Kern geöffnet; im Innern des Nährgewebes n liegt der Keimling k; vergrössert. — Fig. 3, 4, 5, 9, 10 nach Richard, 7 nach Engler-Prantl.
A. J. Nystrom & Co.
EDUCATIONAL MAP PUBLISH
CHICAGO, ILLINO

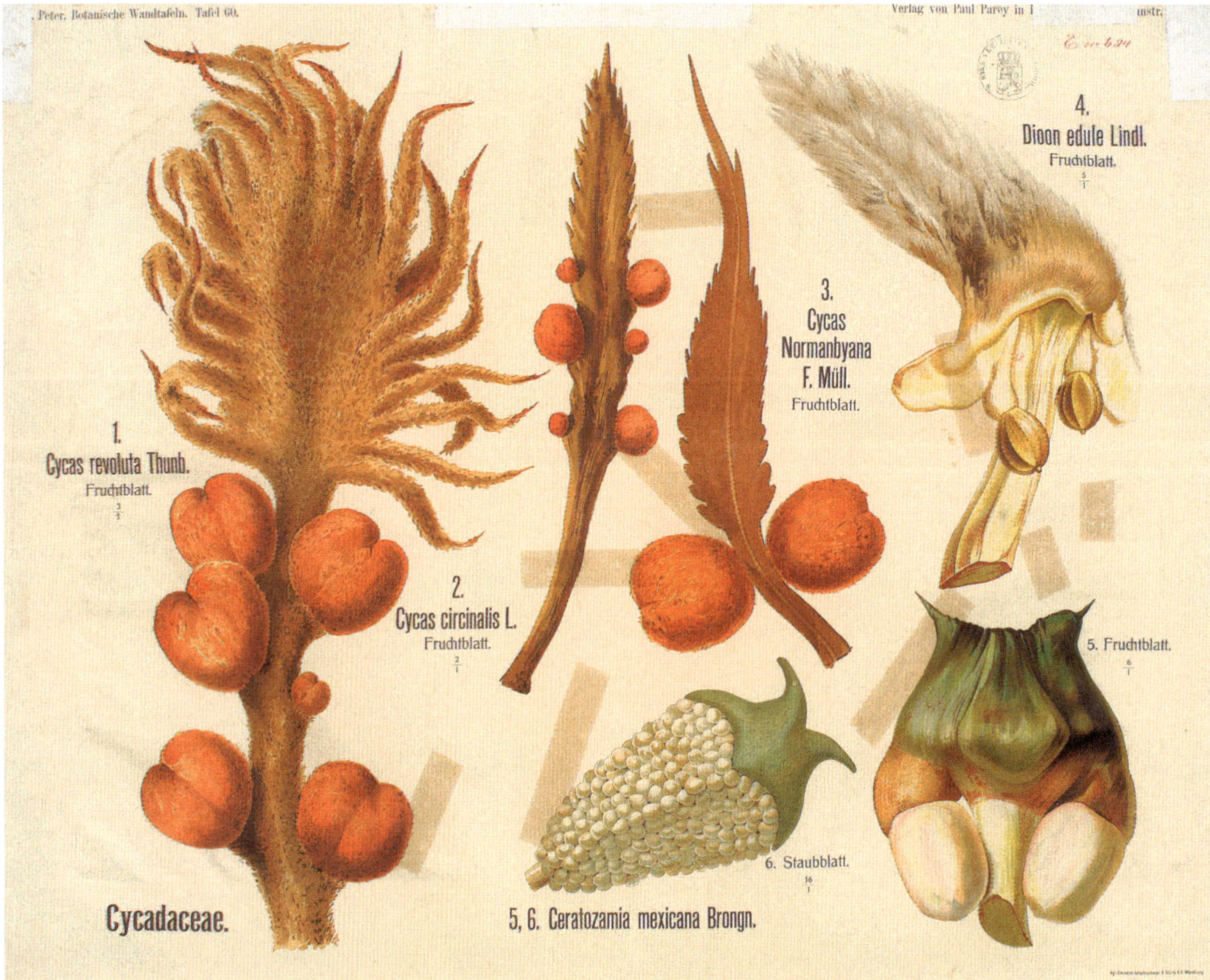
Peter, Botanische Wandtafeln. Tafel 60.
1. Cycas revoluta Thunb.
Fruchtblatt.
2. Cycas circinalis L.
Fruchtblatt.
3. Cycas Normanbyana F. Müll.
Fruchtblatt.
4. Dioon edule Lindl.
Fruchtblatt.
5. Fruchtblatt.
6. Staubblatt.
5, 6. Ceratozamia mexicana Brongn.
Cycadaceae.

OPPOSITE: A. Peter recognized a confusing prevalence of red fruits among cycads; for easier identification, one can look instead to the carpels on which they're borne. First, *Cycas revoluta* and *C. circinalis*. Otherwise known (respectively) as the king and queen sago, the two species are morphologically pretty similar. *Cycas revoluta* can be distinguished either by a tendency to branch from the trunk, or by the woolly carpels that undulate over its fruits. The queen's carpels are more modest and slender. *Cycas normanbyana* (fig. 3), now *Cycas media* subsp., has no royal common name, but, with its feathery crown and corky bark, would be welcome in the sago court. Peter therefore also represents this lesser-known cycad by a carpel – smaller, serrated, and smoother than those of the king and queen.

Dioon edule (chestnut cycad, fig. 4), bears female cones that sprout fibrous silver-haired tips as they mature. Peter represents *D. edule* with a cone scale and two fruits. Any Greek scholars will now understand the binomial name; '*dioon*' means 'two eggs.'

Finally, a *Ceratozamia mexicana* (Mexican horncone) represented by a male cone (fig. 6) and a female scale with two fruits and two horns (fig. 5). When encountered in aggregate, the specimen has the appearance of a long, spiny cucumber.

OPPOSITE:
Title: *Cycadaceae*
Author: A. Peter
Language: German
Country: Germany
Series/Book: *Botanische Wandtafeln*
Plate: 60
Publisher: Paul Parey (Berlin, Germany)
Year: 1901

X DROSERACEAE / SUNDEW FAMILY

A carnivorous family with three genera and around 261 species distributed primarily in tropical and temperate regions. Unlike most angiosperms, which derive a good amount of nutrients through a root system, the *Droseraceae* grow in environments with poor soil conditions, such as fens and sphagnum bogs, and have evolved mechanisms to capture and digest insects and other small prey. Two genera, *Aldrovanda* and *Dionaea*, are monotypic – they both include only one species. *Aldrovanda vesiculosa*, commonly named the waterwheel plant and the only aquatic family member, is a free-floating plant with tiny motion-triggered traps that ensnare prey, especially mosquito larvae. *Dionaea muscipula*, the iconic Venus flytrap, is endemic to a small region in the eastern United States. *Drosera* includes the 259 other species of annuals and perennials that comprise this family.

Characteristics: two distinct trapping mechanisms to entrap insects, and enzymes to absorb nutrients. *Dionaea* and *Aldrovanda* have modified leaves that form a motion-triggered trap; response to stimulus can be within 0.4 seconds and accomplished within one second. *Drosera* are characterized by modified leaves arranged in a basal rosette; prey is entrapped on the upper surfaces, coated with hairs secreting a mucilage or sticky substance resembling a drop of dew. Fruit is a capsule.

RIGHT:
Title: *Drosera rotundifolia/Rundblättriger Sonnentau*
Authors: Heinrich Jung, Dr. Friedrich Quentell
Illustrator: Dr. Gottlieb von Koch
Language: German
Country: Germany
Series/Book: *Neue Botanische Wandtafeln*
Plate: 36
Publisher: Fromann & Morian (Darmstadt, Germany)
Year: 1928

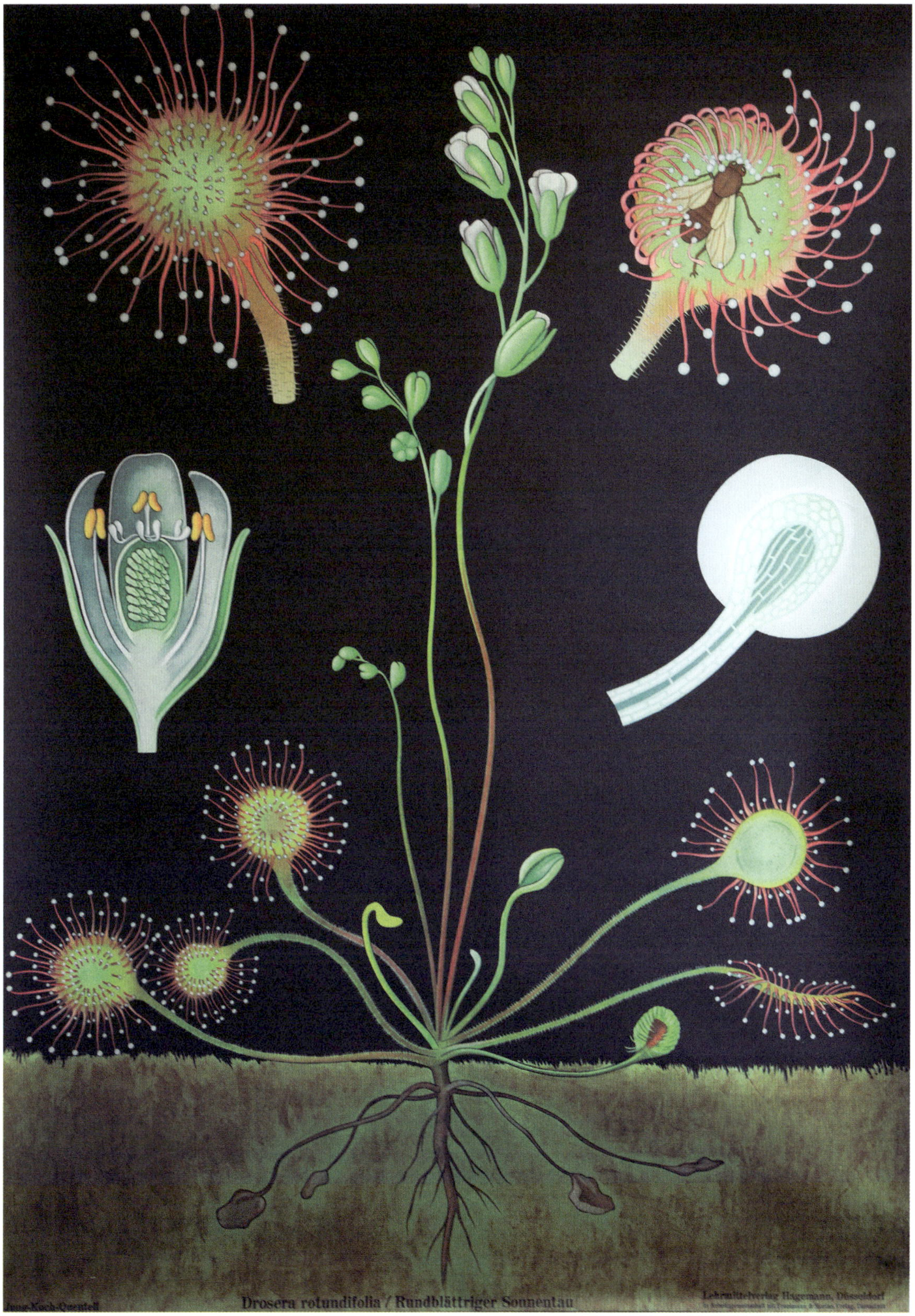

Jung-Koch-Quentell
Drosera rotundifolia / Rundblättriger Sonnentau
Lehrmittelverlag Hagemann, Düsseldorf

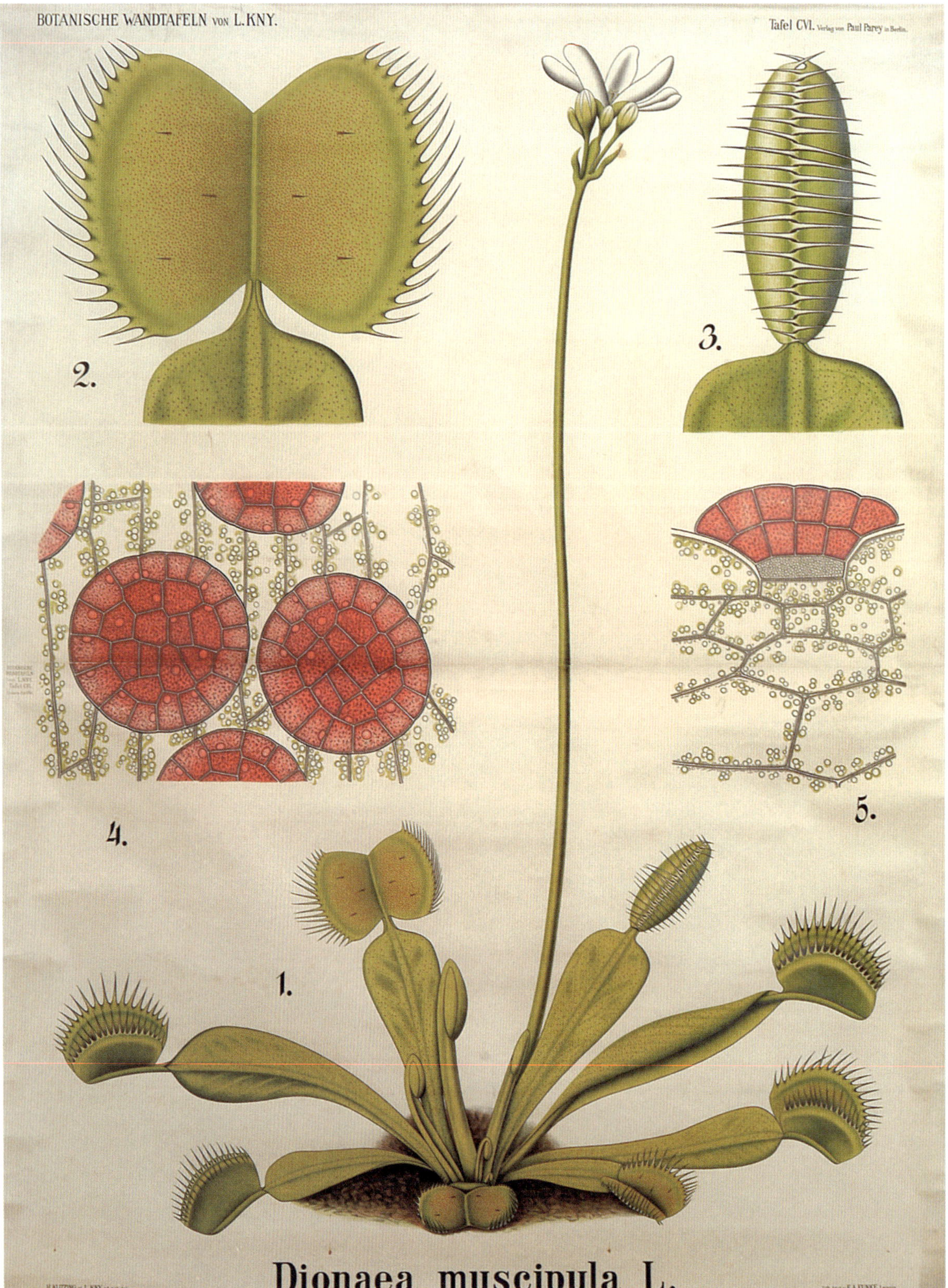
BOTANISCHE WANDTAFELN VON L. KNY.
Tafel CVI. Verlag von Paul Parey in Berlin.
2.
3.
4.
5.
1.
Dionaea muscipula L.

OPENING PAGE: An unusual chart from Jung, Koch, Quentell; the black background is iconic, the depiction of a landscape is rarer. While many other charts by Jung, Koch, Quentell illustrate a root system, this is perhaps the only one to include a ground covering. *Drosera rotundifolia* is portrayed in a genera-appropriate habitat (bog or moss), thus indicating the significance of the soil conditions within which the species has evolved. Within an otherwise black landscape, the plant becomes an extra-terrestrial oddity. Floating around the budding tentacles are enlarged trichomes, a flower cross-section, and a triggered trap; a short root system grounds the curious plant.

LEFT: In the late nineteenth century, the number of known carnivorous species rose sharply with the exploration of the new continents. Botanists marvelled at these curious species, impressively adapted to nutrient-starved environments with behaviours that challenged traditional definitions of a plant. In a 1874 letter to J. D. Hooker, Charles Darwin wrote, 'I do not think any discovery gave me more pleasure than proving a true act of digestion in *Drosera*.' Once their characteristics were known, *Droseraceae* similarly became a popular subject for wall charts. Perhaps more than any other family, it especially makes sense that illustrators would represent the genera in very different styles, as they had yet to adopt a common image.

Here, Leopold Kny portrays *Dionaea muscipula* with a particular emphasis on aesthetic over precision. A Venus flytrap emerges from the suggestion of a landscape (indicating the significance of environment in the evolution of a carnivorous species), with a basal rosette of modified leaves and a flower stalk. The prey is absent, but nonetheless Kny illustrates various stages of entrapment including agape and waiting (enlarged in fig. 2) and snapped tight (fig. 3).

OPPOSITE:
Title: *Dionaea muscipula L.*
Author: L. Kny
Language: German
Country: Germany
Series/Book: *Botanische Wandtafeln*
Plate: 106
Publisher: Paul Parey (Berlin, Germany)
Year: 1874

RIGHT: Unlike other illustrators, A. Peter chooses not to root *Dionaea muscipula* and *Drosera rotundifolia* (common sundew) in the earth. Neither nutrient-deficient soil nor basal rosette are included; instead, Peter rotates the axis and severs the stalks to capture the movement. At the right, a carnivorous head thrusts into frame; many rapacious trichomes (gland-tipped plant hairs) demonstrate *D. rotundifolia*'s particular trapping mechanism, a new capture thus secured. Peter mirrors this action in the lower-left-hand corner, where *Dionaea muscipula* has secured its intake. As always, Peter lists the magnification degree below each illustration.

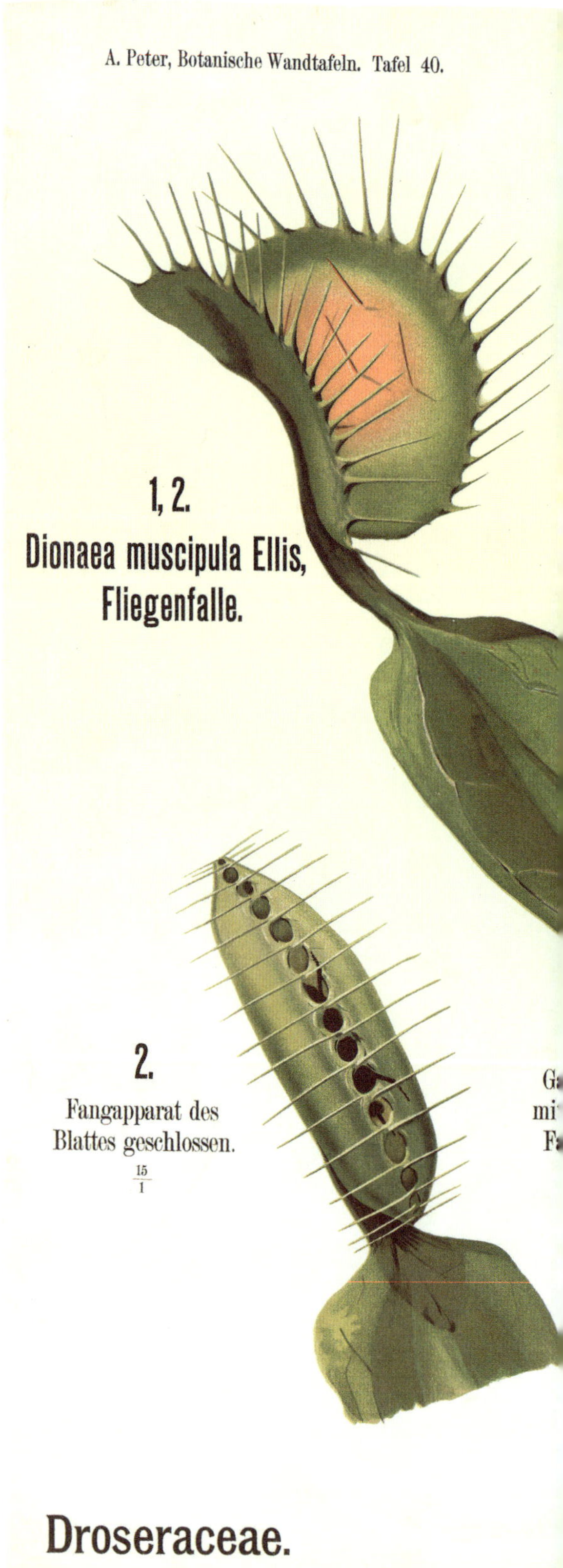

RIGHT:
Title: *Droseraceae*
Author: A. Peter
Language: German
Country: Germany
Series/Book: *Botanische Wandtafeln*
Plate: 40
Publisher: Paul Parey (Berlin, Germany)
Year: 1901

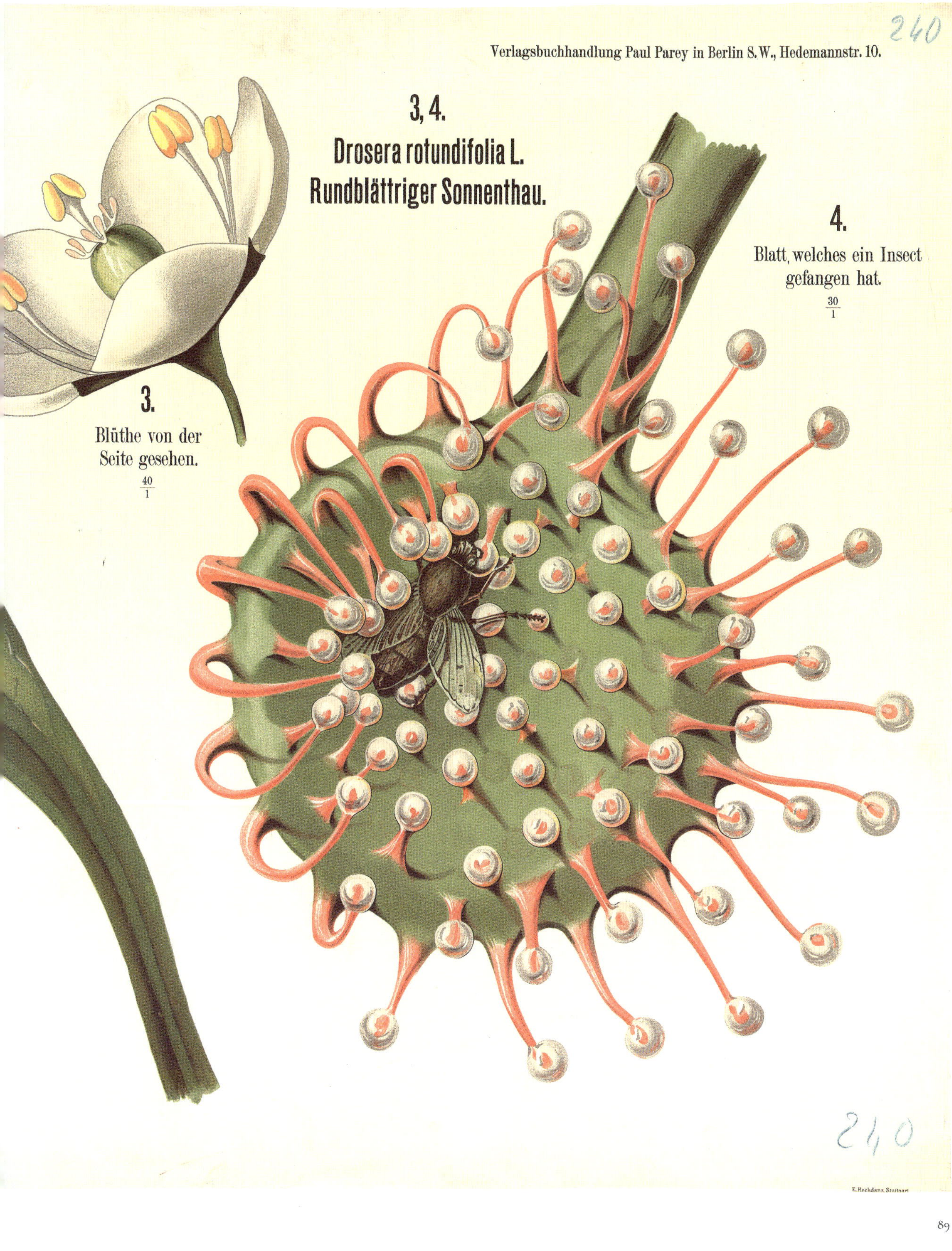
Verlagsbuchhandlung Paul Parey in Berlin S.W., Hedemannstr. 10.
3, 4.
Drosera rotundifolia L.
Rundblättriger Sonnenthau.
4.
Blatt, welches ein Insect gefangen hat.
30/1
3.
Blüthe von der Seite gesehen.
40/1
E. Hochdanz, Stuttgart

Schmeils Botanische Wandtafeln №12: Sonnentau

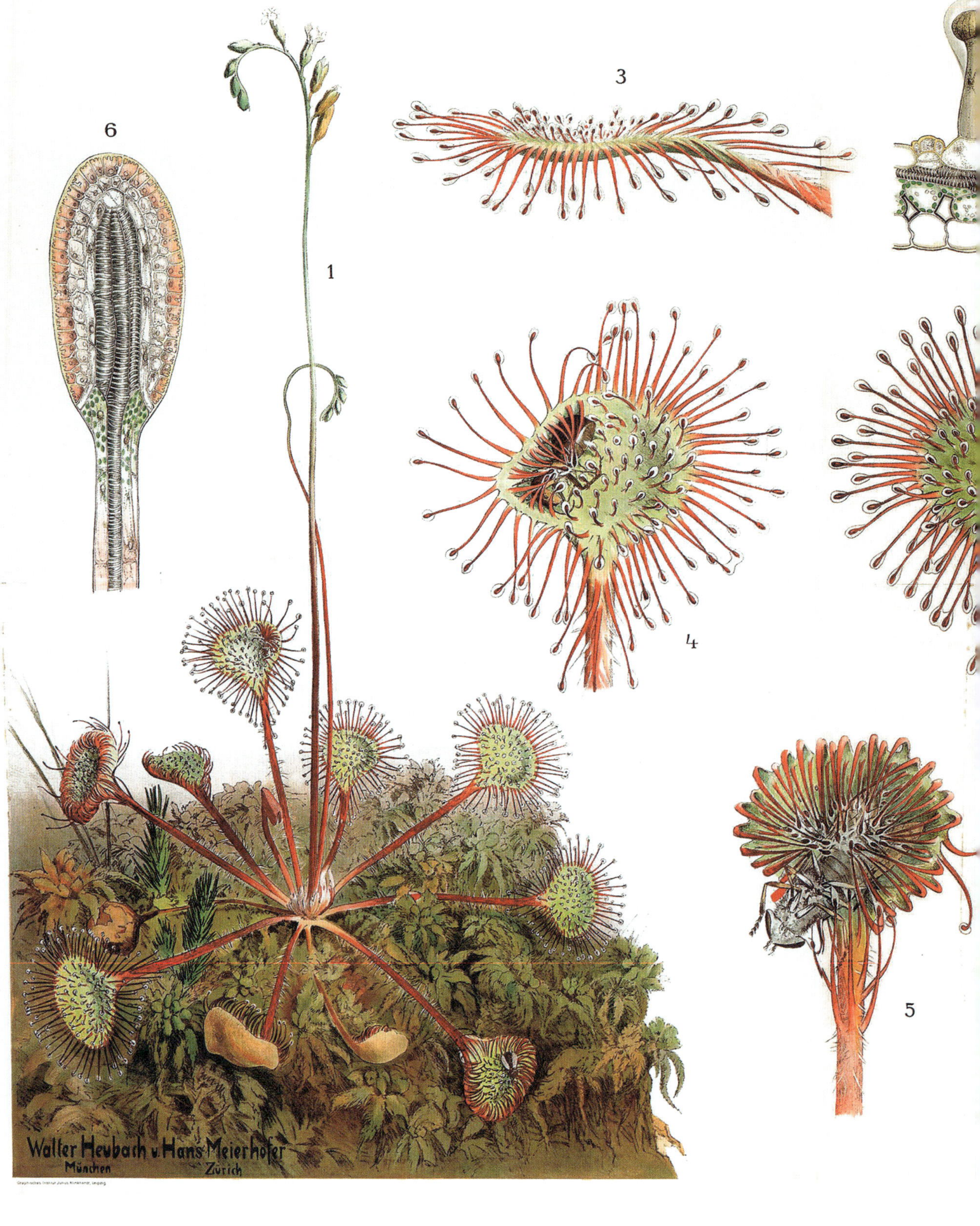

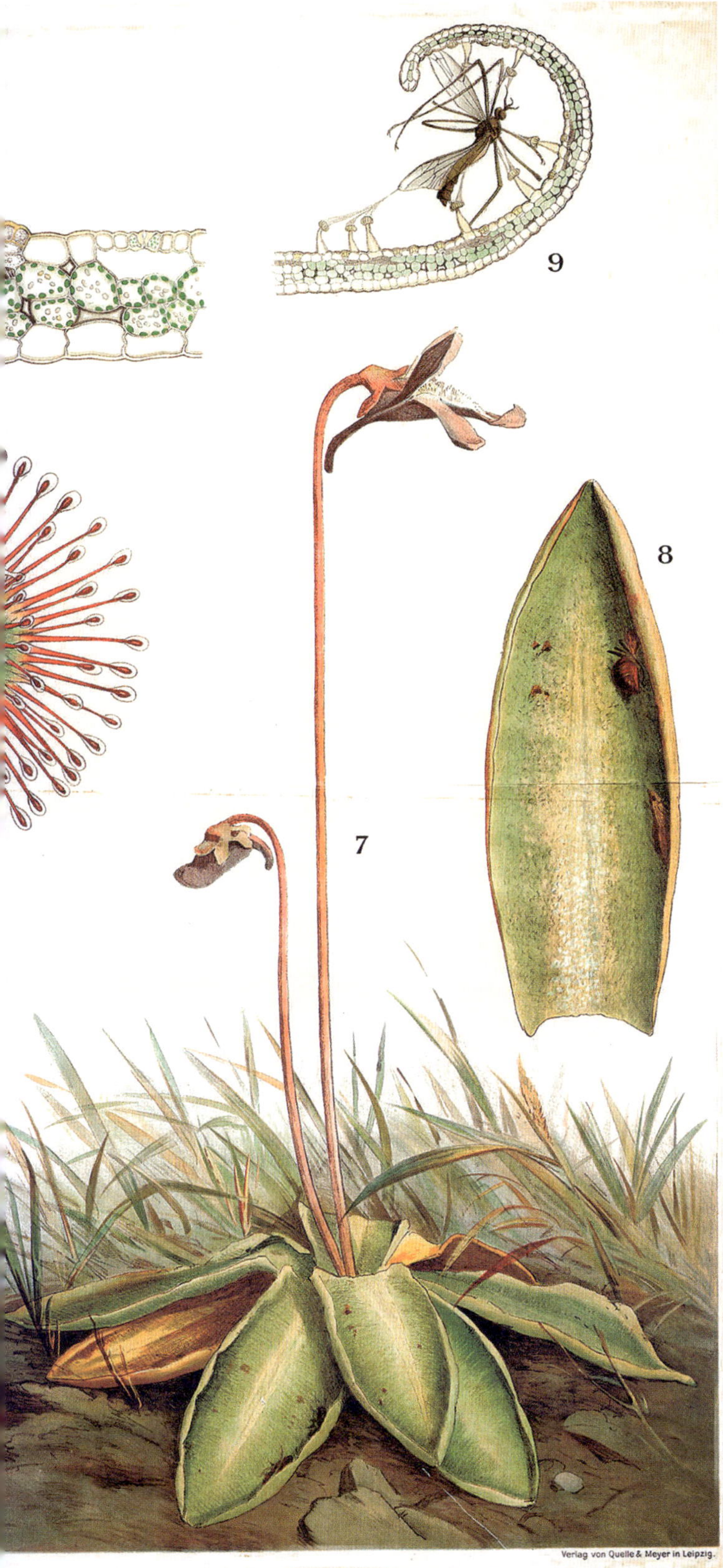

LEFT: In contrast to A. Peter's approach, as both a scientist and an illustrator, Otto Schmeil believed that a plant and its ecology were inseparable; neither one could be properly understood without the other. Thus illustrating a plant within its landscape would facilitate both scientific inquiry, as well as field identification. Here he portrays one species in the *Drosera* genus and one *Pinguicula* (belonging to the *Lentibulariaceae* family), both in their respective habitats: *D. rotundifolia*, a sphagnum bog (fig. 1); *Pinguicula vulgaris* (common butterwort), a peat bog (fig. 7). A drab environmental colour palette provides contrast for the *Drosera*'s red filaments.

Schmeil also includes enlargements of the sundew's fascinating trapping method: *Drosera rotundifolia*'s trichome-covered head, before and after prey (figs. 2, 3, 4), and viewed from the side (fig. 3, upper), and a mucilaginous gland (fig. 6). On the right, he shows *Pinguicula vulgaris*'s leaf (fig. 8), where glands secrete a sticky fluid and entrap insects (figs. 9, 10).

LEFT:
Title: *Sonnentau*
Author: Otto Schmeil
Language: German
Country: Germany
Series/Book: *Botanische Wandtafeln*
Plate: 12
Publisher: Quelle & Meyer (Leipzig, Germany)
Year: 1907

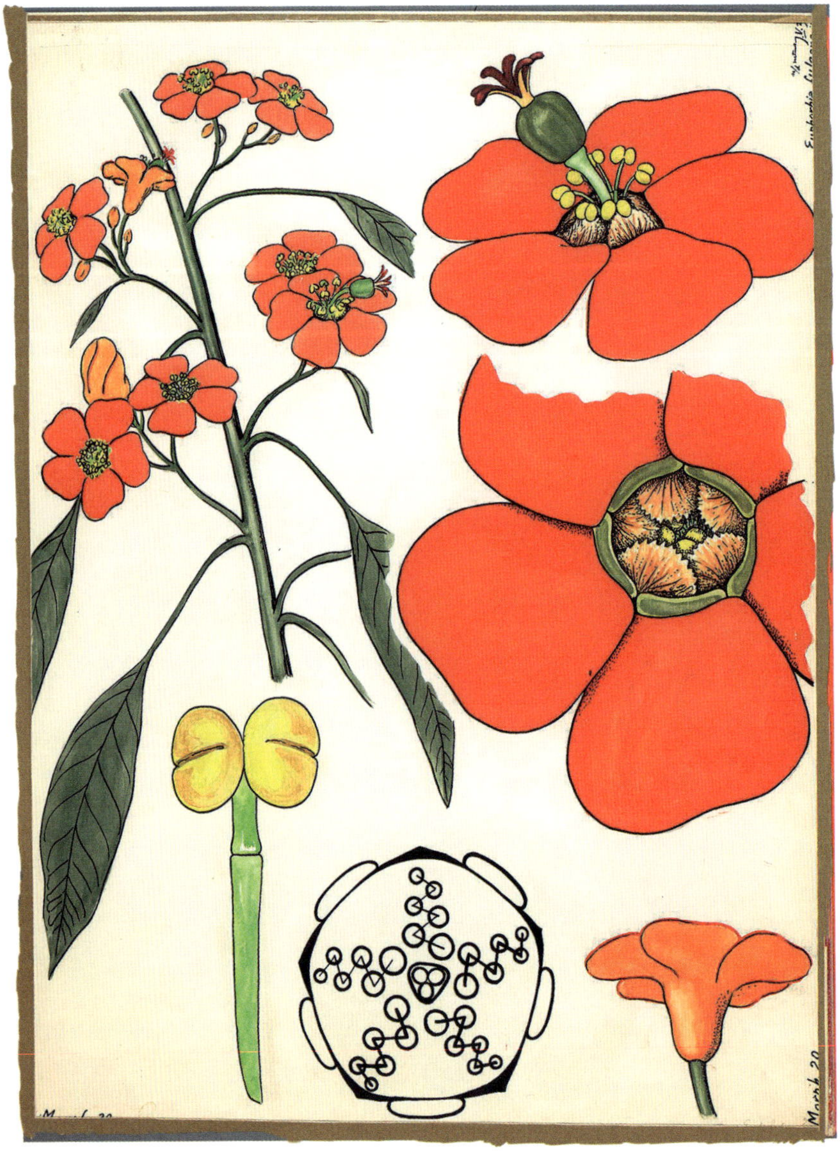
Euphorbia
March 20

XI EUPHORBIACEAE / SPURGE FAMILY

The spurges are a large and variable family: most are herbaceous annuals to perennials, but the 229 genera and roughly 7,000 species also include shrubs, trees and vines. They are found across the globe, especially in the tropics; some are succulents and have evolved to look like cacti, for example *Euphorbia esculenta*. Succulent and non-succulent species are grown horticulturally, such as *Acalypha*, *Codiaeum*, *Croton*, *Euphorbia*, *Jatropha* and *Ricinus*. *Euphorbia pulcherrima* (poinsettia or Christmas flower) is a favourite houseplant for the Christmas season.

Some *Euphorbiaceae* are important commercial crops, such as *Hevea brasiliensis* for rubber, *Manihot esculenta* for cassava, *Triadica sebifera* for candles and soap, and *Ricinus communis* for castor oil. *Euphorbiaceae* have unpleasant aspects – the sap is usually toxic in the *Euphorbioideae*, so plants have been used as a purgatory as far back as the ages of Chaucer and Cleopatra. Today, some spurges are classed as noxious weeds, and the deadly poison ricin is derived from *Ricinus communis*.

Characteristics: leaves sometimes replaced by cladodes; inflorescences of small flowers usually without sepals or petals – in *Euphorbia*, a cyathium (cup of bracts mimicking a single flower); stamens may be single to numerous; fruit a schizocarp. The stem may be spiny and usually has milky, acrid sap. Seeds are carunculate (wart-like surface).

OPPOSITE:
Title: *Euphorbia fulgens*
Author: J. Vuijk
Language: German
Country: Germany
Series/Book: *Morphology*
Plate: 20
Publisher: N/A
Year: 1905

OPENING PAGE: The very caliente *Euphorbia fulgens* (scarlet plume) is a Mexican native with all the best qualities of a euphorbia. The true flowers are small, lacking petals or sepals, composed of a single pistil with three lilting styles, and surrounded by a circle of styles and bracts. Despite a lack of petals, euphorbias do colours very well; the yellows and greens are fluorescent; the vermillions are bright. Here, J. Vuijk, an orchidologist and illustrator with the Botanical Laboratory of Amsterdam illustrated a slender stem of tomato-coloured dabs, more or less evenly paced towards the tip (an inflorescence that distinguishes *E. fulgens* from other euphorbias, most of which display clustered flower heads). A pair of anthers appears at the lower left, next to a diagram of floral anatomy and a view of the goblet-shaped bracts.

OPPOSITE: A rash of the skin, a pain in the stomach, a blindness to the eye: beware the milky latex of the euphorbias. All euphorbias exude a milky latex known for its noxious effects on unassuming gardeners. Generally poisonous if ingested in large amounts, the latex functions to protect the plant from herbivores. However, toxicity varies by species, and a measured application was popular in the days of heroic medicine. The common name for the family, spurge, alludes to its purgative properties. *Euphorbia cyparissias* isn't the most toxic species in the family, but its ubiquity and characteristic white secretions qualified it for Dr. P. Esser's *Poisonous Plants of Germany.*

In his chart, Esser dissects the plant into a gridded anatomy with left- and right-hand columns of whole elements and cross-sections, and a full plant growing up the middle. Especially on a black background, the utilitarian template was a logical way for Esser to disseminate his content. Here, the semi-woody perennial displays its numerous linear leaves, a bustling inflorescence and the hint of an extensive root system (fig. 1). Esser has afforded consideration to a perplexing morphology shared amongst euphorbias, documenting two fundamental points: First, the yellow leaves aren't petals, they are bracts (fig. 2). Second, the true flowers are tiny, and clustered in a cyathium (figs. 4, 5), a choreography of one female and several surrounding males. Third, this pistil and her stamens develop (fig. 4) within an involucre with four horned yellow glands (fig. 3). Two pairs of anthers (fig. 6) peer at a three-lobed fruit (figs. 7, 8) and a seed (fig. 9).

OPPOSITE:
Title: *Anprellen-Wolfsmilch, Euphorbia cyparissias*
Author: Dr. P. Esser; Illustrator: Carl Bollmann
Language: German
Country: Germany
Series/Book: *Die Giftpflanzen Deutschlands (Poisonous Plants of Germany)*
Plate: 11
Publisher: Friedrich Vieweg & Sohn (Braunschweig, Germany)
Year: 1910

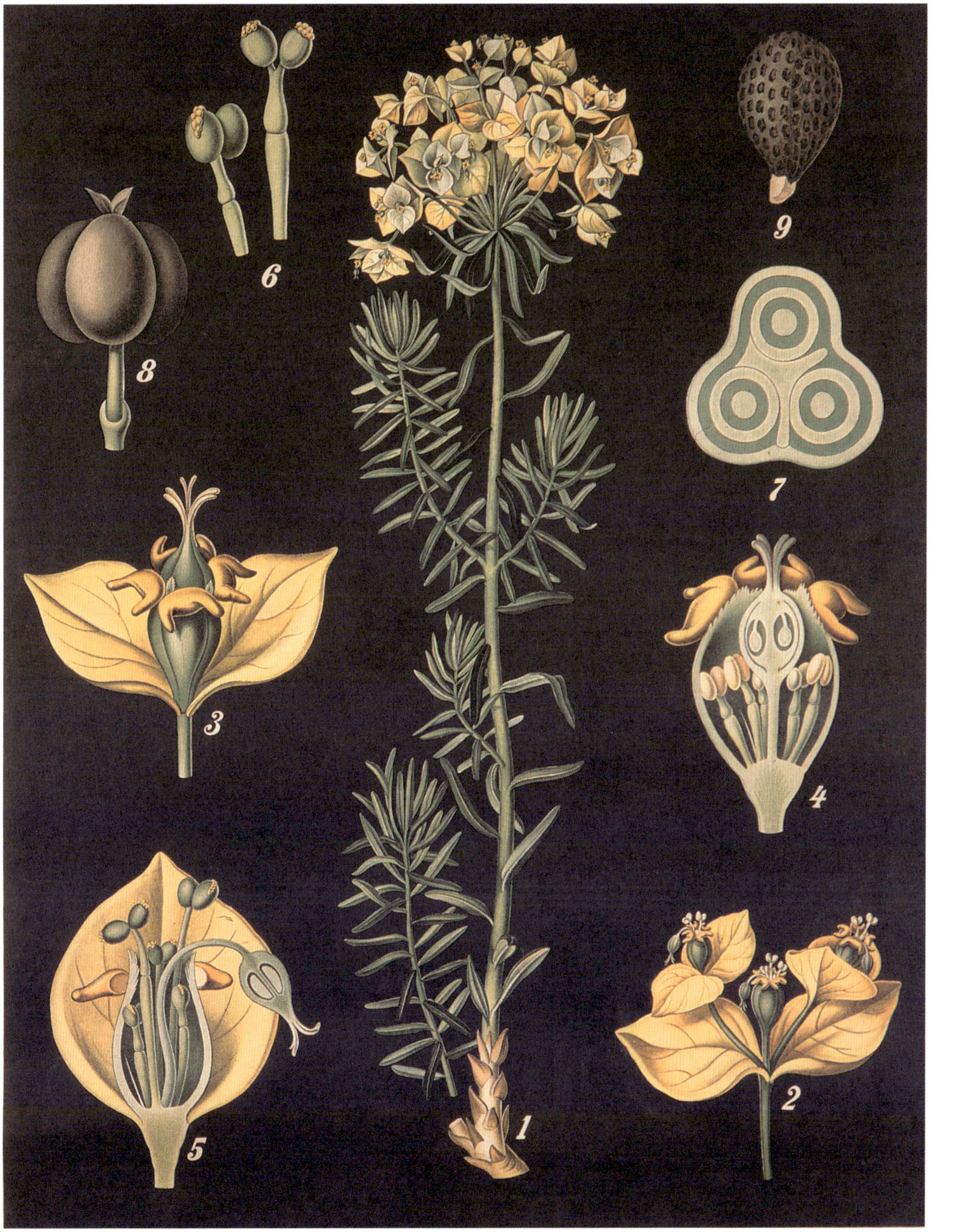
6
9
8
7
3
4
5
1
2

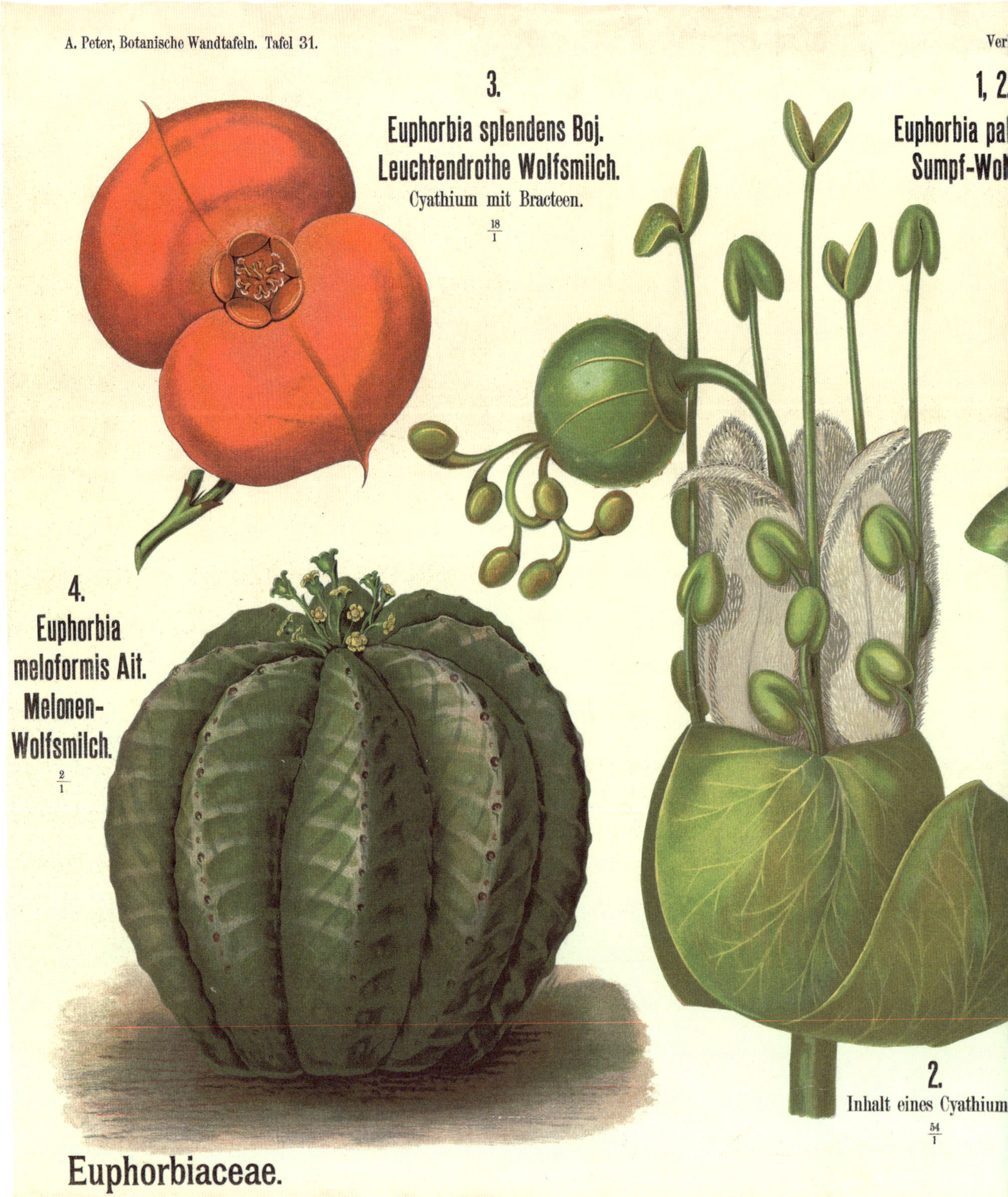
A. Peter, Botanische Wandtafeln. Tafel 31.
3.
Euphorbia splendens Boj.
Leuchtendrothe Wolfsmilch.
Cyathium mit Bracteen.
18/1
4.
Euphorbia
meloformis Ait.
Melonen-
Wolfsmilch.
2/1
2.
Inhalt eines Cyathium
54/1
Euphorbiaceae.

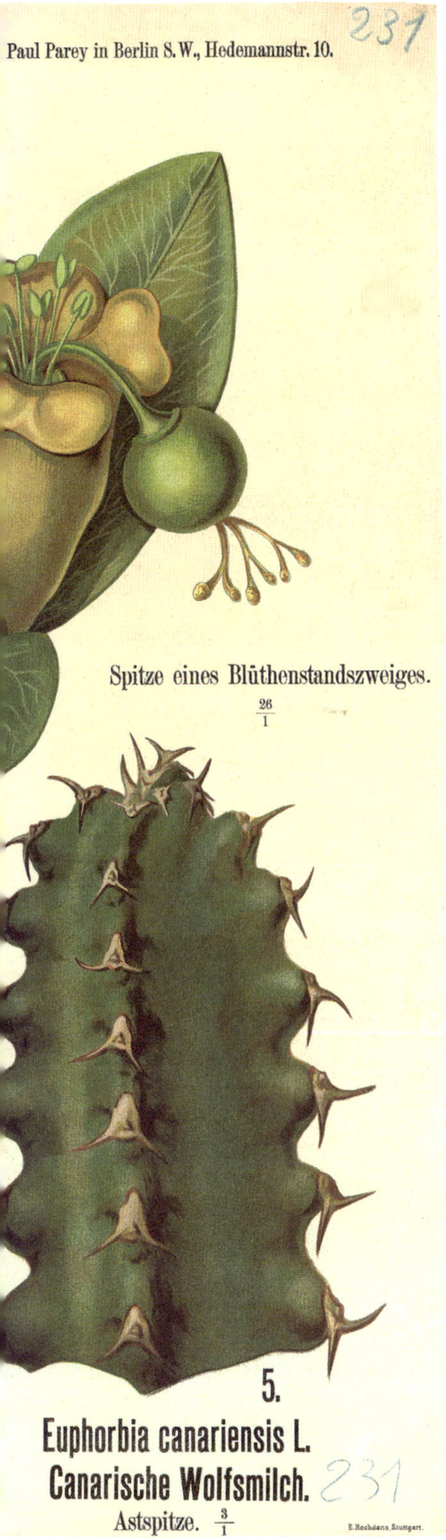

LEFT:
Title: *Euphorbiaceae*
Author: A. Peter
Language: German
Country: Germany
Series/Book: *Botanische Wandtafeln*
Plate: 31
Publisher: Paul Parey (Berlin, Germany)
Year: 1901

LEFT: Euphorbias may lack petals, but they can be colourful; they may be entirely unrelated to cacti, but their forms can be as succulent. Finally, they may be angiosperms, but their flower structure is unlike any other. A. Peter illustrates all three of these euphorbia truths. To elucidate the unusual reproductive anatomy, he selected *Euphorbia palustris* (marsh spurge, figs. 1, 2). A magnified cyathium commands centre stage on the chart, a pistillate female in coy repose and her coterie of staminate males. To the right, a lemon involucre and a pair of chartreuse bracts.

For their bright colours, Peter chose *Euphorbia splendens*, the symmetric flaming beauty in the upper-left corner (fig. 3). A Madagascar native with a long history, its common (ecclesiastic) name is 'crown of thorns,' or 'Christ plant,' referring to the belief that Jesus Christ wore *E. splendens* at his crucifixion (historical evidence indicates the Madagascar native could have found its way to the Middle East prior to the birth of Christ).

The next two illustrate *Euphorbiaceae*'s succulent tendencies. Aptly named the 'melon spurge,' *E. meloformis* (fig. 4) swells into a rotund body with a tuft of flowers. The South African native was introduced to England in 1774 by Francis Masson, a Scottish botanist and Kew garden's first plant hunter, appointed by its new director Sir Joseph Banks. *E. meloformis* was first described in 1789 by William Aiton, gardener to King George III, in the first edition of *Hortus Kewensis*, a catalog of Kew's plants, and thus most species cultivated in England. A nod in the book earned *E. meloformis* the interest of many renowned botanists.

Euphorbia canariensis (Canary Island spurge, fig. 5), is a columnar species with succulent arms that may reach as high as 12 feet (3.65m). Each stem is indented with four to five vertical ridges; each undulating ridge is furnished with pairs of small spines. Ephemeral leaves and reddish-green flowers sprout near the ends of the stems (where Peter's priorities were not).

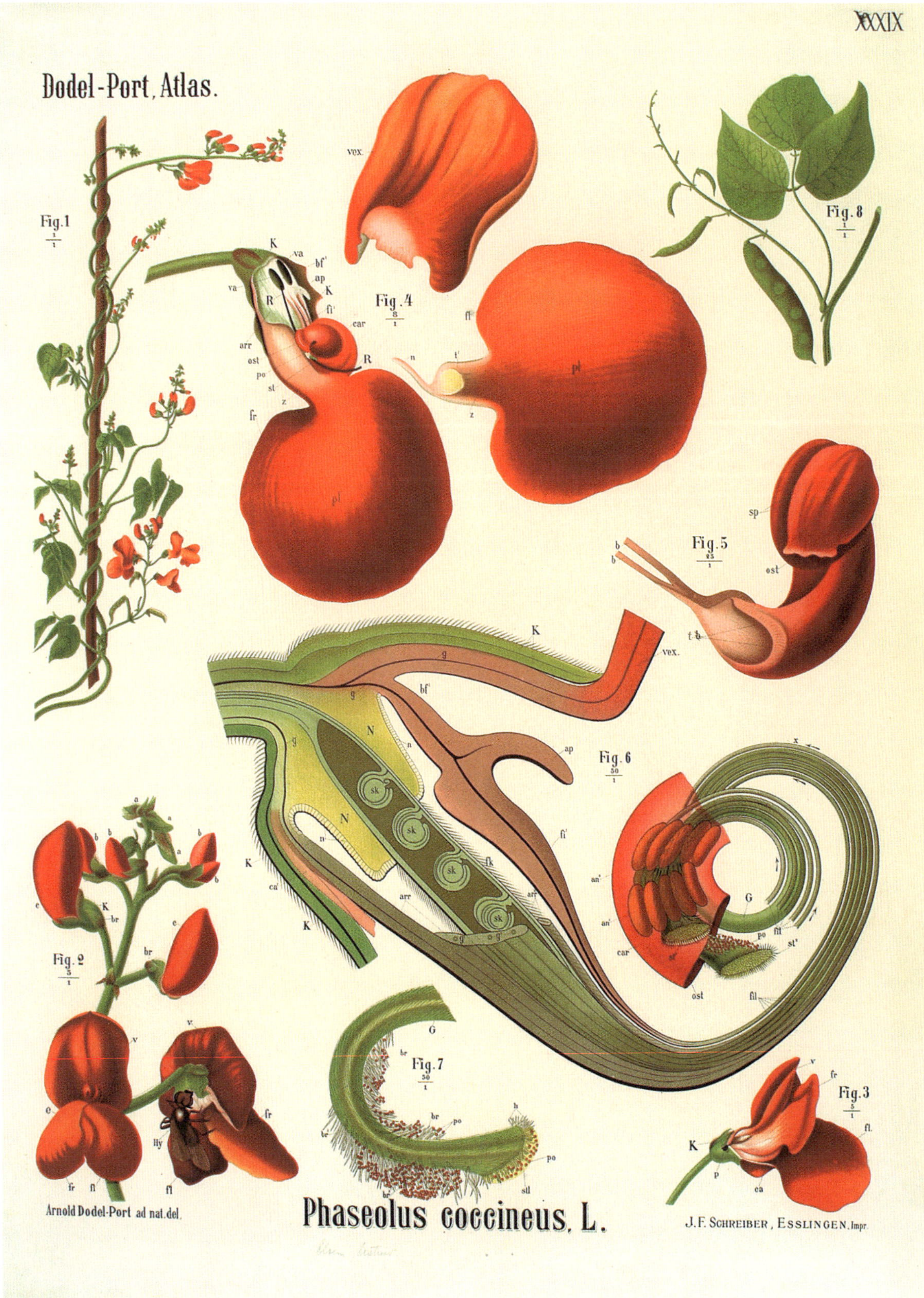
XXXIX
Dodel-Port, Atlas.
Fig. 1
Fig. 2
Fig. 3
Fig. 4
Fig. 5
Fig. 6
Fig. 7
Fig. 8
Arnold Dodel-Port ad nat. del.
Phaseolus coccineus, L.
J. F. Schreiber, Esslingen, Impr.

XII FABACEAE / BEAN FAMILY

Also known as the *Leguminosae*, this variable bean family is one of the largest among angiosperms and encompasses formerly separate families *Caesalpiniaceae*, *Mimosaceae*, and *Papilionaceae*, now treated as subfamilies *Caesalpinioideae*, *Mimosoideae* and *Papilionoideae*. The first two subfamilies are mainly shrubs and trees, but the *Papilionoideae* also includes herbaceous annuals and perennials, some climbing. The nearly 800 genera and 20,000 species are widespread, from tropical to temperate climates.

Many legumes have root nodules containing *Rhizobium* bacteria that fix nitrogen from the air, so they are able to thrive in poor soils, as well as producing highly nutritious seeds. These properties help to make this family one of the most important to agriculture and industry. As well as the many pea and bean species familiar to the vegetable gardener, legumes provide us with fodder crops, green manures, gums and resins, insecticides, medicines, tannins, and timber. *Indigofera* yields a blue dye, *Ceratonia siliqua* carob, and *Glycyrrhiza glabra* liquorice. Garden ornamentals include *Acacia*, *Albizia*, *Cercis*, *Cytisus*, *Laburnum*, *Lathyrus*, *Lupinus*, *Robinia*, *Sophora*, and *Wisteria*.

Characteristics: fruit is a legume, a pod with large seeds; leaves are stipulate, often pinnate or bipinnate, some with tendrils; the flowers bear five sepals and five petals. Inflorescence: raceme or cyme with variable, zygomorphic flowers (*Caesalpinioideae*); capitulum with many prominent stamens (*Mimosoideae*); zygomorphic flowers with petals differentiated into large, upper 'standard,' two lateral 'wings' and lower two fused into 'keel' (*Papilionoideae*).

OPPOSITE:
Title: *Phaseolus coccineus L.*
Authors: Arnold and Carolina Dodel-Port
Language: German
Country: Germany
Series/Book: *Anatomisch physiologische Atlas der Botanik* (*The Anatomical & Physiological Atlas of Botany*)
Plate: 39
Publisher: J. F. Schreiber (Esslingen, Germany)
Year: 1878–93

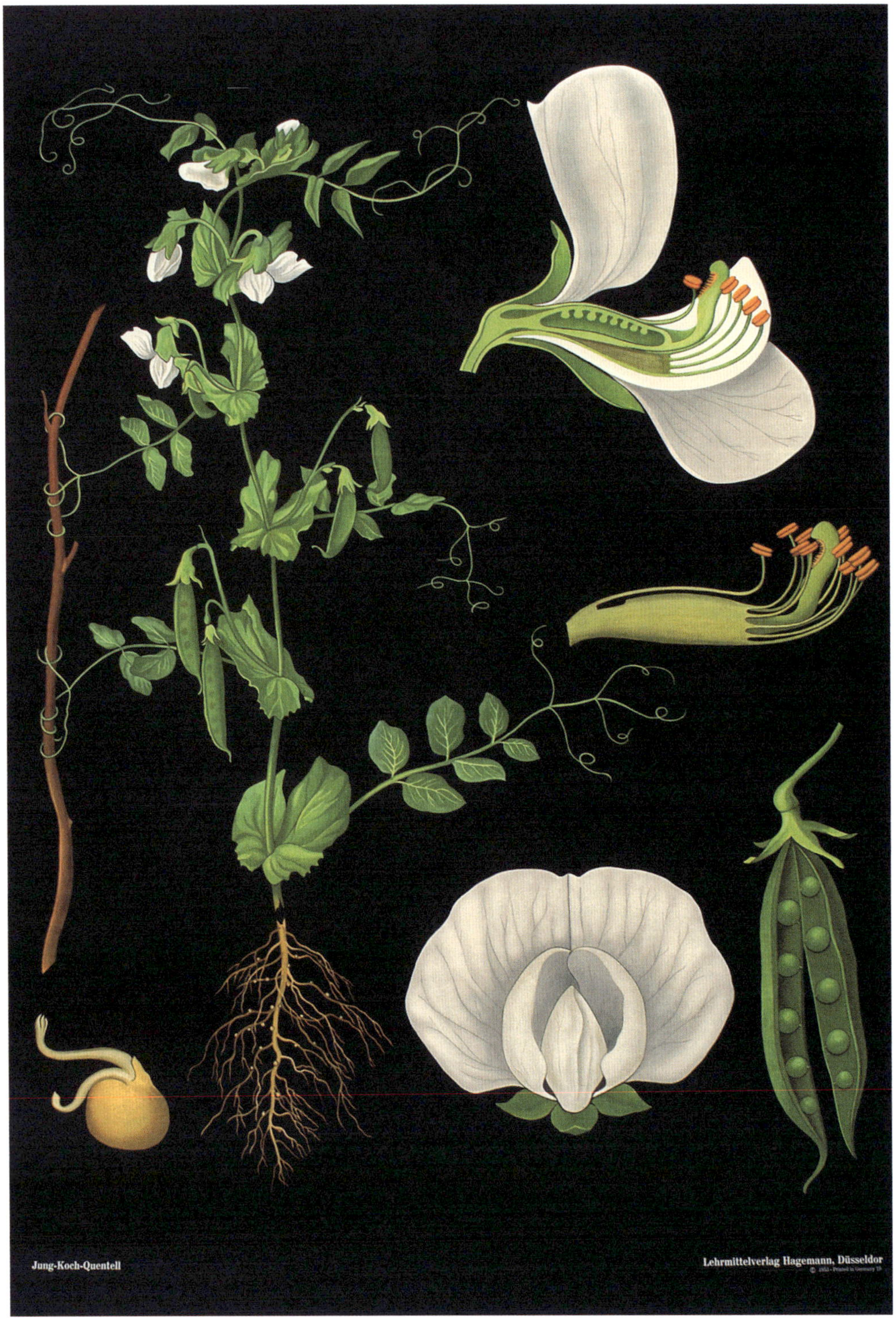
Jung-Koch-Quentell
Lehrmittelverlag Hagemann, Düsseldor

OPENING PAGE: *Phaseolus coccineus* (runner bean) has known many admirers. Thomas Jefferson planted this vining *Fabaceae* in 1812, noting how it brightened Monticello with 'arbor beans of white, crimson, scarlet, purple ... on long walks in the garden,' and gardeners have praised the vigorous ornamental for centuries. To Arnold and Carolina Dodel-Port, however, the scarlet runner bean is a specimen exhibiting impressive adaptation to its preferred pollinators, namely bumblebees and hummingbirds. Morphologically complicated blossoms (fig. 6) offer a wealth of nectar only accessed when a pollinator of sufficient weight lands on a petal. A stylus is 'rocketed forth' and the nectar appears, but not before the bee jostles through strategically positioned stigmas and pollen. In return for its pickiness, the flower can hope for seeds of whose quality and number would not be possible if it settled from a visit from any other pollinator.

OPPOSITE: Gardeners know a common pea for its beans. A pollinator, however, will recognize *Pisum sativum* by the structure of its flowers, which Jung, Koch, Quentell have shown as brilliant white forms on their standard black canvas: a tiny puzzle comprised of five petals, each with a different name and function. The large oyster-shell petal at the rear is called the 'standard,' and is a welcoming flag for pollinators; a little below the standard is a pair called 'wings'; inside those, a second pair form the 'keel.' Their nestled arrangement keeps pollinators from navigating inside, where the fused anthers lie, without considerable pressure. A very small apical opening at the top of the two keel petals will reveal the style and anthers when the keel is depressed. Pollen is released just before the flower opens, which means that the small pouch is soon filled, and dusts the style as well. When pollinators arrive, a small amount of pollen is received and released; usually, however, *Pisum sativum* is self-pollinated.

OPPOSITE:
Title: *Pisum sativum / Erbse*
Authors: Heinrich Jung, Dr. Friedrich Quentell;
Illustrator: Dr. Gottlieb von Koch
Language: N/A
Country: Germany
Series/Book: *Neue Botanische Wandtafeln*
Plate: 13
Publisher: Fromann & Morian (Darmstadt, Germany); Hagemann (Düsseldorf, Germany)
Year: 1928; 1951–63

OPPOSITE: Many illustrators and authors focused on the flower structure of *Fabaceae* species. Eugen Warming, however, dominates his chart with highly magnified illustrations of the seed germinating (figs. 1, 2, 3) and developing leaves. Only one figure illustrates the blossoms (fig. 6), yet at too small a size to explore the legume's peculiar morphology. Instead, the pistils and stamens are included as reference, with no visual explanation of their relationship to the keel petals (see previous page). While he omits a dissection of the flower, Warming did include a dehiscent pod and a sprawl of vining tendrils.

OPPOSITE:
Title: *Fabaceae*
Author: Eugen Warming; Illustrator: Vilhelm Balslev
Language: Danish
Country: Denmark
Series/Book: *Onderwijsplaat botanie* (*Educational Botany Plates*)
Plate: 4
Publisher: Chr. Cato (Copenhagen, Denmark)
Year: c.1910

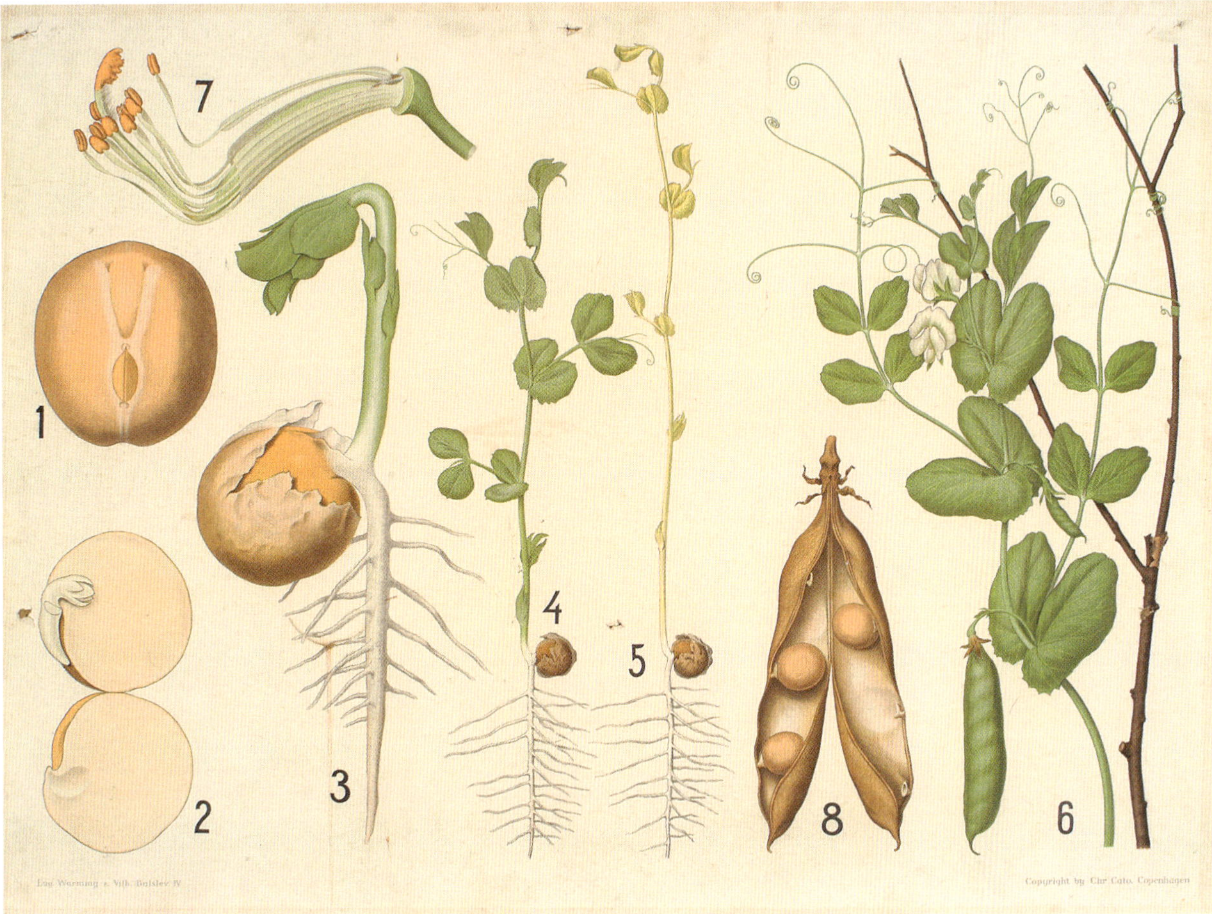
7
1
3
4
5
8
6
2
Copyright by Chr Cato. Copenhagen

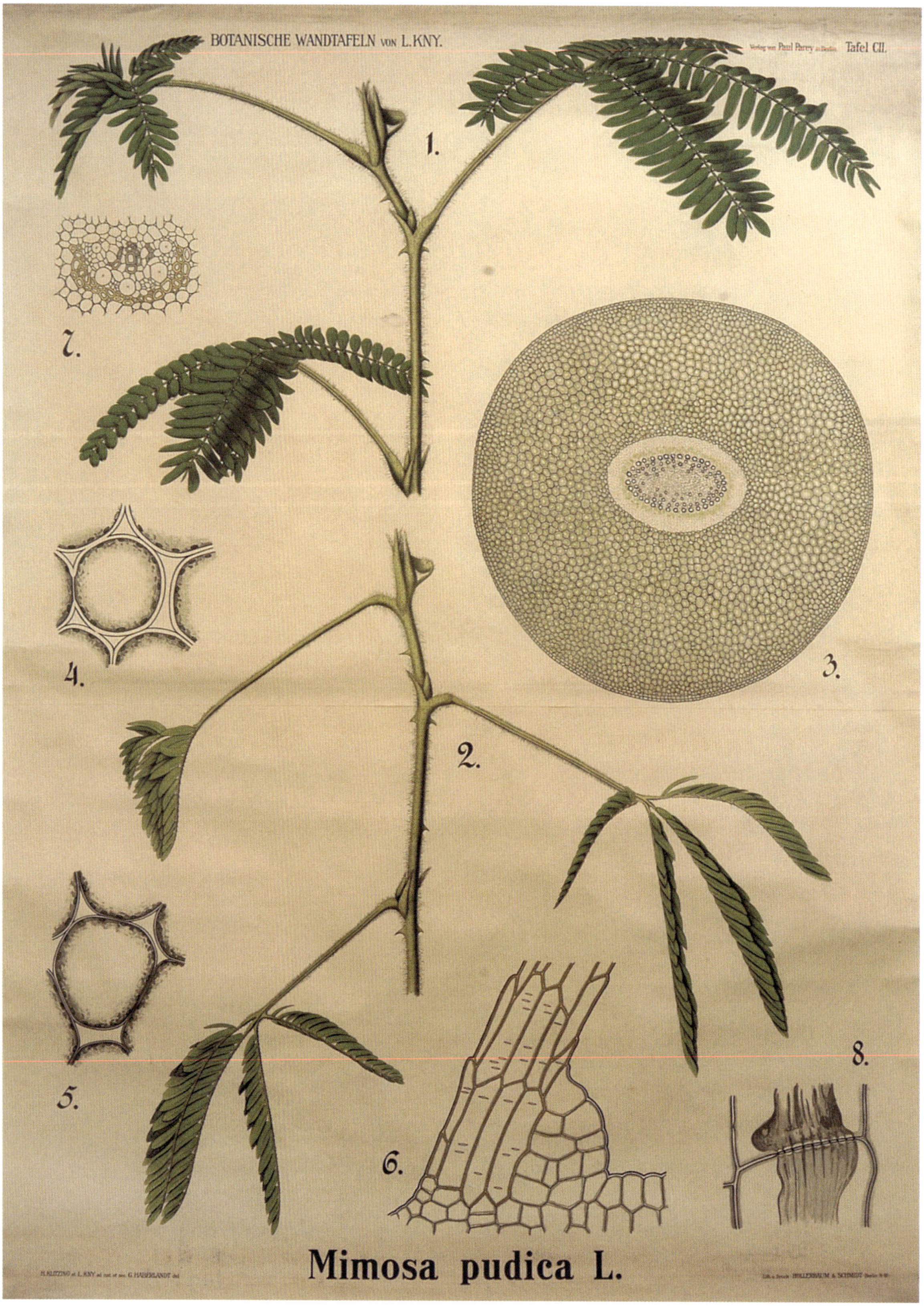
BOTANISCHE WANDTAFELN VON L. KNY.
Verlag von Paul Parey in Berlin
Tafel CII.
1.
2.
3.
4.
5.
6.
7.
8.
Mimosa pudica L.

OPPOSITE: *Mimosa pudica*, or the 'sensitive plant', has drawn the curiosity of horticuralists, writers and botanists for centuries. With delicate fringe leaves that fold in on themselves at the brush of a finger or an insect's wing, the scientific world marvelled at how to explain their motion. Was it controlled by spirits communicating with us? Or was a plant's movement merely an automatic response to stimuli? In the nineteenth century, *M. pudica* flourished in literary writing as authors philosophically personified the species; various epithets such as weak, fearful, shameful all associated the plant with the feminine gender. Eventually, with centuries of debate, new tools offered clarity and conversations moved away from personifying *M. pudica* as a nervous lady in the court of flora, to study its mechanistic processes. Robert Hooke, the English scientist known for his seventeenth-century microscopy work, was one of the first to discover that plant movement is a result of the internal movement of water, whereby water moves out of cell vacuoles and leads to cellular collapse. Here, Leopold Kny offers several stages of its compound leaves (figs. 1, 2) and then microscopic details of the cellular mechanisms that mobilize the movement (figs. 3–8).

OPPOSITE:
Title: *Mimosa Pudica*
Author: L. Kny
Language: German
Country: Germany
Series/Book: *Botanische Wandtafeln*
Plate: 102
Publisher: Paul Parey (Berlin, Germany)
Year: 1874

OPPOSITE: Q. Haslinger's chart is bisected with a vigorous *Vicia sativa* (common vetch); its tendrils and leaves divide the landscape into pockets for flower, fruit, seed and floral diagrams. The hollow vine is four-sided, as shown by a severed central stem; including the roots allows for instruction in the classroom about the plant's nitrogen-fixing capabilities. Other notable features are a dehiscent fruit that has dispersed its seeds ballistically, a common mechanism that many *Fabaceae* have evolved to fling seeds at a significant distance. The full plant also includes a single shrivelled purple flower – a fascinating adaptation that some plants have developed to indicate when a blossom has been pollinated. Bees are attracted to pink flowers, so flowers shift to blue in post-pollination, redirecting attention to younger pink blossoms.

OPPOSITE:
Title: *Vicia sativa L.*
Author: Q. Haslinger
Language: N/A
Country: Germany
Series/Book: *Schoolplaat: Haslinger Botanische Wandtafeln*
Plate: Unknown
Publisher: Dr. te Neues (Kempen, Germany)
Year: 1950

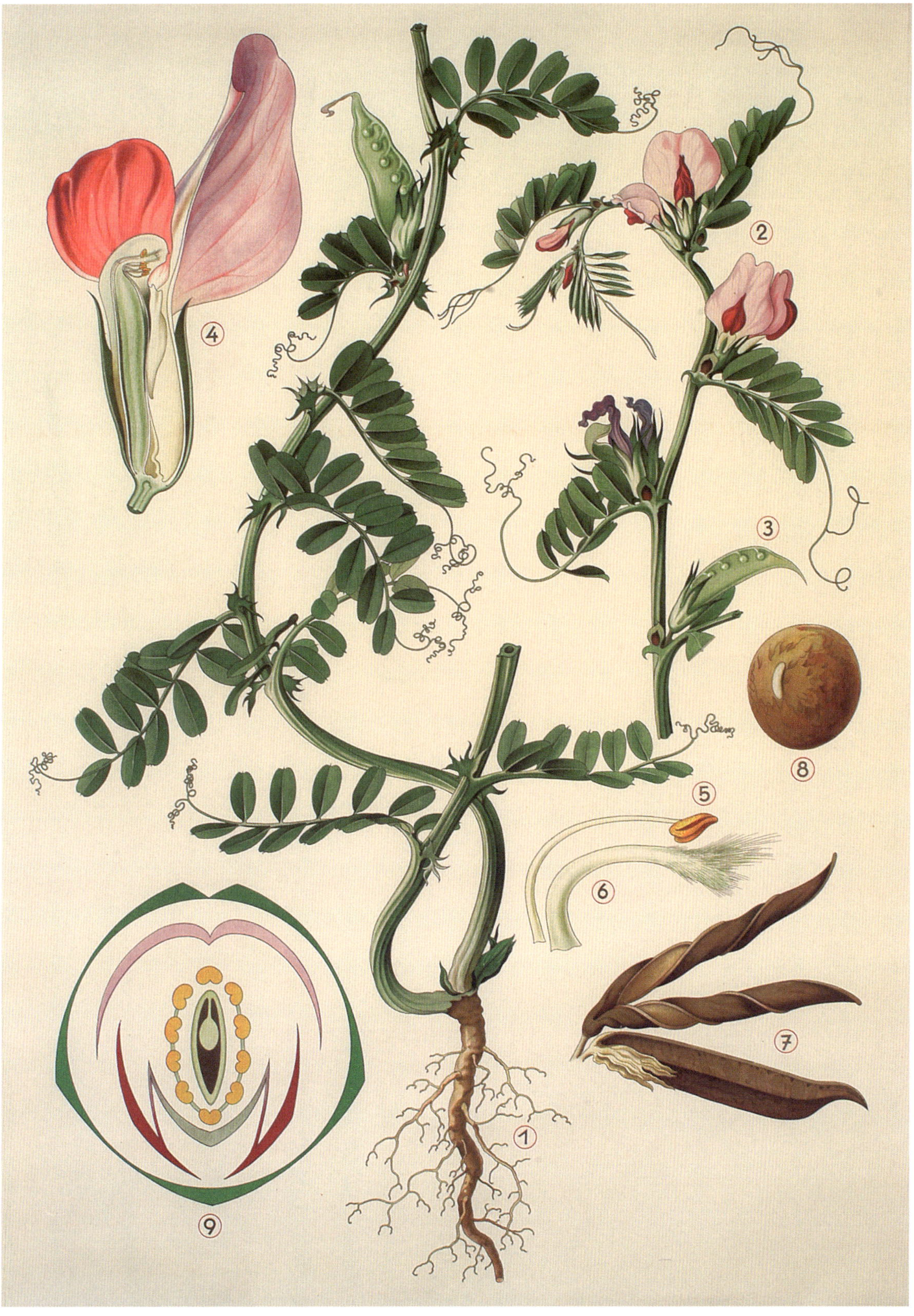
4
2
3
8
5
6
7
9
1

III. Abteilung. **Ausländische Kulturpflanzen in farbigen Wandtafeln.** Tafel 11.

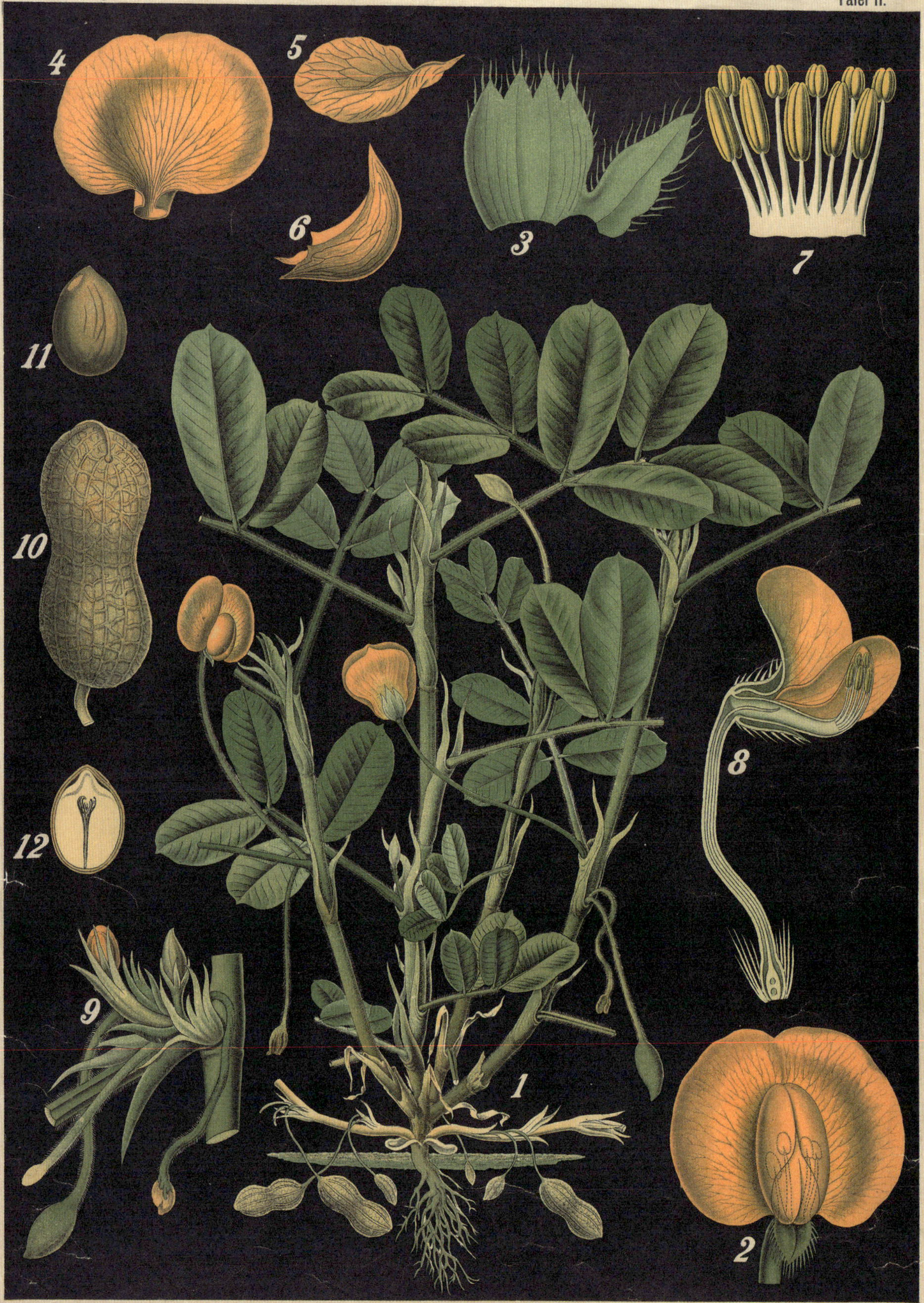

Verlag von FRIEDR. VIEWEG & SOHN, Braunschweig. Nach H. ZIPPEL bearbeitet von O. W. THOMÉ, gezeichnet von CARL BOLLMANN. Lith.-art. Inst. von CARL BOLLMANN, Gera, Reuss j. L.

Erdnuss (Arachis hypogaea Linné). Nach der Natur.

1. Pflanze; *Vergr. 2.* — 2. Oberer Teil der Blüte; *Vergr. 10.* — 3. Saum des Kelches; *Vergr. 12.* — 4. Fahne, 5. Flügel, 6. Kiel der Blumenkrone; *Vergr. 10.* — 7. Oberes Ende der geöffneten Staubblattröhre; *Vergr. 20.* — 8. Blüte im Längsschnitte; *Vergr. 10.* (Nach Taubert in Engler-Prantl.) — 9. Stengelknoten nach Entfernung der Nebenblätter; rechts neben dem rinnenförmigen Blattstiele sind drei stielartige verlängerte Blütenachsen, welche an ihrer Spitze eine junge Frucht tragen; an der Frucht rechts ist der Kelch gesprengt, an der Frucht links abgeworfen; *Vergr. etwa 8.* (Nach Bentham.) — 10. Frucht; *Vergr. 10.* — 11. Same; *Vergr. 10.* — 12. Same im Längsschnitte; zahlreiche Blattanlagen sind bereits erkennbar; *Vergr. 10.*

OPPOSITE: Rather than emphasize the peanut's fruits (which unusually develop both above and underground), or flowers (papilionaceous), Zippel and Bollmann have curiously chosen to emphasize its leaves, which aren't especially distinctive. The decision to minimize the blossoms isn't particularly odd. The keel, standard and wing structure is not unique to *Arachis hypogaea* (it is represented in other members of the subfamily *Papilionoideae*), nor is it the part of the plant for which the peanut is known. But one can't help but wish to see more of the peanuts, especially because *A. hypogaea* demonstrates a fascinating behaviour in which the legume originally forms above ground, after fertilization. The flower stalk then curves downward, and the developing pod is pushed into the ground by the proliferation and growth of cells below the ovary (figs. 1, 9).

OPPOSITE:
Title: *Erdnuss (Arachnis hypogaea Linné)*
Author: Hermann Zippel; Illustrator: Carl Bollmann
Language: German
Country: Germany
Series/Book: *Ausländische Kulturpflanzen in farbigen Wandtafeln (Foreign Crops in Coloured Wall Panels)*
Plate: III Abteilung, 11
Publisher: Friedrich Vieweg & Sohn (Braunschweig, Germany)
Year: 1897

XIII HYDROCHARITACEAE / FROGBIT FAMILY & NYMPHAEACEAE / WATERLILY FAMILY

These families of submerged, floating, or emergent, annual or perennial, aquatic plants are found globally – *Hydrocharitaceae* especially in tropical regions – and mostly in freshwater habitats, though there are also a few marine species. The 16 *Hydrocharitaceae* genera include 114 species; the 5 *Nymphaeaceae* genera comprise nearly 140 species. *Nymphaeaceae* are grown principally as ornamentals for their showy flowers, such as *Nymphaea* and *Nuphar*, and huge leaves, as with *Euryale*, and *Nymphaea nounchali* var. *caerulea* (blue lotus) was commonly featured in ancient Egyptian art as a sacred flower. Some *Hydrocharitaceae* are grown as oxygenating plants, but escapees have been spread by boats to become invasive weeds, clogging up waterways and water habitats – they include *Elodea canadensis*, *E. densa*, *Hydrilla verticillata* and *Lagarosiphon major*.

Characteristics of *Hydrocharitaceae*: leaves varied but with distinct petiole and lamina; three sepals and three petals, perianth often in two whorls; one to many stamens; inflorescence usually emergent, spathe; fruit is a capsule, rarely a berry.

Characteristics of *Nymphaeaceae*: rhizomatous; ovate to circular, peltate or leathery leaves, long petiole; floating or emergent, solitary flower, with usually four sepals, poorly differentiated, and many, often brightly coloured petals and many stamens; fruit is a berry.

RIGHT:
Title: *Hydrocharitaceae*
Author: A. Peter
Language: German
Country: Germany
Series/Book: *Botanische Wandtafeln*
Plate: 34
Publisher: Paul Parey (Berlin, Germany)
Year: 1901

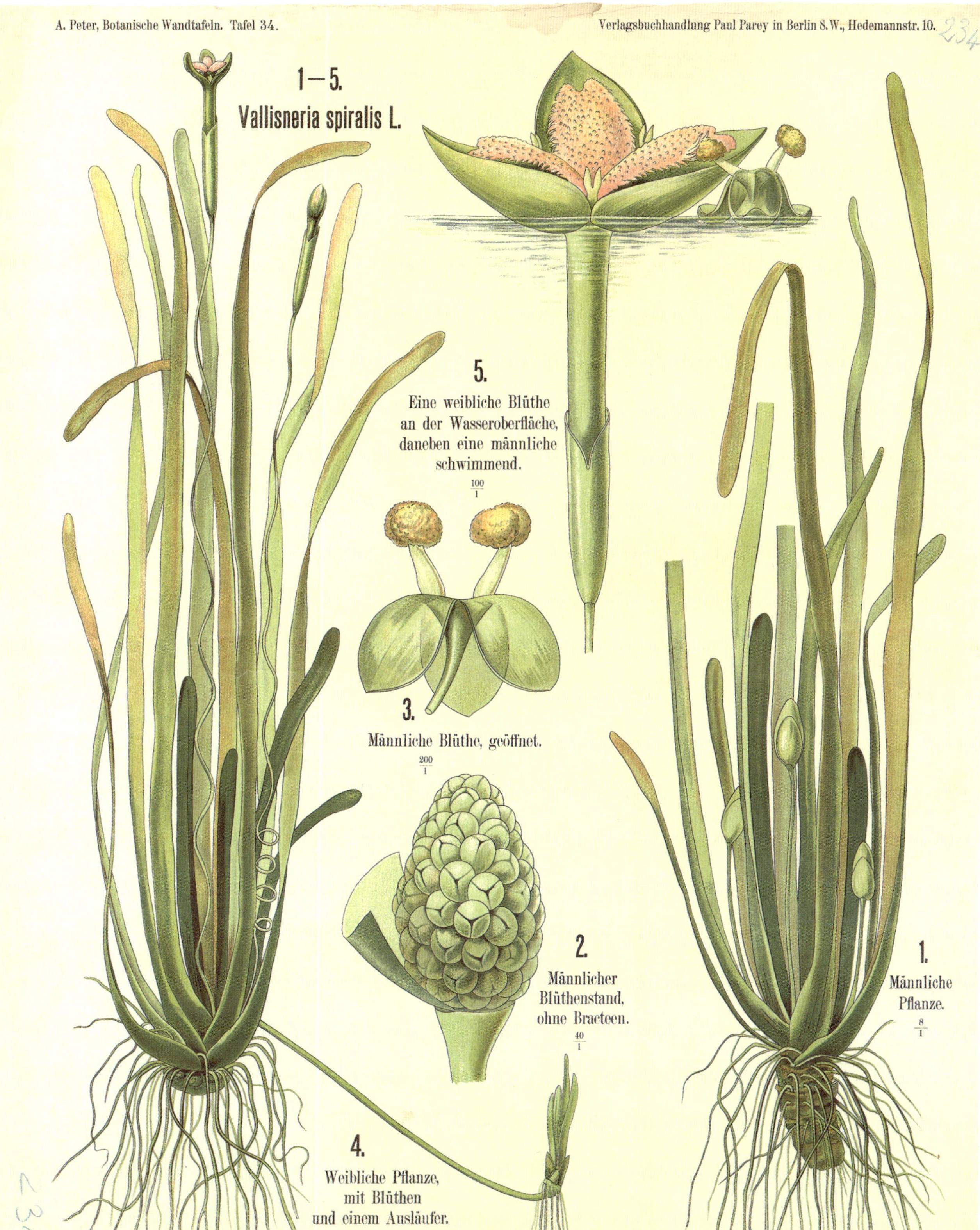
A. Peter, Botanische Wandtafeln. Tafel 34.
Verlagsbuchhandlung Paul Parey in Berlin S.W., Hedemannstr. 10.
1—5.
Vallisneria spiralis L.
5.
Eine weibliche Blüthe an der Wasseroberfläche, daneben eine männliche schwimmend.
100/1
3.
Männliche Blüthe, geöffnet.
200/1
2.
Männlicher Blüthenstand, ohne Bracteen.
40/1
1.
Männliche Pflanze.
8/1
4.
Weibliche Pflanze, mit Blüthen und einem Ausläufer.
8/1
Hydrocharitaceae.

OPENING PAGE: *Vallisneria spiralis* has evolved a curious method of pollination, the mechanisms of which were debated by Darwin and his colleagues. Commonly known as eel grass, the tropical and sub-tropical underwater aquatic species is dioecious; male (pistillate) and female (staminate) flowers are borne on individual plants, growing toward the water's surface. Both develop as clustered rosettes, rooted in the soil. A. Peter illustrates a male (fig. 1) and female (fig. 2) on mirroring halves of the chart, a divergence from his characteristic asymmetry. The single female flower develops underwater and is brought to the surface by its slender peduncle. When mature, it floats on the water's surface. Male inflorescences, however, are produced underwater. At maturity, these male buds detach and rise to the water's surface where the three sepals open to expose the stamens (fig. 3). While a female flower rests, anchored to its parent plant, male flowers are dispersed, ultimately aggregating around the female flower. Here, at the water's surface, pollination occurs when a wandering stamen settles against a patient stigma.

RIGHT: Taxonomy has always been a slippery, contentious, confusing discipline, even for botanists, and especially for students. For example, the two species depicted here, morphologically similar, are in different families. *Hydrocharis morsus-ranae* (frogbit) is a small aquatic plant with floating leaves in the *Hydrocharitaceae* family. *Lemna minor* (common duckweed or brandy bottle) is a small aquatic plant with floating leaves in the *Araceae* family (see page 30). Zippel and Bollman also remind students that, beyond superficial similarities of ecology and structure, the two species have little in common; *Hydrocharis morsus-ranae* has small white flowers and kidney-shaped leaves; *Lemna minor* rarely flowers and has oval leaves. One can understand why Zippel and Bollmann paired the two species; it's certainly useful to compare plants with similar structures and habitats.

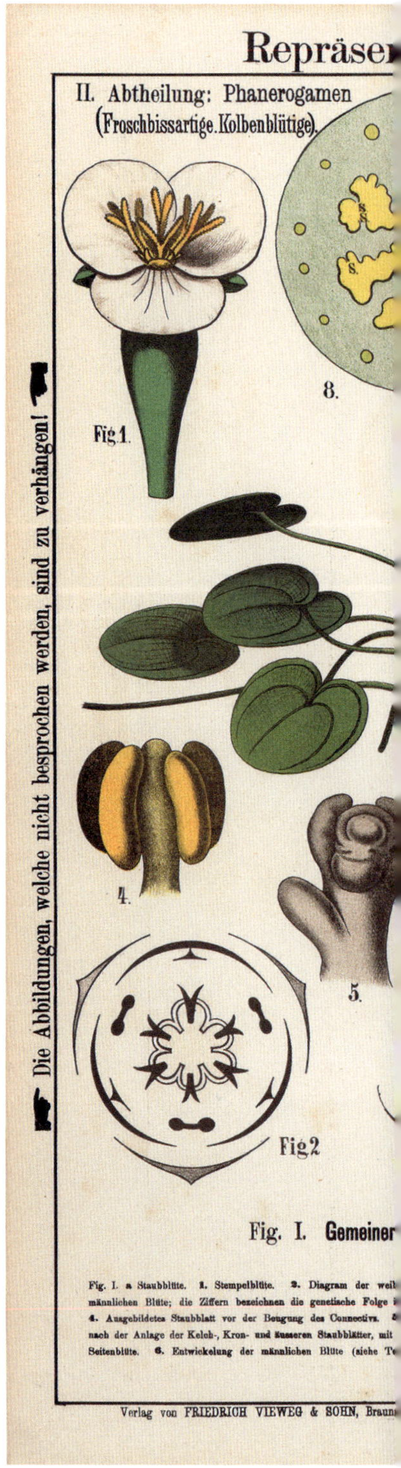

RIGHT:
Title: *Phanerogamen (Froschbissartige, Kolbenblütige) – Phanerogams (Frogbit, Arecidae Family)*
Author: Hermann Zippel; Illustrator: Carl Bollmann
Language: German
Country: Germany
Series/Book: *Repräsentanten einheimischer Pflanzenfamilien (Representatives of Indigenous Plant Families)*
Plate: II Abteilung, 5
Publisher: Friedrich Vieweg & Sohn (Braunschweig, Germany)
Year: 1879–82

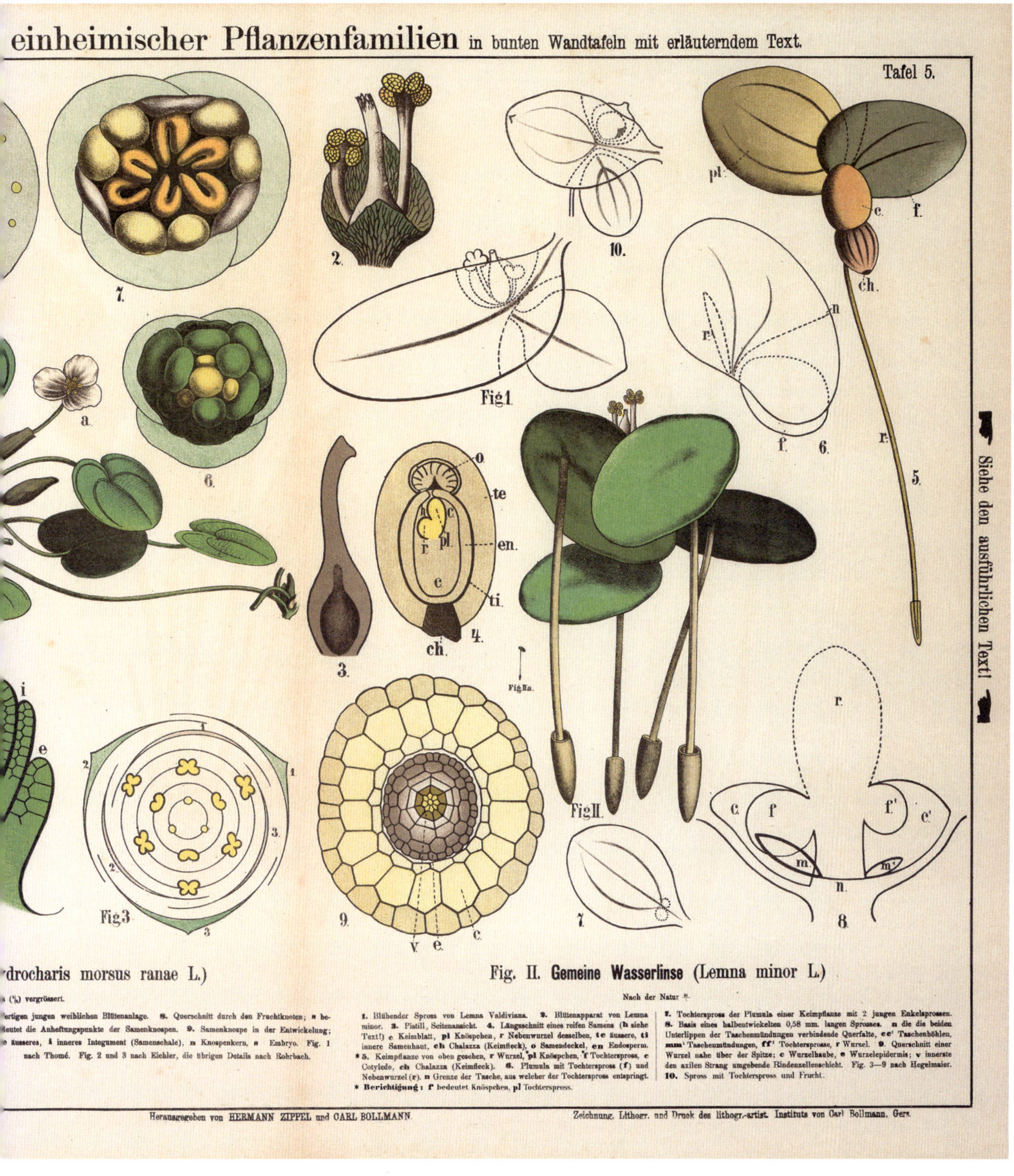

einheimischer Pflanzenfamilien in bunten Wandtafeln mit erläuterndem Text.
Tafel 5.
Siehe den ausführlichen Text!
Fig1.
FigII.
Fig3
drocharis morsus ranae L.)
(³/₁) vergrössert.
ertigen jungen weiblichen Blütenanlage. 8. Querschnitt durch den Fruchtknoten; s bedeutet die Anheftungspunkte der Samenknospen. 9. Samenknospe in der Entwickelung; äusseres, i inneres Integument (Samenschale), n Knospenkern, e Embryo. Fig. 1 nach Thomé. Fig. 2 und 3 nach Eichler, die übrigen Details nach Rohrbach.
Fig. II. Gemeine Wasserlinse (Lemna minor L.)
Nach der Natur
1. Blühender Spross von Lemna Valdiviana. 2. Blütenapparat von Lemna minor. 3. Pistill, Seitenansicht. 4. Längsschnitt eines reifen Samens (h siehe Text!) c Keimblatt, pl Knöspchen, r Nebenwurzel desselben, te äussere, ti innere Samenhaut, ch Chalazza (Keimfleck), o Samendeckel, en Endosperm. *5. Keimpflanze von oben gesehen, r Wurzel, *pl Knöspchen, *f Tochterspross, c Cotyledo, ch Chalazza (Keimfleck). 6. Plumula mit Tochterspross (f) und Nebenwurzel (r). n Grenze der Tasche, aus welcher der Tochterspross entspringt.
* Berichtigung: f bedeutet Knöspchen, pl Tochterspross.
7. Tochterspross der Plumula einer Keimpflanze mit 2 jungen Enkelsprossen. 8. Basis eines halbentwickelten 0,58 mm. langen Sprosses. n die die beiden Unterlippen der Taschenmündungen verbindende Querfalte, cc' Taschenhöhlen, mm' Taschenmündungen, ff' Tochtersprosse, r Wurzel. 9. Querschnitt einer Wurzel nahe über der Spitze; c Wurzelhaube, e Wurzelepidermis; v innerste den axilen Strang umgebende Rindenzellenschicht. Fig. 3—9 nach Hegelmaier. 10. Spross mit Tochterspross und Frucht.
Herausgegeben von HERMANN ZIPPEL und CARL BOLLMANN.
Zeichnung, Lithogr. und Druck des lithogr.-artist. Instituts von Carl Bollmann, Gera.

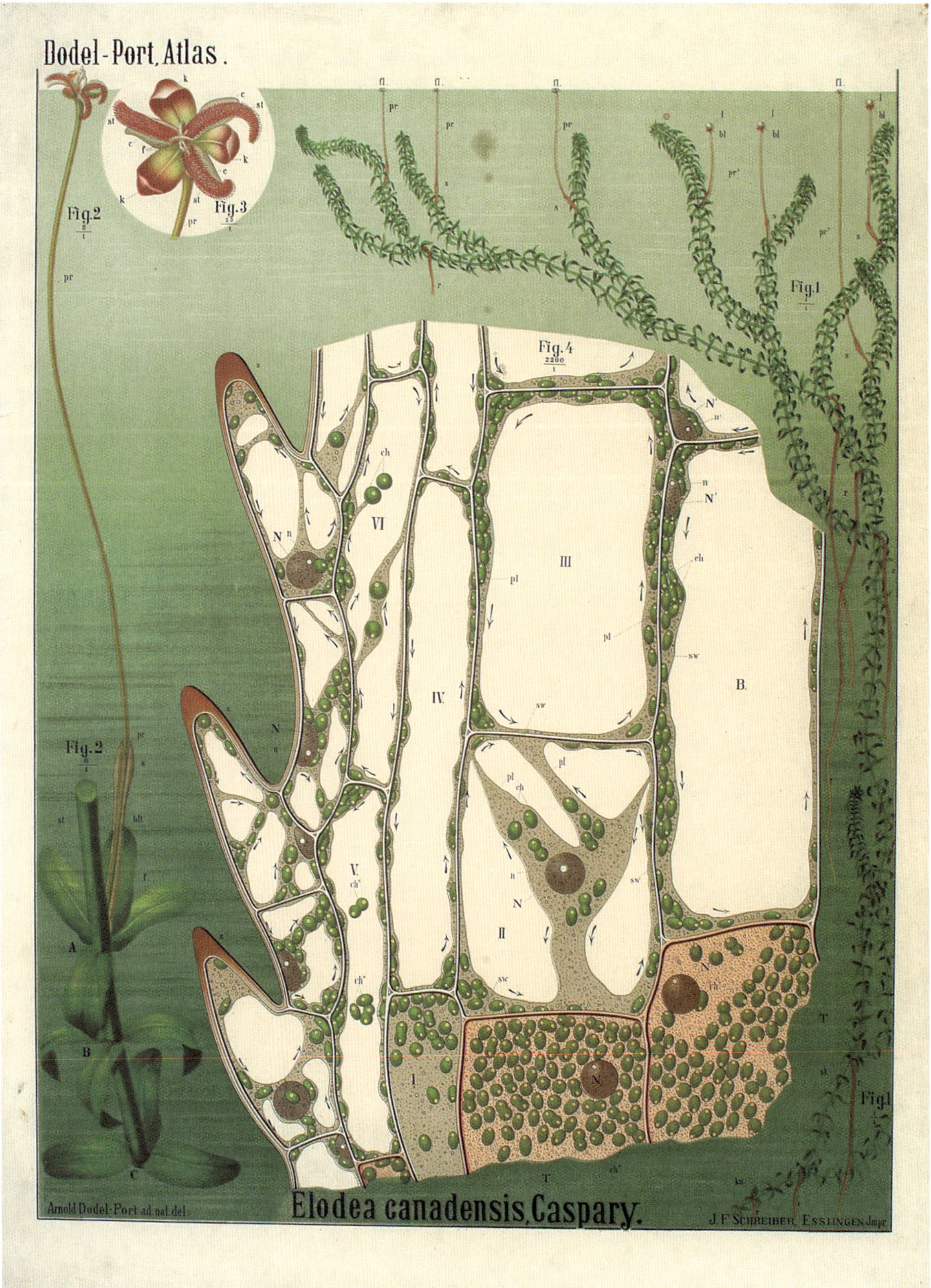
Dodel-Port, Atlas.
Fig. 1
Fig. 2
Fig. 3
Fig. 4
Elodea canadensis, Caspary.
Arnold Dodel-Port ad. nat. del.
J. F. Schreiber, Esslingen Impr.

OPPOSITE:
Title: *Elodea canadensis, Caspary.*
Authors: Arnold and Carolina Dodel-Port
Language: German
Country: Germany
Series/Book: *Anatomisch physiologische Atlas der Botanik* (*The Anatomical & Physiological Atlas of Botany*)
Plate: 28
Publisher: J. F. Schreiber (Esslingen, Germany)
Year: 1878–93

OPPOSITE: *Elodea canadensis* was a popular pondweed for nineteenth-century botanists to study. In their *Botanic Atlas*, Carolina and Arnold Dodel-Port explain why: 'we offer the Canadian waterweed ... as the best living demonstration of transitions between circulating and rotating plasma masses.' In other words, the plasmic movements behind translocation, or the movement of soluble nutrients. from one part of the plant to another. Unlike other species, in *E. canadensis*, 'the whole plasmatic cell content, chlorophyll grains and nuclei included, perform the movement.'

Native to North America, this perennial aquatic plant was introduced to the British Isles in the mid-1800s, and quickly multiplied throughout the European continent. *Elodea canadensis* produces few viable seeds, so the species relies on asexual vegetative reproduction, allowing populations to be exclusively male or female. Dodel-Port note that 'the male has never been observed in Europe. Therefore [we rely on our] American colleagues to determine the pollination and fruiting process of *Elodea canadensis*.'

The Dodel-Ports don't concern themselves with the male. Instead, they illustrate a full and segmented female (figs. 1, 2) and flower (fig. 3), and a fragment of a serrated leaf, where the aforementioned plasma movement can be observed. Arrows within the cells indicate directions of movement. They cite *Vallisneria spiralis* (see previous page) as a species with comparable cellular movement, while recommending *Elodea canadensis* as a preferable study for its ubiquity: 'Without much effort and cost ... they should be easily available for all colleges in which our *Atlas* has been found.'

RIGHT: As always, A. Peter's chart poses intriguing questions surrounding his oft-inscrutable aesthetic and scientific logic. Rather than represent each of the three species with a common element (e.g. flower, leaf, fruit), equal space within the chart, and a comparable scale, he treats each plant very differently and gives no overview of either the individual plants or the family as a whole.

The only plant to be represented in full is *Cabomba peltata* (now called *Brasenia schreberi*, water shield), presumably to represent its feathery submerged leaves (fig. 4). A *Nymphaea nouchali* var. *caerulea* (blue lotus) blossom is halved, revealing a convergence of tall anthers with purplish-blue tips – a distinguishing feature of the species (fig. 3). Finally, the dazzlingly bright *Nuphar lutea* (yellow waterlily) blossom blooms improbably large, beaming beyond the chart. Below, *N. lutea's* notably large, many-seeded fruit is magnified to an equal degree (7.5 times).

Brasenia schreberi has since moved to the *Cabombaceae* family, though the other two still float among the *Nymphaeaceae*.

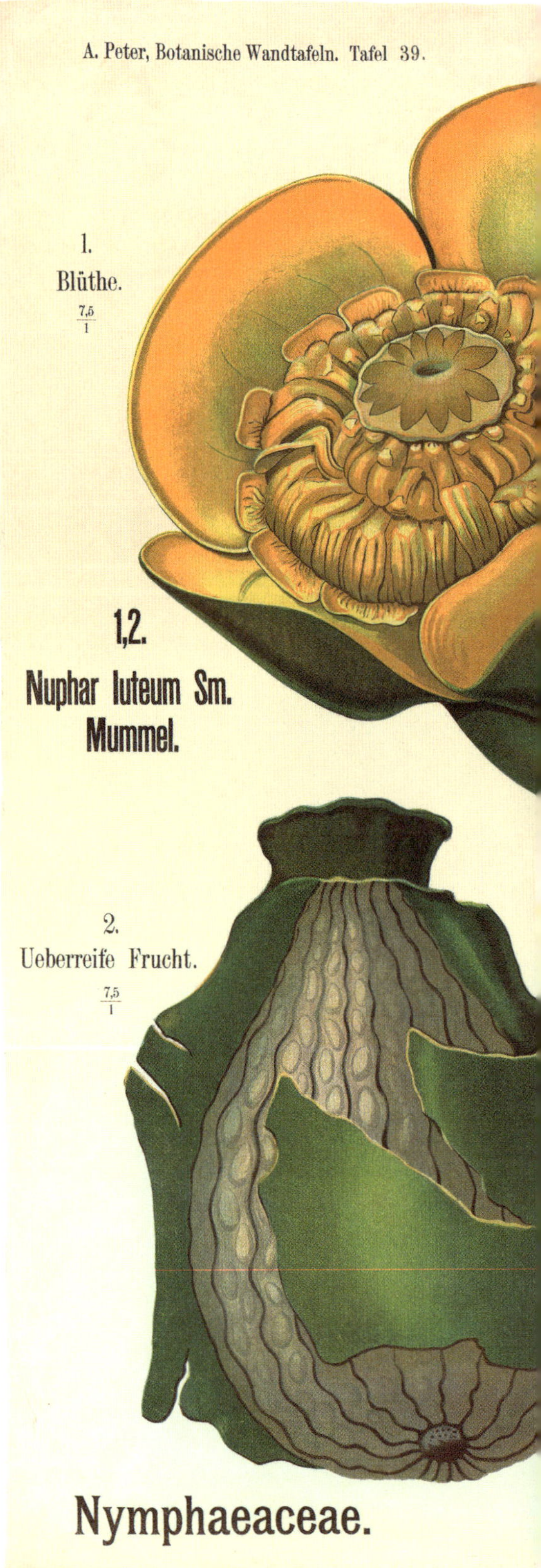

RIGHT:
Title: *Nymphaeaceae*
Author: A. Peter
Language: German
Country: Germany
Series/Book: *Botanische Wandtafeln*
Plate: 39
Publisher: Paul Parey (Berlin, Germany)
Year: 1901

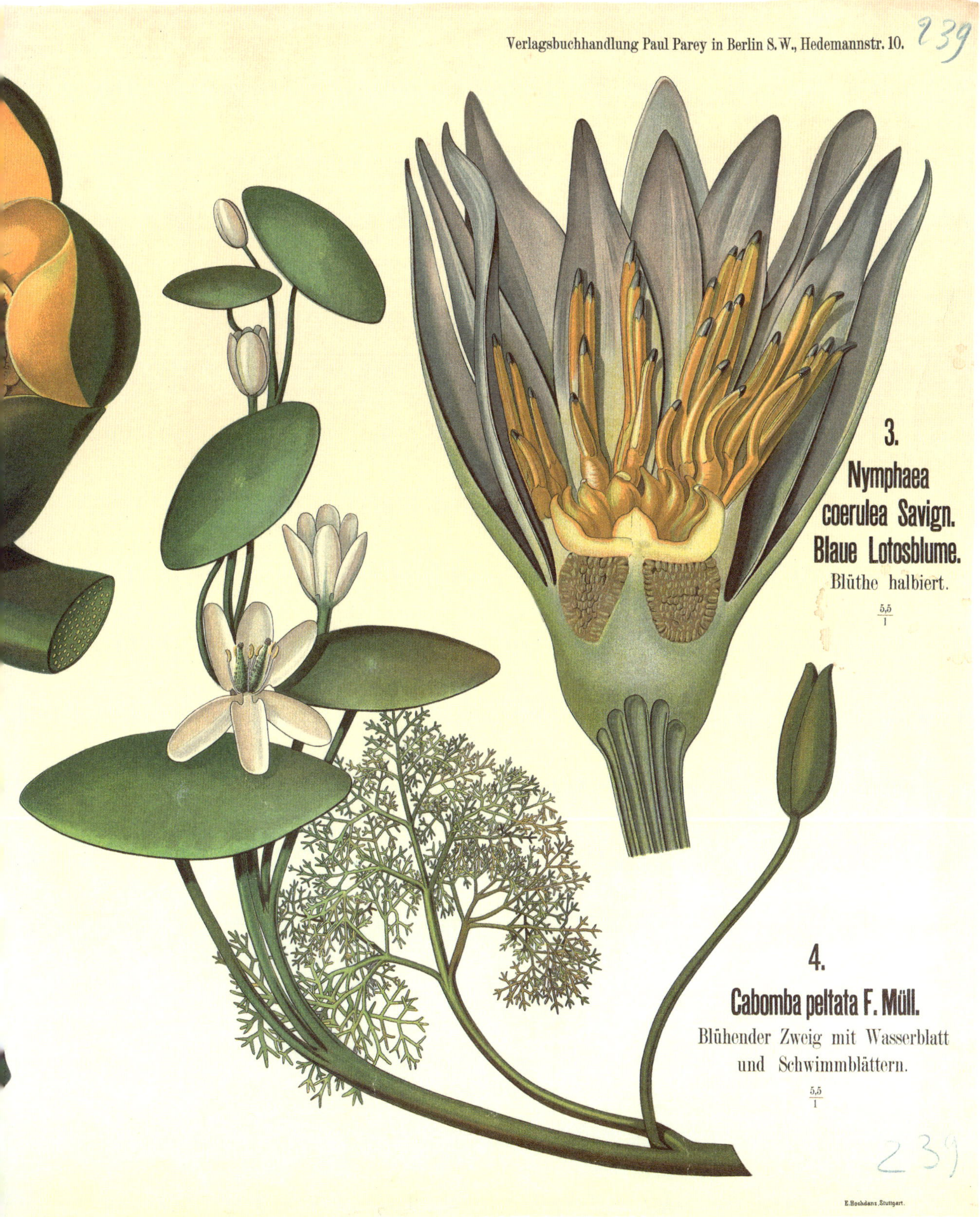
Verlagsbuchhandlung Paul Parey in Berlin S. W., Hedemannstr. 10.
3.
Nymphaea coerulea Savign.
Blaue Lotosblume.
Blüthe halbiert.
5,5/1
4.
Cabomba peltata F. Müll.
Blühender Zweig mit Wasserblatt und Schwimmblättern.
5,5/1
E. Hochdanz, Stuttgart.

OPPOSITE: A comprehensive monograph of wall charts by Jung, Koch, Quentell would primarily be an album of anthers and cross-sections against a black backdrop. Here is one of the few to push the palette. And it's lovely. *Nymphaea alba* and *Nuphar lutea* float in an ecru pond with a surface of pale blue (the use of an alternate colour to suggest a dimensional landscape is equally unorthodox within the Jung, Koch, Quentell style guide).

At the left, *Nymphaea alba* stems from a large rhizome renowned for its medicinal properties. For hundreds of years, the root has been crushed, decocted, extracted, dissected and prescribed for various maladies (monks and nuns ingested the root as an aphrodisiac). Submerged leaves are acutely arrow-shaped; lily pads and white blossoms float above. The fruit (at left, with horizontal and vertical cross-sections at right), a large capsule, matures beneath the water and ruptures, releasing seeds buoyed by bubbles of air confined in their arils. After a couple days, the aril detaches, the air is released, and the seed (right, with cross-section) sinks.

A smaller *Nuphar lutea* sways beyond the white lotus, bearing two flowers and a developing fruit; beneath it a tiny seedling.

At the very bottom of the chart, we witness one of botany's curiosities: the ability of certain plants to modify their organs. Here, we see the petal of a *Nymphaea alba* morphing into a stamen, beginning with the telltale yellow pigmentation at the top of the petal.

OPPOSITE:
Title: *Seerose/Nymphaea alba; Teichrose/Nuphar luteum*
Authors: Heinrich Jung, Dr. Friedrich Quentell
Illustrator: Dr. Gottlieb von Koch
Language: N/A
Country: Germany
Series/Book: *Neue Botanische Wandtafeln*
Plate: 7
Publisher: Fromann & Morian (Darmstadt, Germany); Hagemann (Düsseldorf, Germany)
Year: 1928; 1951–63

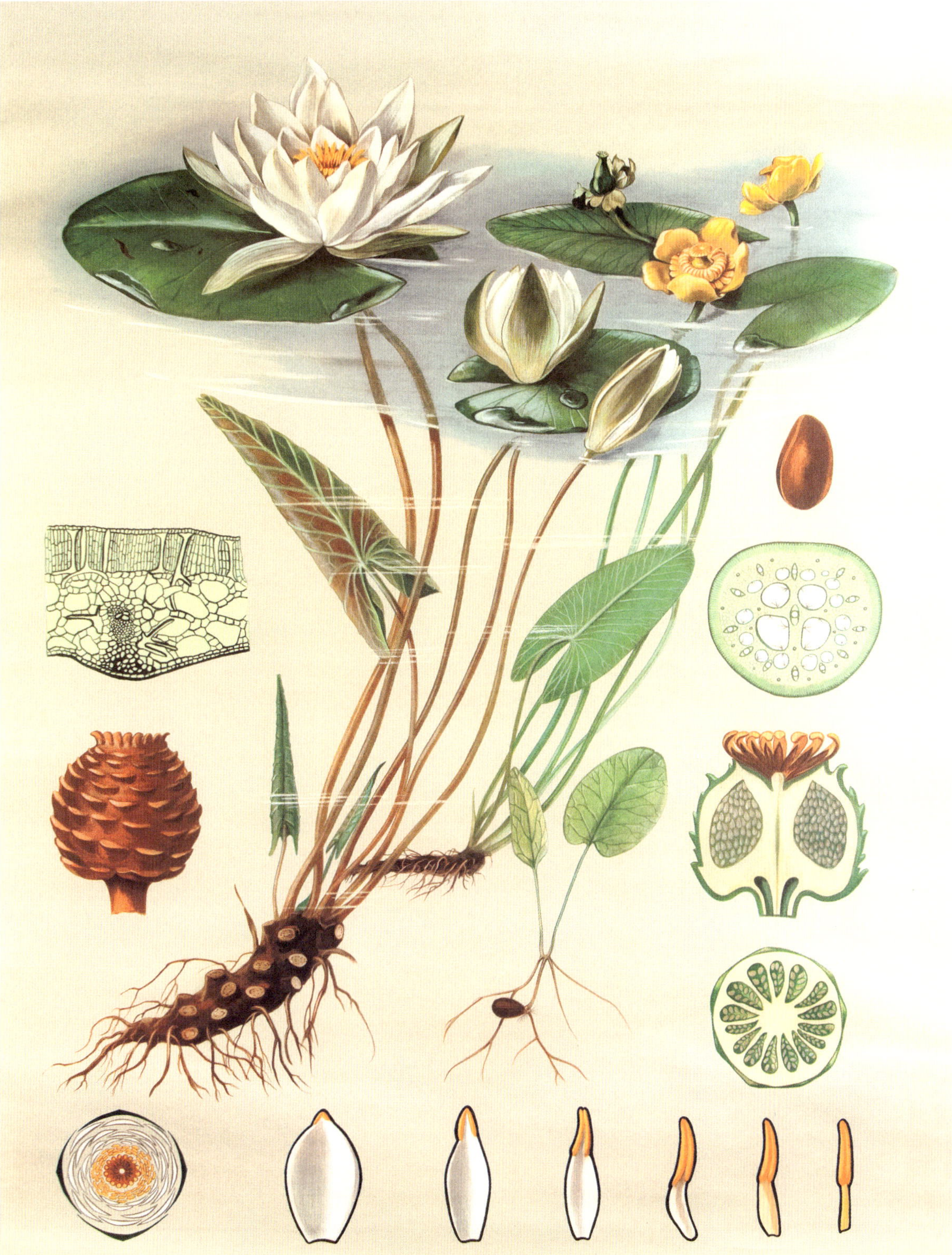
Jung-Koch-Quentell
Lehrmittelverlag Hagemann, Düsseldorf

RIGHT: At left, a *Nymphaea alba* soars to the surface, in almost identical form to Jung, Koch, Quentell's specimen as seen on the previous page – we witness blossoms in three stages, truncated rhizome, severely angled lower leaves, and a sunning lilypad. In figure 2, the illustrator has also included the transition of petal to stamen. Perhaps this is coincidence, perhaps an example of an illustrator and his reference material (naturally, most charts were not drawn from life. Instead, authors sought precedent illustrations as reference). *Nymphaea alba* is featured with a very young fruit, in full and cross-section, both angled for perspective's sake (figs. 3, 4).

Annals of early nineteenth-century botanic expeditions are filled with poetic accounts of a flower with superlative size, intoxicating fragrance and mysterious beauty. Plant hunters wrote of harrowing encounters on the wild Amazonian waters while on a quest for one waterlily, the *Victoria regia* (now called *V. amazonica*), the largest in the world and a darling of Victorian horticulturists since its discovery in 1801. As one gardener wrote, 'It would not be extravagant to call the beauties of this plant unsurpassable. It is everything to be wished for.' And little wonder the world marvelled: her nocturnal blossoms are white for the first moon, pink with the second, with a sweet aroma that feathers through the night; her leaves span 9 feet (3m) in diameter, with intersecting veins and a ribbed underside, were 'a natural feat of engineering,' and inspiration for architect Sir Joseph Paxton's Crystal Palace, designed for the Great Exhibition in 1851. Naturally, the architect tested his blueprints by floating his daughter on a leaf. Here, lacking a scale or reference to indicate the size of the leaf, the illustrator presumably reasoned that three would convey their improbable circumference. Floating on a haze of blue, the three leaves are a formidable backdrop for the blossom, no less extraordinary (though not quite as large).

RIGHT:
Title: *Nymphaeaceae*
Author: V. G. Chrzhanovskii
Language: Russian
Country: Russia
Series/Book: *Sistematika Rasteniy Komplekt Plakatov iz 53 Listov (Systematics of Plants – Posters from 53 Sheets)*
Plate: 4
Publisher: Kolos Publishing
Year: 1971

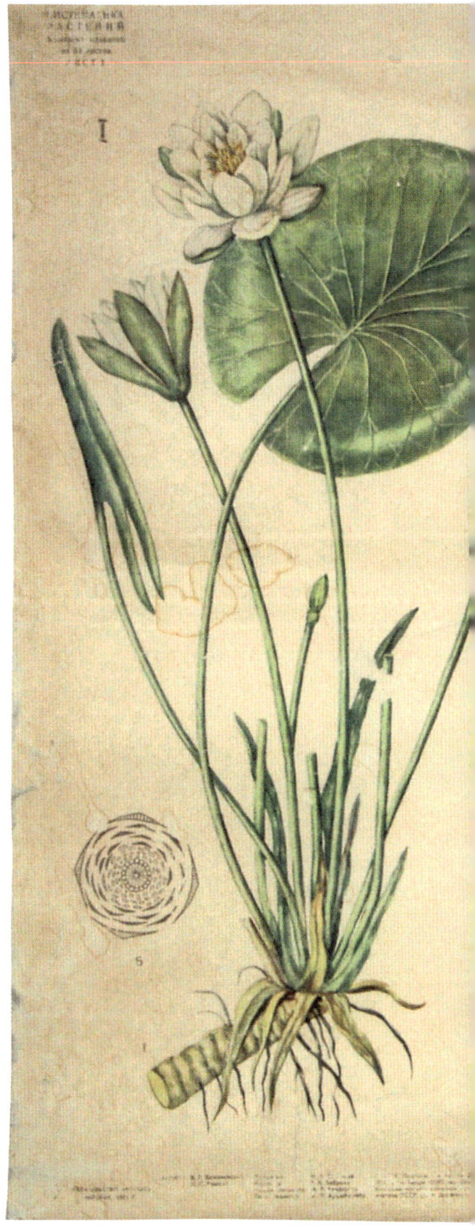

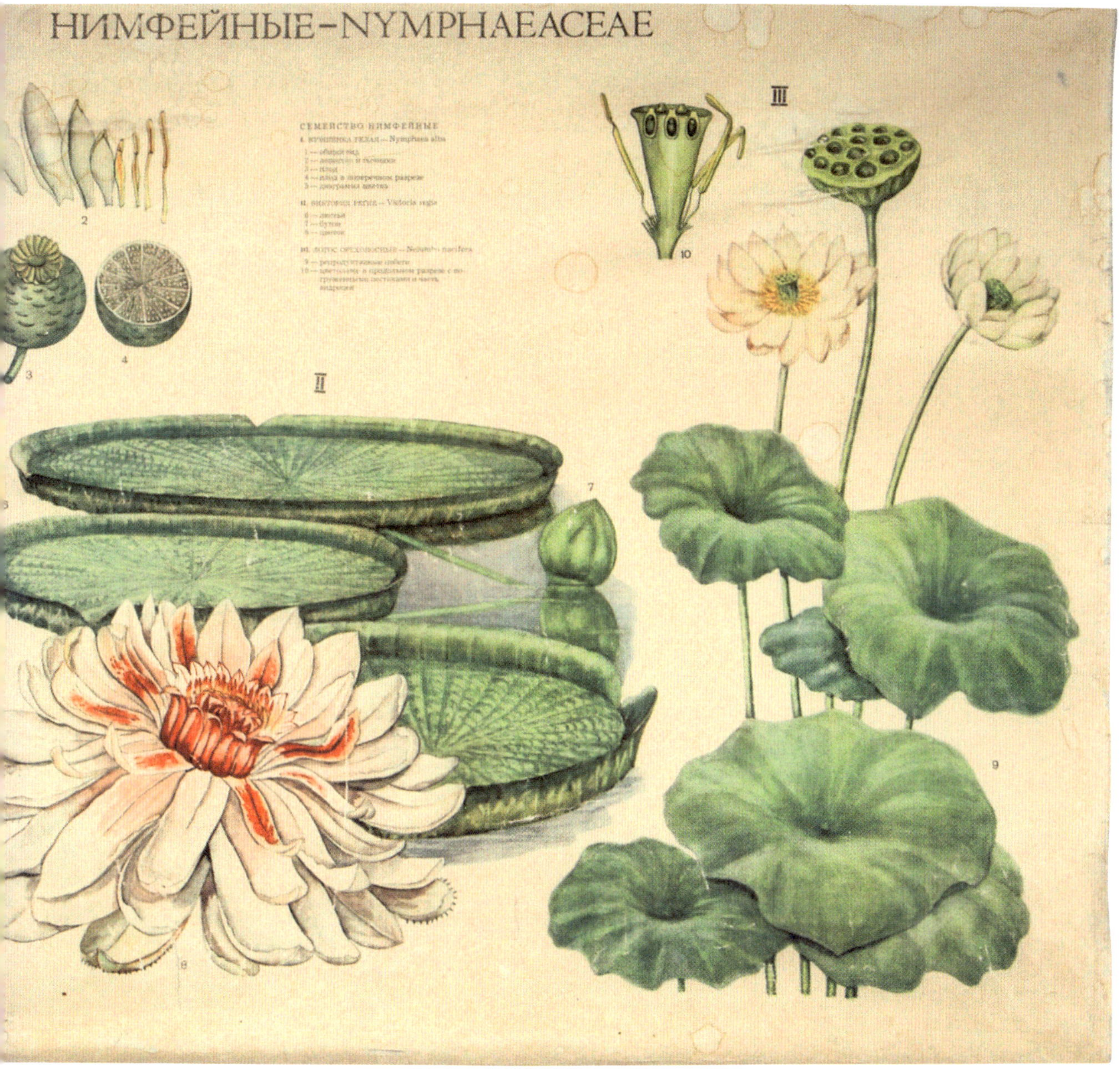
НИМФЕЙНЫЕ–NYMPHAEACEAE
СЕМЕЙСТВО НИМФЕЙНЫЕ

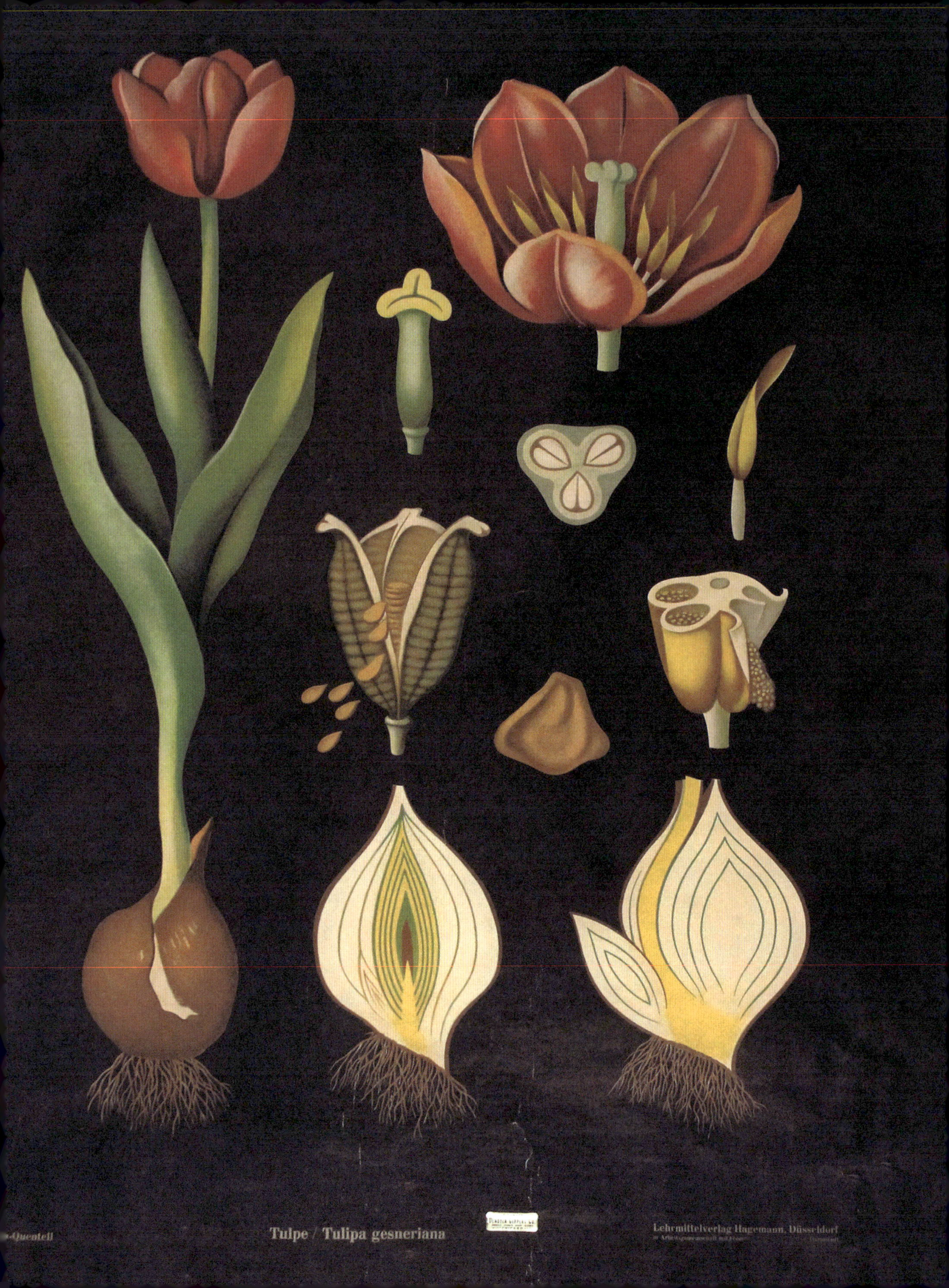

Quentell
Tulpe / Tulipa gesneriana
Lehrmittelverlag Hagemann, Düsseldorf

XIV LILIACEAE / LILY FAMILY

There are currently only 15 genera and around 780 species in the lily family, although there used to be many more. Largely perennial, but variable herbaceous plants with rhizomes, bulbs, corms or tubers; some are fibrous-rooted while others have woody stems or a climbing habit. They are widespread, but especially common in temperate climates in the northern hemisphere.

Liliaceae such as *Calochortus*, *Erythronium*, *Fritillaria*, *Lilium*, *Tricyrtis* and *Tulipa* are very popular ornamental plants in the garden and also mainstays of the floristry industry, for their large, beautiful flowers. Lilies and tulips especially have been revered in sacred texts, literature and art for many centuries: lilies often were potent religious symbols, while a single tulip bulb, at the height of the flower's popularity in the seventeenth century, could cost as much as a house.

Characteristics: leaves linear, simple, sometimes basal or whorled on stem, parallel veins; six tepals, free or fused, in two whorls in perianth; six stamens; fused style with three stigmas; solitary flower or in umbels, clusters, racemes or panicles; fruit often a three-chambered capsule or berry.

OPPOSITE:
Title: *Tulpe/Tulipa gesneriana*
Authors: Heinrich Jung, Dr. Friedrich Quentell
Illustrator: Dr. Gottlieb von Koch
Language: N/A
Country: Germany
Series/Book: *Neue Botanische Wandtafeln*
Plate: 34
Publisher: Fromann & Morian (Darmstadt, Germany); Hagemann (Düsseldorf, Germany)
Year: 1928; 1951–63

OPENING PAGE: Tulips have been cultivated as ornamental plants from at least the tenth century in Persia, but it was under the Ottoman Empire in Turkey that their legacy truly began. In the mid-sixteenth century they were introduced to western Europe, where unusual variegated patterning in the petals caused them to be greatly sought after. This effect was in fact caused by a virus which weakened the plant, causing supply numbers to fall. The combined effect was the 'tulip mania' that took over the Netherlands between 1634 and 1637; a speculative craze wherein tulips became so valuable that people traded their land and their life savings for a single bulb.

Jung, Koch, Quentell let us peer into coyly unfolding petals of *Tulipa gesneriana* (Didier's tulip) to illustrate the tulip's reproductive cycle. Like all bulbous plants, the tulip has two methods of reproduction: asexual reproduction via bulbs, and sexual reproduction via pollination to produce fruits. Here, the authors show both: underground, a bulb and bulblet that will cleave to produce a new plant; and at the centre of the chart the seeds are shown neatly escaping their withered capsule.

OPPOSITE: Otto Schmeil chooses the fragrant *Tulipa suaveolens* to represent the *Liliaceae* family, the species from which most contemporary varieties have been cultivated. More than two centuries had passed since tulip mania, so the tulip isn't included for its flash in the spotlight as a symbol of wealth – quite the opposite. By the nineteenth century, tulips were ubiquitous in European gardens, accessible and appealing to all; as Schmeil writes in the companion text, 'pleasurable to even a man indifferent about plants.'

Schmeil's chart deals with the same subject as Jung, Koch, Quentell's, but the execution is quite different. While Jung, Koch, Quentell tend toward the abstract, Schmeil gives us a comprehensive timeline of tulip reproduction.

At the bottom of the chart, rooted in soil, a new bulb (fig. 1) leans toward its future: modified leaves, or scales, surround a wreath of fibrous roots and tissue, out from which pushes a forthcoming stalk and flower (figs. 2–4). After blooming, the head forms a fruit (fig. 4), while the bulb produces a new bulbil (fig. 5). Aboveground, figures six to ten give us the tulip's sexual method of reproduction in explicit detail, from tight young bloom (fig. 8) to the pollinator at work (fig. 6), and finally the dehiscent capsule, casting seeds back to the soil (fig. 10), below which the bulb has drawn nutrients from the leaves to produce a daughter bulb.

OPPOSITE:
Title: *Tulp*
Author: Otto Schmeil
Language: German
Country: Germany
Series/Book: *Botanische Wandtafeln*
Plate: 1
Publisher: Quelle & Meyer (Leipzig, Germany)
Year: 1907

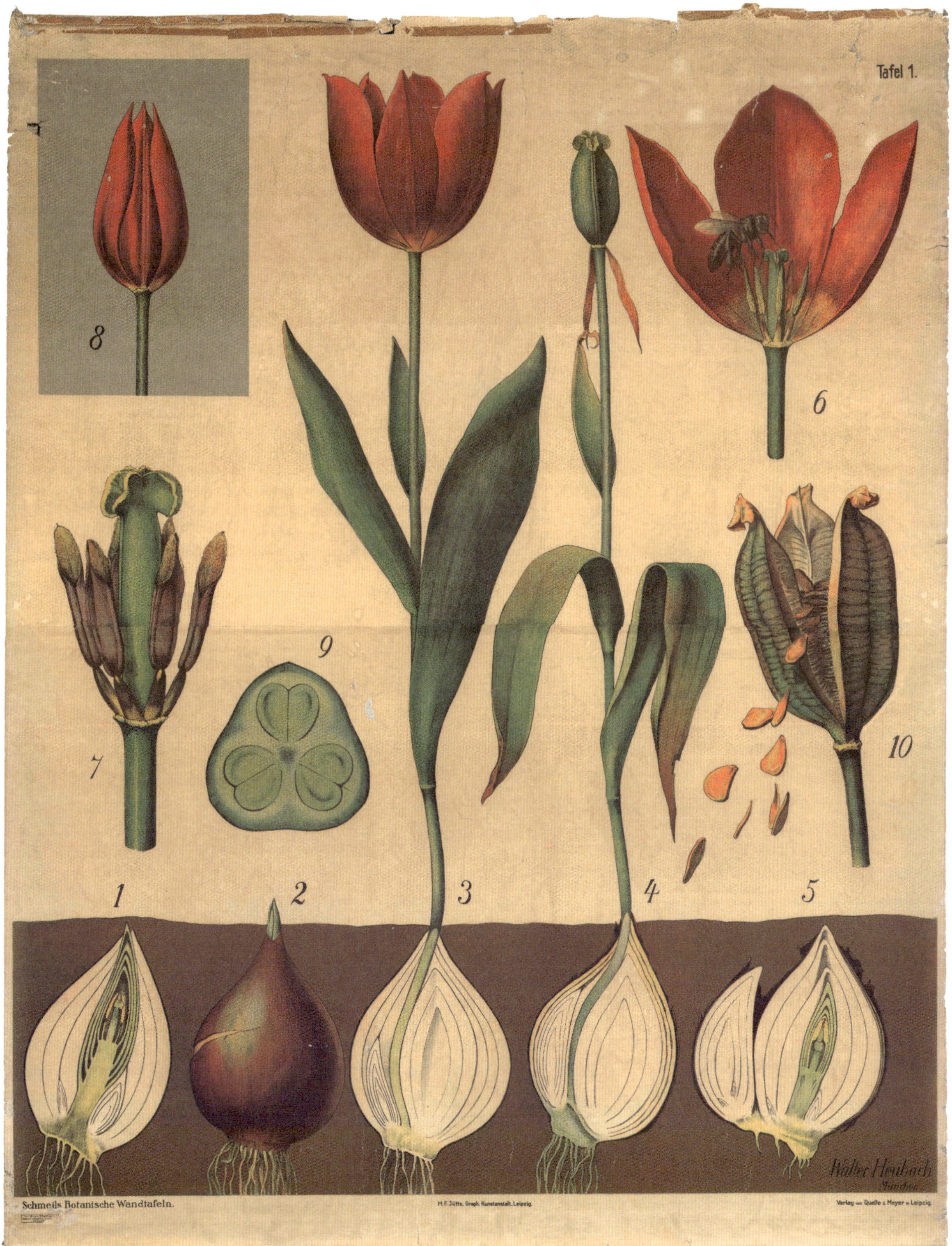
Tafel 1.
8
6
7
9
10
1
2
3
4
5
Walter Heubach
München
Schmeils Botanische Wandtafeln.
H.F. Jütte, Graph. Kunstanstalt, Leipzig.
Verlag von Quelle & Meyer in Leipzig.

OPPOSITE: The Dodel-Ports lavish as much praise on *Lilium martagon* as any other species in their atlas – as always, with a specific intent: 'She is a hallmark example of an angiosperm excelling at reproduction. She both offers conveniences to honey-seeking insects, and, failing that, can alternately self-pollinate with stunning success.' It was also one of the first lilies to be grown in British gardens: it was listed by the gardener John Gerard in his *Catalogue of Plants* in 1596. The common name comes from a style of turban adopted by a Turkish ruler, known as a martagon, with a similarly pendulous shape. Hence, too, the alternative alias, 'Turk's cap lily.'

As with most angiosperms, the flower of *L. martagon* can be understood by its reproductive mission: it must attract a pollinator, provide a navigation map to its sexual organs, and offer irresistible splendours to reward a leisurely visit. *Lilium martagon* is highly evolved to tempt particular pollinators through its shape, colour, fragrance, texture and a rare golden treasure within. Take the petals: when the anthers are mature, the petals arch their necks backwards to display a gradient of dark purple markings that concentrate at the base, funnelling a pollinator's attention towards the pale green tunnels, offering nectar and honey – accessible only to a select class of pollinators. Like other species with hidden honey repositories, *L. martagon*'s sweet troughs are positioned such that the pollinator must hover at the stamens to access with a proboscis. *Macroglossum stellatarum*, the hummingbird hawk-moth, is the preferred pollinator, illustrated here, tapping the chamber while trembling at the anthers, brushing pollen onto its legs. The anthers chambers are lined with an oily substance that causes pollen grains to cluster, such that a raft of pollen is dispatched when the anther is jostled.

OPPOSITE:
Title: *Lilium Martagon*
Authors: Arnold and Carolina Dodel-Port
Language: German
Country: Switzerland
Series/Book: *Anatomisch physiologische Atlas der Botanik (The Anatomical & Physiological Atlas of Botany)*
Plate: 33
Publisher: J. F. Schreiber (Esslingen, Germany)
Year: 1878–93

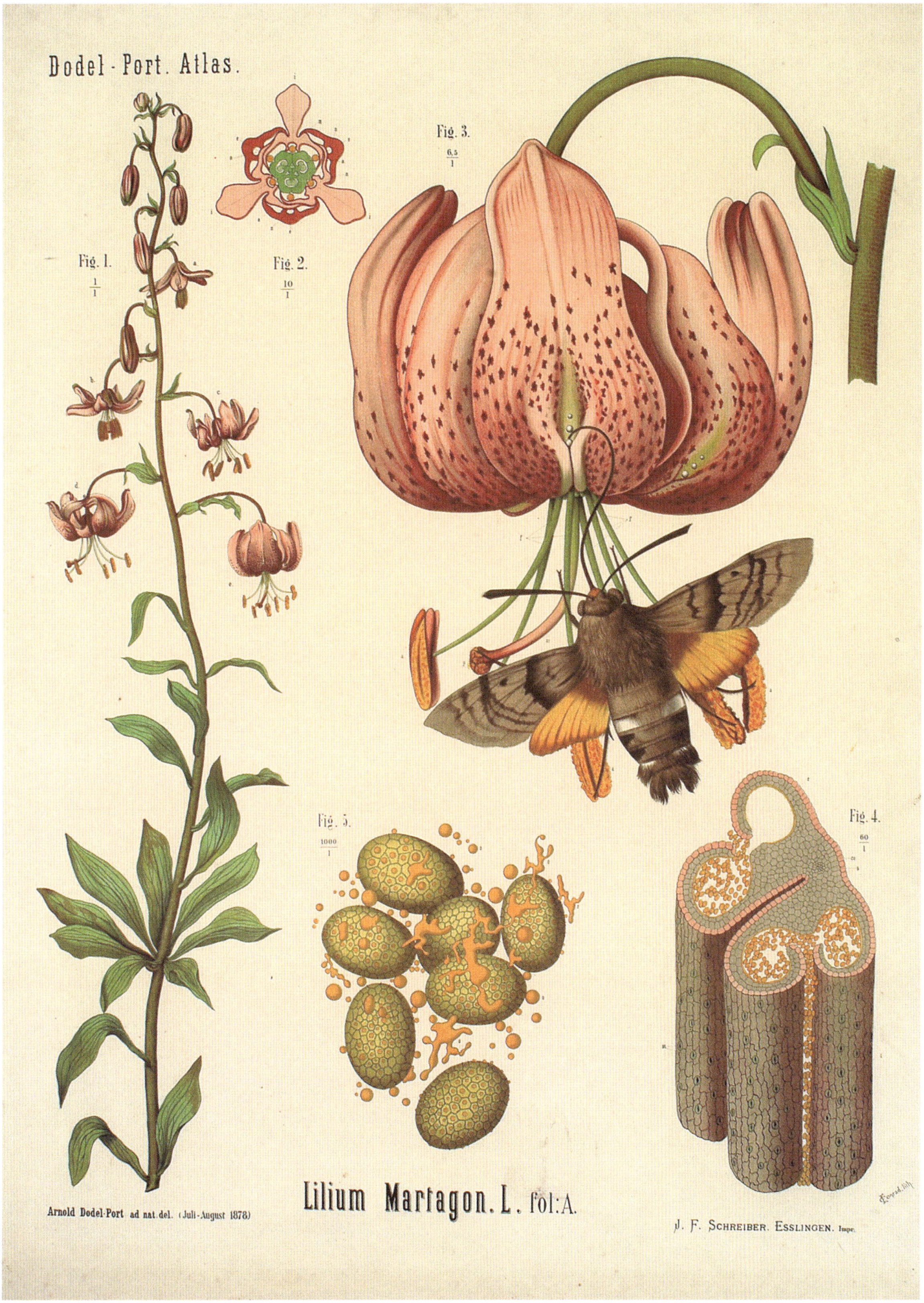
Dodel-Port. Atlas.
Fig. 1.
Fig. 2.
Fig. 3.
Fig. 4.
Fig. 5.
Lilium Martagon. L. fol: A.
Arnold Dodel-Port ad nat. del. (Juli-August 1878)
J. F. Schreiber. Esslingen. Impr.

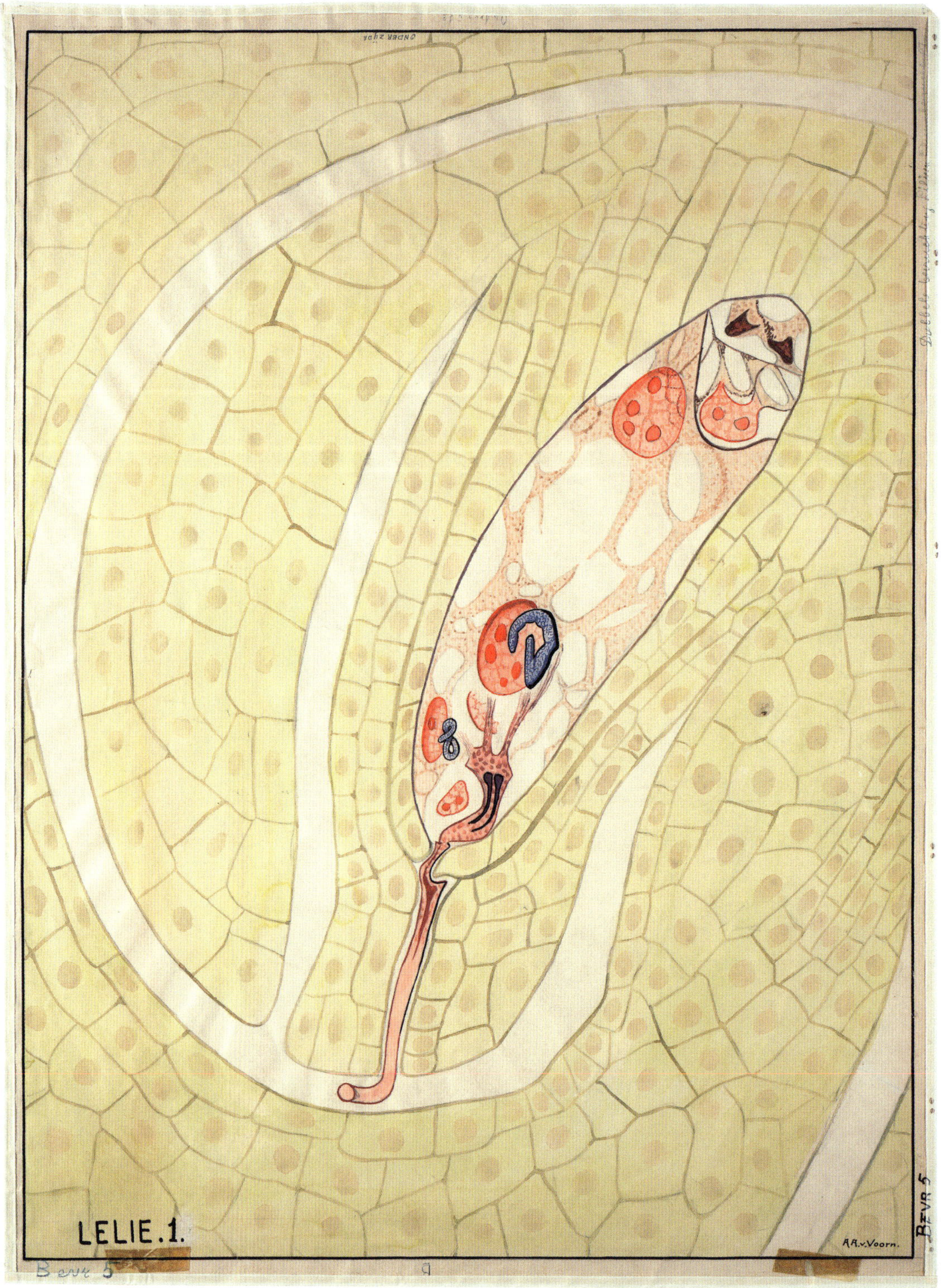
LELIE.1.
A.R.v.Voorn.

OPPOSITE: In 1898 a Russian biologist named Sergei Navashin discovered a curious feature of angiosperm reproduction – double fertilization, wherein two sperm cells must reach the ovary. The journey begins at the sticky stigma, where a pollen grain (a protective structure containing sperm cells) adheres and grows a pollen tube that reaches the ovary, where there is one ovule. A. A. van Voorn illustrates the next stages of the process in a *Lilium martagon* specimen. *Lilium martagon* and *Fritillaria orientalis* (also a member of the *Liliaceae* family) were the species first studied with the classical light microscope to observe the phenomenon of double fertilization. In van Voorn's chart we can see the two sperm cells that have entered the ovary through the pollen tube. One combines with an egg to form the zygote, and the other fertilizes the two polar nuclei to form the endosperm that will eventually envelop the growing zygote. The mature seed comprises the embryo, which will become a seedling; the protective seed coat; and the endosperm, which nourishes the seedling – as well as being the source of nutrients in cereal crops grown as food and fodder.

OPPOSITE:
Title: *Dubbele bevruchting Lilium Martagon (Double fertilization of Lilium Martagon)*
Author: A. A. van Voorn
Language: Dutch
Country: The Netherlands
Series/Book: N/A
Plate: N/A
Publisher: N/A
Year: c.1900s

RIGHT: Kazakhstan has a rich collection of tulips, about 34 wild species, and is possibly the birthplace of the tulip. Three of these Kazakh species have played a particularly important role in the establishment of the cultivated tulip, one of which was *Tulipa greigii*, seen on the left of the chart, named after Samual Greig, known as 'the Father of the Russian Navy' after a tour in battle against the Turks and Swedes in 1764. Here, this venerable species is illustrated by a mature bloom (fig. 1), a clipped stem and bulb (fig. 2), a chorus of stamens and a pistil, the cross-section of an ovary (fig. 4), and a floral diagram (fig. 5).

Now classified in the *Asparagaceae* family, *Convallaria majalis* (lily of the valley), in the centre of the chart, was first identified as a *Liliaceae*. Despite being highly poisonous, the sweet-smelling species has been popularly used in May flower celebrations, as well as homeopathic remedies – in Russia, *C. majalis* was used to treat epilepsy. A homeopathy journal published in 1884 describes how to prepare and administer the remedy: 'A quart bottle is filled with the flowers, covered with liquid, and then macerated in the sun for one week. The bottle is refilled and both tinctures are mixed. The dose is as many drops as the patient counts years, to be given in a spoonful of table wine mornings, afternoon, and in the evening.'

Finally, *Lilium lancifolium* – another Russian native. Similar to other true lilies, the flowers are borne on an erect stem. Unlike other lilies, it produces aerial bulblets, known as 'bulbils,' in the leaf axils along the stems. Cultivated in Europe since 1804, it also grows wild among the meadows, rocky slopes, and river valleys in northeastern China, Japan, and Korea. It can grow up to 6 feet (1.8m) tall, with nodding flowers and bowed pedicels. The species is beloved of butterflies and bees and – unfortunately for the lily – also deer and rodents.

RIGHT:
Title: *Liliaceae*
Author: V. G. Chrzhanovskii
Language: Russian
Country: Russia
Series/Book: *Sistematika Rasteniy Komplekt Plakatov iz 53 Listov (Systematics of Plants – Posters from 53 Sheets)*
Plate: 39
Publisher: Kolos Publishing
Year: 1971

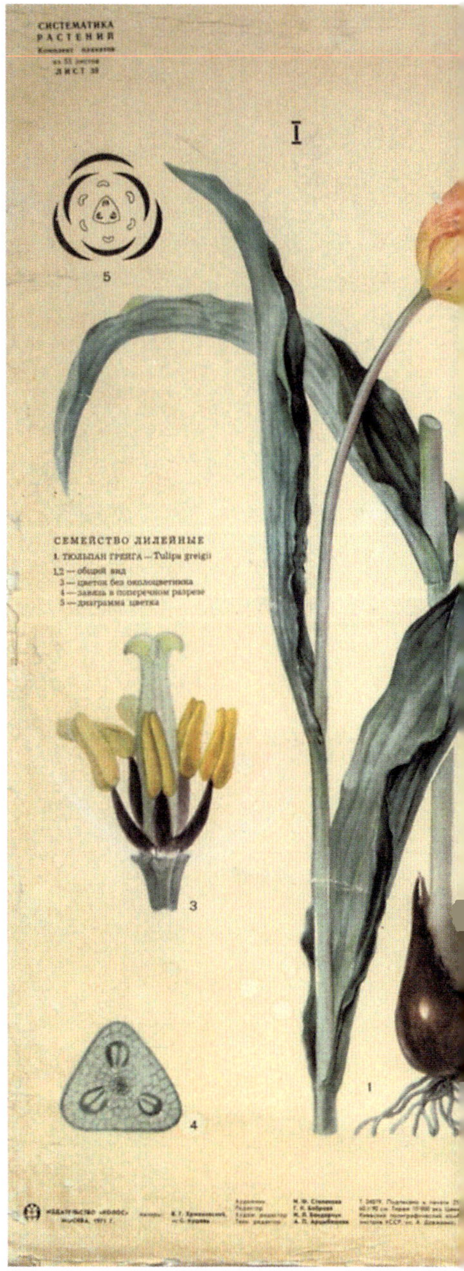

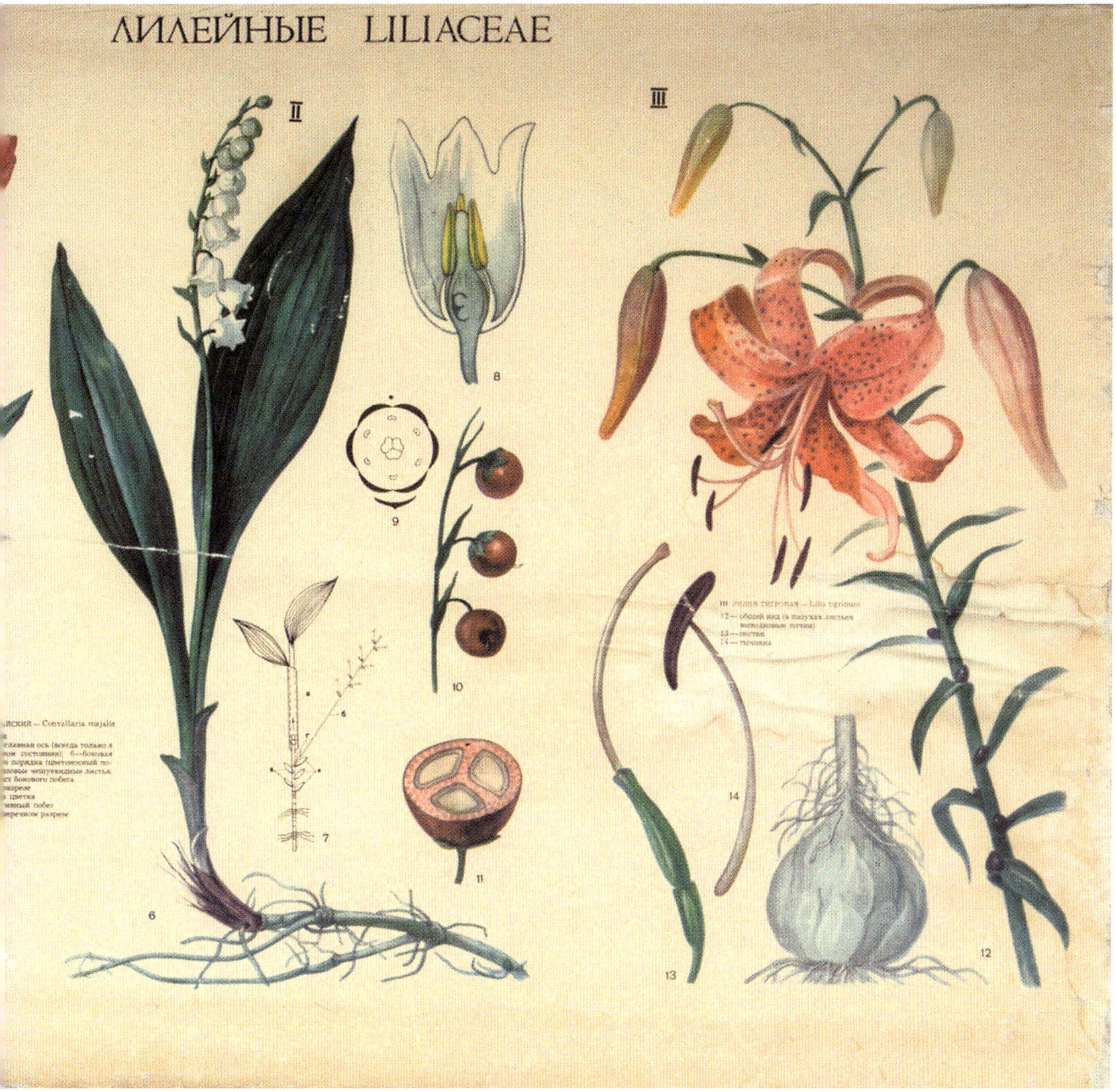
ЛИЛЕЙНЫЕ LILIACEAE
II
III
8
9
10
Convallaria majalis
7
11
6
14
13
12

XV MUSACEAE / BANANA FAMILY & BROMELIACEAE / PINEAPPLE FAMILY

There are only three genera in the *Musaceae* (banana) family, encompassing just over 90 species from tropical Africa and Asia. The genera *Ensete* is grown as an ornamental banana, and *Musa* as an important commercial crop, while *Musella* has just one species, the Chinese dwarf banana. Bananas are not trees, but monocarpic, evergreen, perennial herbs.

The *Bromeliaceae* are more numerous and varied, with 3,700 species in 76 genera, including the pineapple. Also perennial herbs, bromeliads exhibit various adaptations to difficult environments, being in the main epiphytic (air plants), with some terrestrial species. Their often dramatic forms and hues make them valued as ornamental plants, especially as houseplants in temperate climates. Unfortunately, their popularity has led to many species becoming endangered, such as *Puya raimondii*. This species from the Andes produces a spectacular flower spike, up to 40 feet (12m) tall, and is thought to be one of the most ancient of plants.

Characteristics of *Musaceae*: pseudostem of overlapping leaf sheaths; huge leaves with long petioles and entire, conspicuously veined laminas; large, pendent spike of whorled flowers in bract axil, with female flowers at base and male flowers at top. Fruit is a cylindrical, fleshy berry; sometimes a capsule.

Characteristics of *Bromeliaceae*: leaves usually in rosettes, often forming a water and nutrient reservoir or 'tank,' but variable – spiny or toothed, with peltate scales or leathery; often conspicuous, colourful floral bracts; three sepals, three petals, six stamens; variable inflorescence, solitary or in raceme, spike, panicle, or capitula (condensed head); fruit is a berry or capsule; often specialized, epiphytic roots.

RIGHT:
Title: *Ananas (Ananas sativus Schult.)*
Author: Hermann Zippel; Illustrator: Carl Bollmann
Language: German
Country: Germany
Series/Book: *Ausländische Kulturpflanzen in farbigen Wandtafeln (Foreign Crops in Coloured Wall Panels)*
Plate: II Abteilung, 7
Publisher: Friedrich Vieweg & Sohn (Braunschweig, Germany)
Year: 1897

Ausländische Kulturpflanzen in farbigen Wandtafeln.

II. Abteilung. Tafel 7.

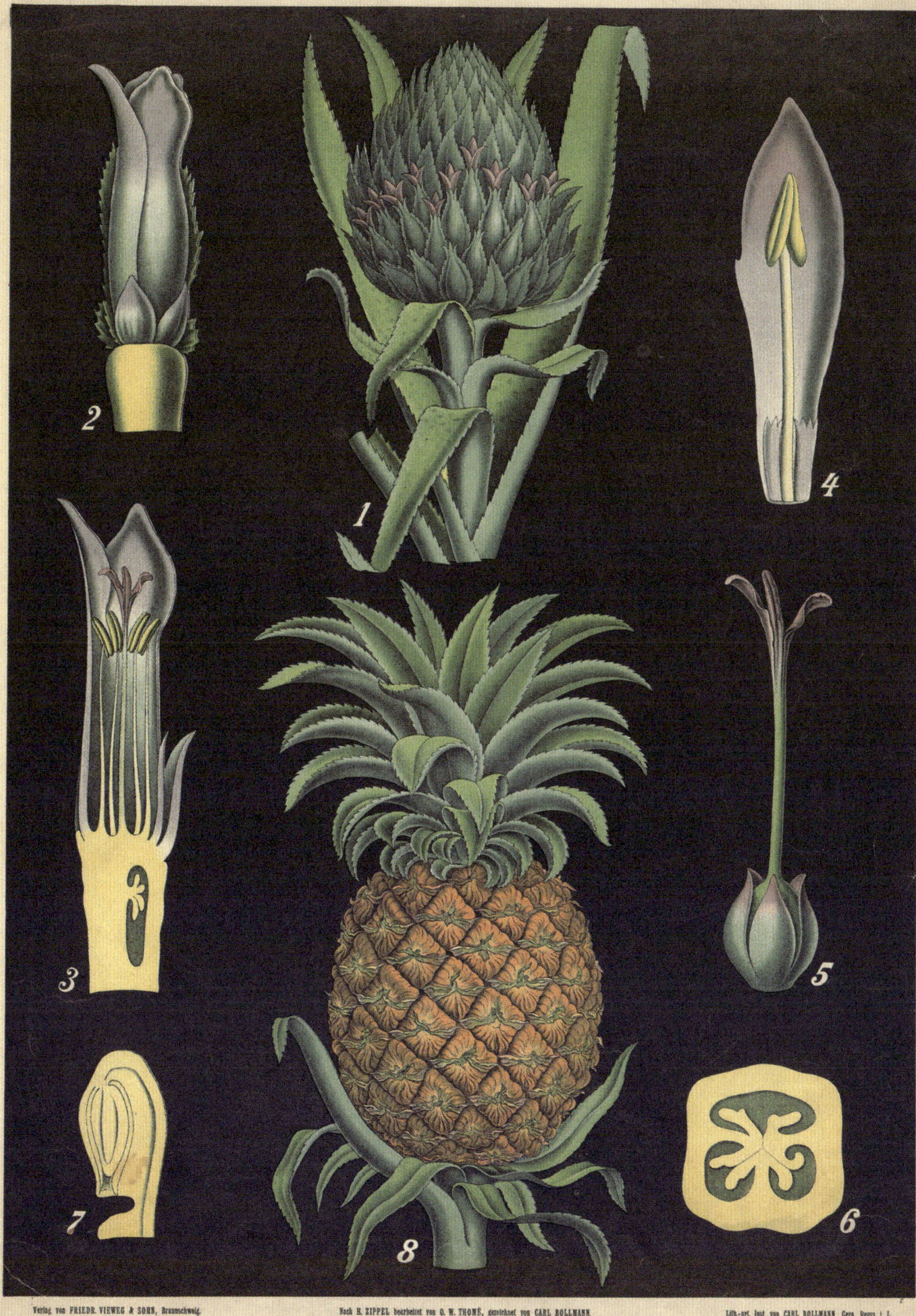

Verlag von FRIEDR. VIEWEG & SOHN, Braunschweig. Nach H. ZIPPEL bearbeitet von O. W. THOMÉ, gezeichnet von CARL BOLLMANN Lith.-art. Inst. von CARL BOLLMANN, Gera, Reuss j. L.

Ananas (Ananas sativus Schult.).

1. Blütenstand; *etwas vergr.* — 2. Blüte; *Vergr.* 5. — 3. Blüte im Längsschnitte; *Vergr.* 6. — 4. Blumenkronblatt mit den beiden Schüppchen und dem an seinem Grunde angewachsenen Staubblatte; *Vergr.* 8. — 5. Kelch und Stempel; *Vergr.* 8. — 6. Querschnitt durch den Fruchtknoten; *vergr.* — 7. Samenanlage; *vergr.* — 8. Fruchtstand; *natürl. Größe.* — Figur 1 nach K. Koch in Engler-Prantl; 2 bis 7 nach Le Maout und Decaisne.

OPENING PAGE: While the pineapple (*Ananas*) was considered an exotic oddity in Europe for several centuries, and a symbol of wealth and royalty (Louis XV grew them at Versailles), by the late nineteenth century, it was not unusual to see sliced and sugared pineapple slices as a treat. As always, Zippel and Bollmann include a few paragraphs about the social history of the species: 'in recent years, their consumption in Europe has increased significantly thanks to the fast steamer connection.' Their illustrations begin with an inflorescence that will swell to a fruit (fig. 1), flower (fig 2), bloom in longitudinal sections (fig. 3), a stamen (fig. 4), a style and calyx (fig. 5), cross-section of the ovary (fig. 6), ovule (fig. 7), and a mature pineapple (fig. 8). The fruit is a cumulation of flowers called an infructescence.

RIGHT: A. Peter illustrates the *Bromeliaceae* with four species. In the lower right (fig. 1), a *Vriesea carinata* (lobster claw) native to Brazil, explodes in reds and yellows, shooting stamens and styles across the chart.

Tillandsia usneoides (Spanish moss, fig. 2), is neither Spanish nor a moss. Spanish moss was given its name by French explorers, who were reminded of the Spanish conquistadors' long beards, so they called it *Barbe Espagnol*, or 'Spanish Beard.' An epiphytic angiosperm, it is adept at draping trees with silver-grey blankets. Its flowers are inconspicuous, and so Peter offers students a magnified view of the stamens, stigma and ovary. While it can reproduce sexually, *Tillandsia usneoides* is more likely to propagate by fragmented pieces of plant called festoons. When a festoon is broken off and carried away by wind or birds (using it for nest material), it will begin to grow into a full plant if it lands in an acceptable place – ideally a healthy tree in a tropical swampland.

Billbergia bakeri (now called *B. distachya*) is a Mexican native and a firecracker of a bromeliad, erupting in chartreuse and pink tassels of tubular bracts. Instead of illustrating its signature feature, Peter instead dissects the ovary to show ten ovules that will become a multi-seeded fruit.

Finally, Peter illustrates a luscious *Ananas sativus* (now called *A. comosus* var. *comosus*, fig. 4). He gives equal emphasis to the pineapple's exterior topography, allowing a study of the individual fruits, and the interior, where multiple fruits have been fused into a fibrous flesh interrupted by a few seeds.

RIGHT:
Title: *Bromeliaceae*
Author: A. Peter
Language: German
Country: Germany
Series/Book: *Botanische Wandtafeln*
Plate: 17
Publisher: Paul Parey (Berlin, Germany)
Year: 1901

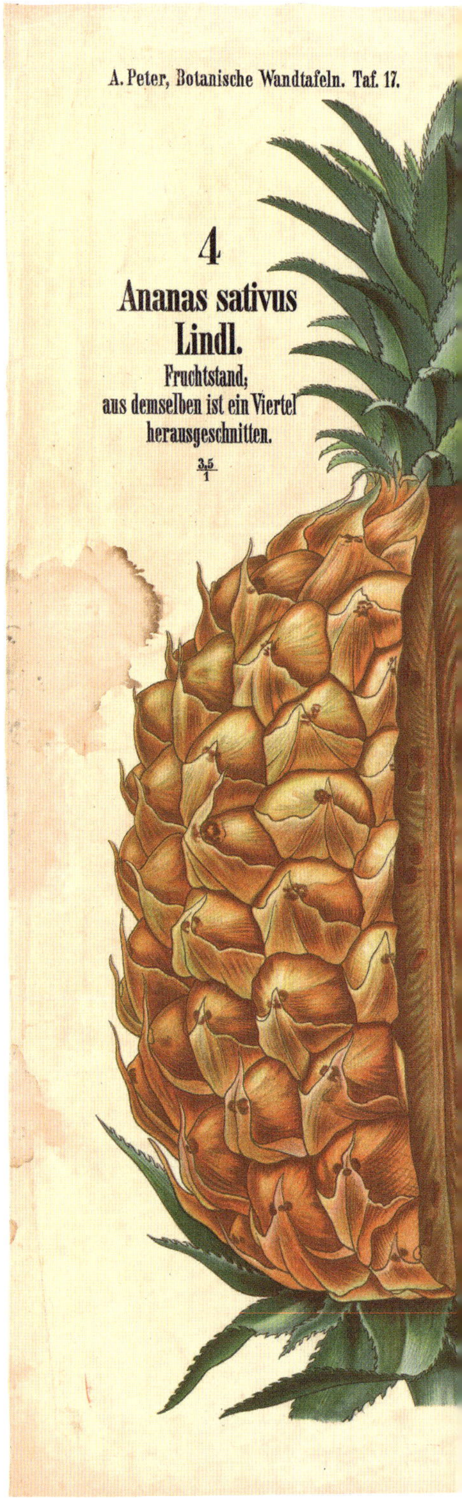

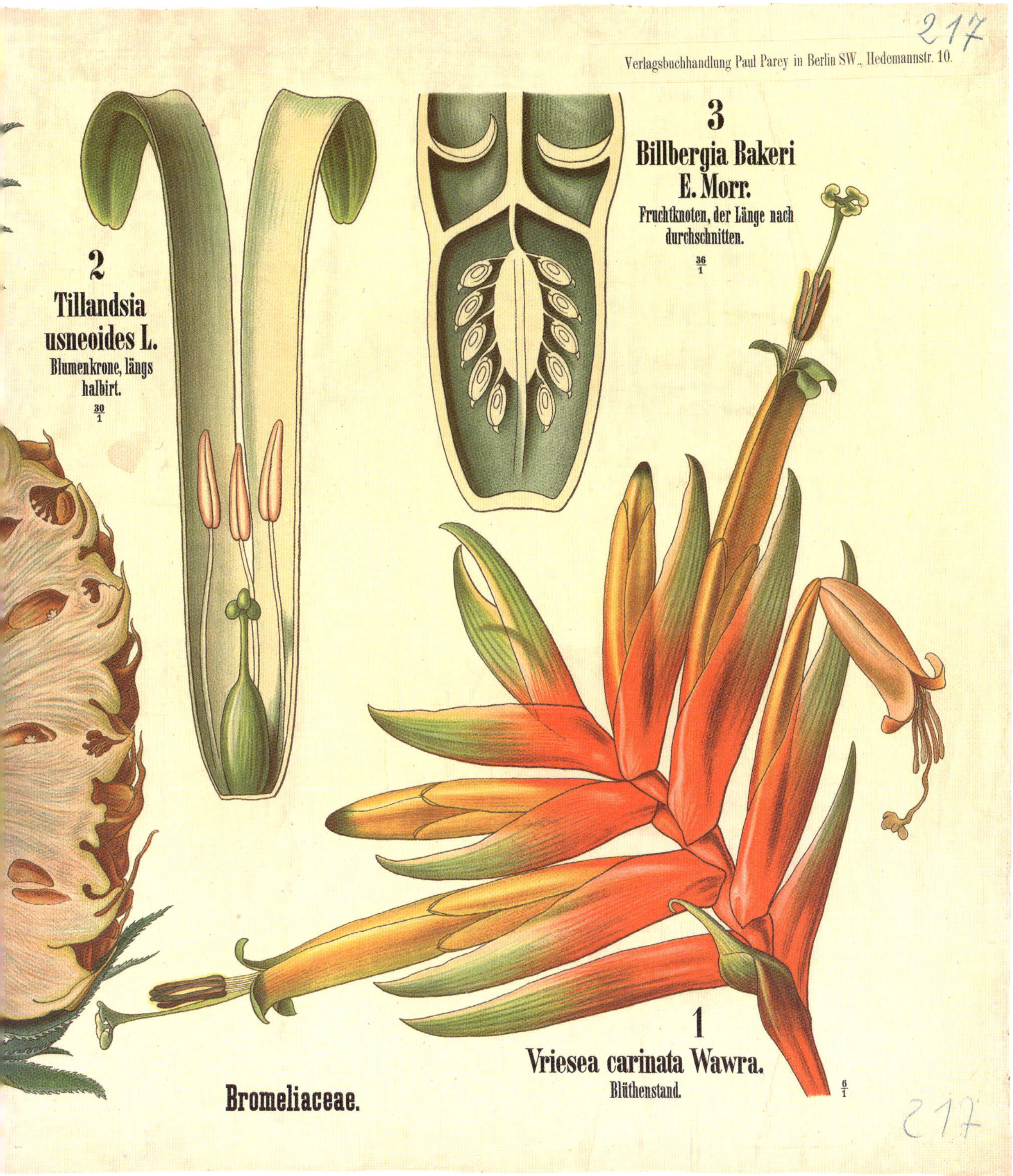
217
Verlagsbuchhandlung Paul Parey in Berlin SW., Hedemannstr. 10.
3
Billbergia Bakeri
E. Morr.
Fruchtknoten, der Länge nach
durchschnitten.
36/1
2
Tillandsia
usneoides L.
Blumenkrone, längs
halbirt.
30/1
1
Vriesea carinata Wawra.
Blüthenstand.
6/1
Bromeliaceae.
217

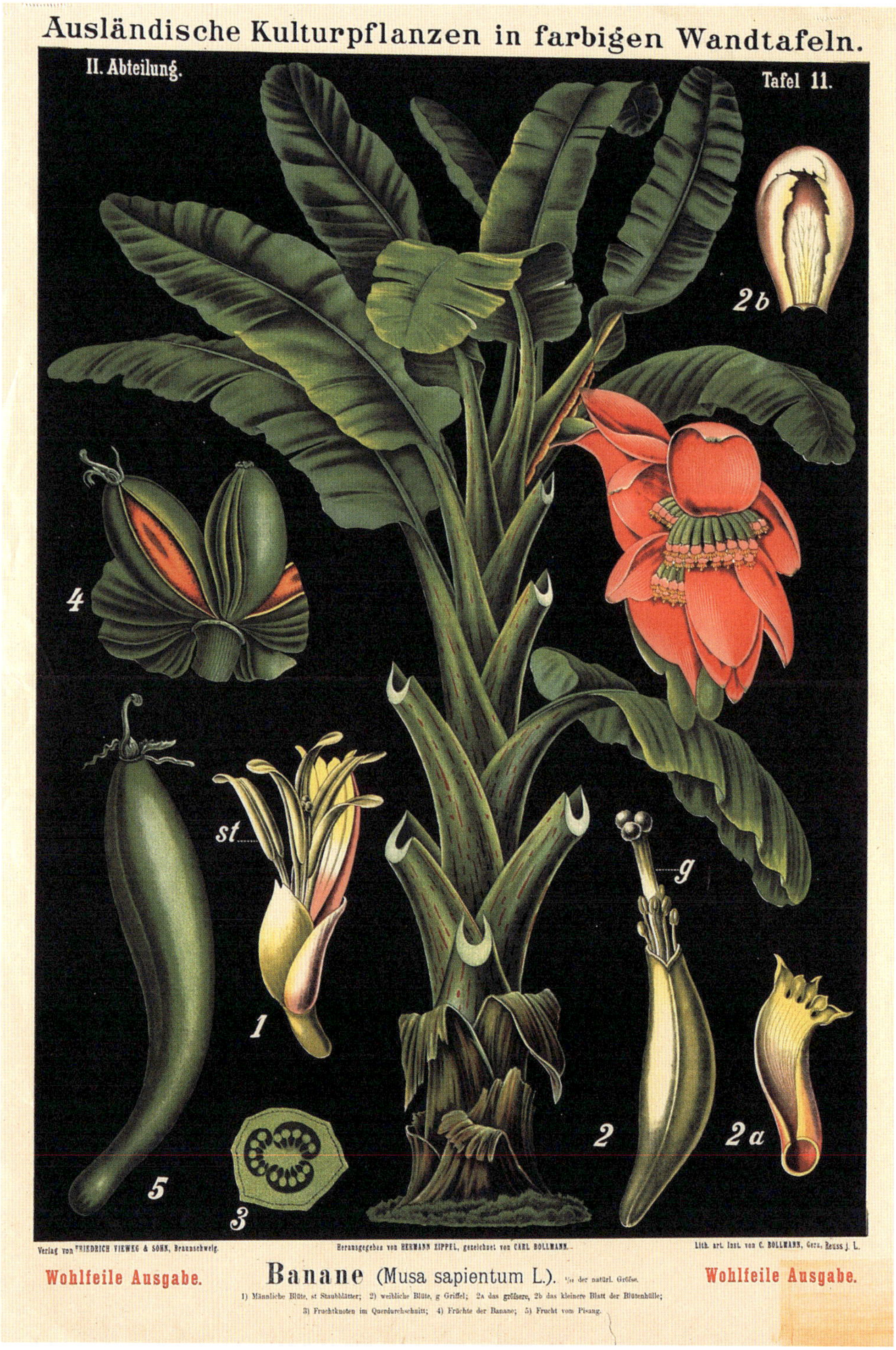
Ausländische Kulturpflanzen in farbigen Wandtafeln.
II. Abteilung.
Tafel 11.
2b
4
st
1
g
2
2a
5
3
Verlag von FRIEDRICH VIEWEG & SOHN, Braunschweig.
Herausgegeben von HERMANN ZIPPEL, gezeichnet von CARL BOLLMANN.
Lith. art. Inst. von C. BOLLMANN, Gera, Reuss j. L.
Wohlfeile Ausgabe.
Banane (Musa sapientum L.).
Wohlfeile Ausgabe.
1) Männliche Blüte, st Staubblätter; 2) weibliche Blüte, g Griffel; 2a das größere, 2b das kleinere Blatt der Blütenhülle;
3) Fruchtknoten im Querdurchschnitt; 4) Früchte der Banane; 5) Frucht vom Pisang.

LEFT: Oddly enough, Zippel and Bollmann's chart of a banana plant is dominated not by the fruit, but by the leaves, which, they observe, 'are arranged in a right-handed spiral.'

It's useful to consider the context of the chart and the species; first, *Musa × sapientum* (now called the *M. × paradisiaca*, the Latundan banana), is smaller and not quite as sweet as the *M. acuminata* (Cavendish banana) which wouldn't become ubiquitous for several more decades. At the time, the banana plant was considered an important economic crop in tropical regions. Indeed, Zippel and Bollmann write prolifically on the many useful parts of the plant: First, the fruit, which is edible cooked or raw, and makes an acceptable beverage when mixed with water. Second, the terminal bed of the flower bulb, which is eaten as a vegetable in southern China. Third, the leaf sheaths and rootstock, which are consumed in Ethiopia. Fourth, the leaves that, when warmed, become 'soft and pliable like paper and waterproof, and provide an excellent packing material,' or can be used to make ceilings, umbrellas, or rope. Fifth, the leaf sheath fibres, which can be woven into mats or wickerwork; and sixth, the shafts, which contain finer fibres known as 'Manila hemp', can be spun with silk to make 'luxury goods' or paper. When the whole plant has so many various uses, does the fruit not become incidental?

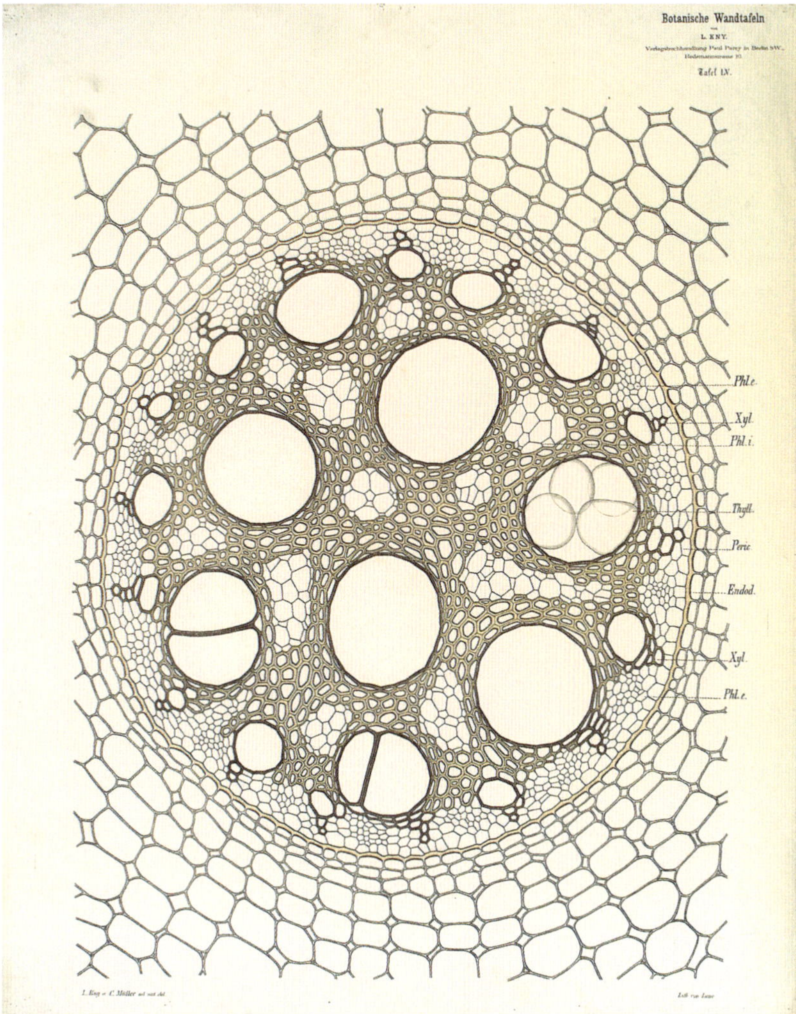

ABOVE RIGHT: As a counterpoint to Zippel and Bollmann's overview of the whole plant and its various practical uses, consider Leopold Kny's microscopic view of the vascular bundle of its root. The biologist has selected *M. × sapientum* to represent a primary transport system of all vascular plants (that is to say, all seed-bearing plants, as well as those that reproduce via spores, such as ferns; examples of non-vascular plants include mosses and algae). Most important of the tissues that constitute this system are the xylem and phloem (marked Xyl. and Phl., respectively, on Kny's diagram). The xylem is responsible for the transportation of water and minerals throughout a plant, while the phloem conducts organic nutrients, known as photosynthates, such as sucrose, to where they are needed. Kny's chart could be used to illustrate a finer point of the anatomy of *M. × sapientum*, or an example of a transportation map of vascular plants.

OPPOSITE:
Title: *Banane (Musa sapientum L.)*
Author: Hermann Zippel; Illustrator: Carl Bollmann
Language: German
Country: Germany
Series/Book: *Ausländische Kulturpflanzen in farbigen Wandtafeln (Foreign Crops in Coloured Wall Panels)*
Plate: II Abteilung, 11
Publisher: Friedrich Vieweg & Sohn (Braunschweig, Germany)
Year: 1897

ABOVE:
Title: *Musa sapientum*
Author: L. Kny
Language: German
Country: Germany
Series/Book: : *Botanische Wandtafeln*
Plate: 55
Publisher: Paul Parey (Berlin, Germany)
Year: 1874

XVI OLEACEAE / OLIVE FAMILY

This family of forest trees, shrubs and a few woody climbers comprises 27 genera and approximately 700 species. They grow in many areas of the world, but are most numerous in East and Southeast Asia and Australia. The deciduous plants tend to originate in northern temperate zones and the evergreens in warmer climates. They are of course named after *Olea europaea*, the olive tree, valued since ancient times as a staple crop for its fruits and the oil they yield. Plants in this family have hard, long-lasting wood, such as *Fraxinus* (ash), so provide timber; another industrial product is oil of jasmine, used in perfumes, from *Jasminum grandiflorum*. Garden *Oleaceae* often have showy, fragrant blooms or are grown as hedges, such as *Forsythia*, *Ligustrum*, *Osmanthus*, *Phillyrea* and *Syringa*.

Characteristics: leaves usually opposite, exstipulate, simple or pinnately compound. Flowers often white, in cymose panicles, actinomorphic; usually four each of united sepals and petals, two stamens on corolla tube; fruit may be a long capsule, samara, drupe or berry.

OPPOSITE:
Title: *Olea europaea*
Author: Alois Pokorny
Language: N/A
Country: Germany
Series/Book: *Botanische Wandtafeln*
Plate: N/A
Publisher: Smichow (Neubert, Germany)
Year: 1894

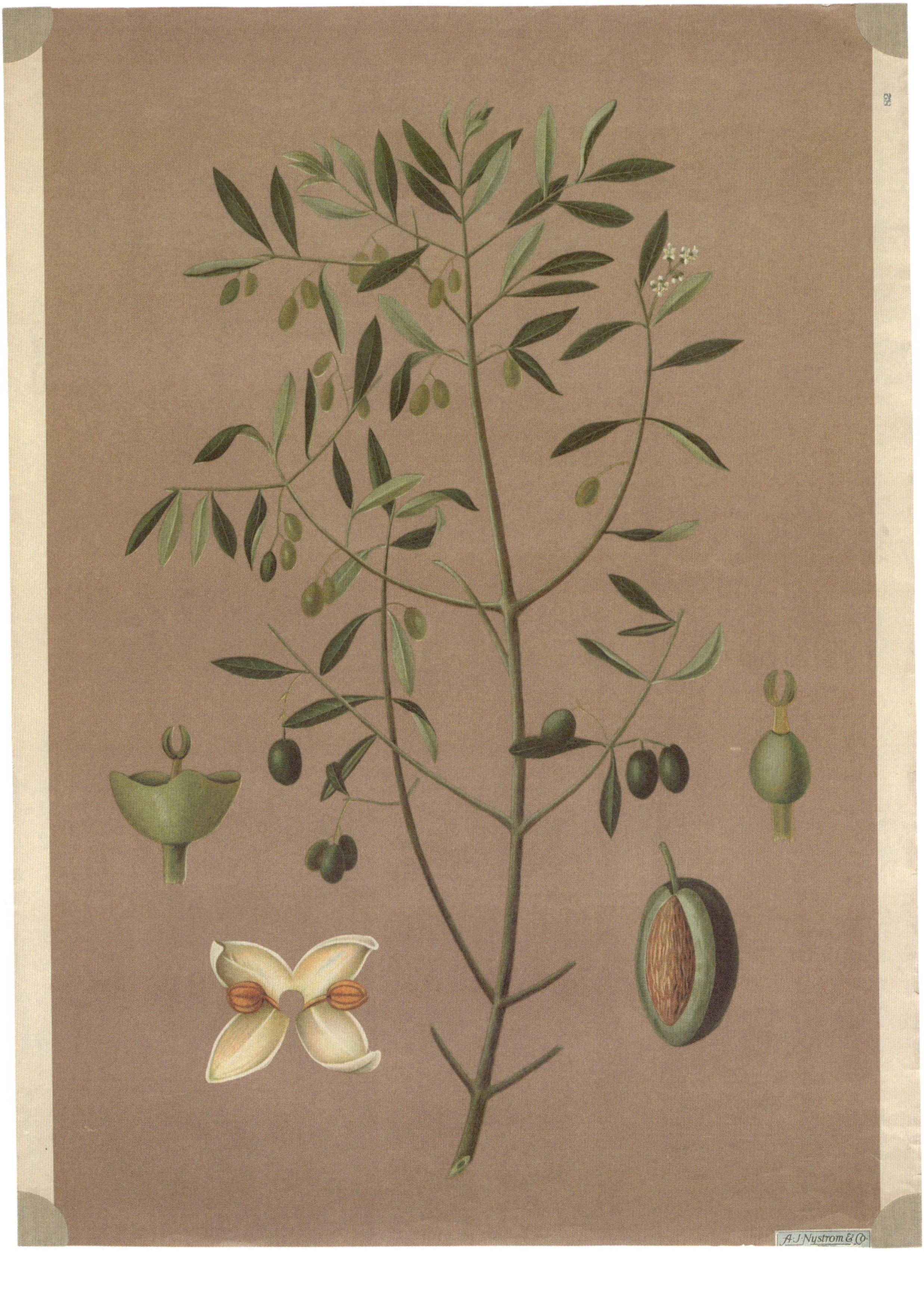
A·J·Nystrom & Co·

OPENING PAGE: As well as authoring educational wall charts, Alois Pokorny was known for his work in nature printing, a technique that was popular in Austria in the mid-nineteenth century. Working with botanic specimens, Pokorny developed an aesthetic that is reflected in the charts he authored. His style was minimalist and natural, evoking an herbarium specimen. Wall charts, however, allowed varying degrees of magnification, which Pokorny has applied to the elements surrounding an *Olea europaea* branch. The chart includes a calyx and pistil and a petal with anthers at left, and a developing fruit and the cross-section of a mature fruit at right. The branch itself demonstrates all these elements to scale, from the delicate white flowers to the young fruit, and the more mature fruit at the bottom, which will eventually ripen to glossy black.

RIGHT: What is the olive without its grove? Here, an Italian chart by an unknown author puts *Olea europaea* in its environment; a sun-kissed, rocky grove by the gentle waters of the Mediterranean. The scene is idyllic by any standards, including those of the European olive, a species that favours craggy, coastal landscapes and a warm, sunny climate. Also in the scene are the olive pickers, with sacks full of fruit. Olive trees are of great economic importance in the Mediterranean, particularly for olive oil – about 90 percent of all harvested fruit will be made into oil. If the economic intention of the chart were not already clear, the author has illustrated a fruiting bough, flower and cross-section of the fruit. At the very top is also an illustration of the hanging fruit of *Ficus carica* (fig) – another popular fruit, belonging to the mulberry family – mirroring the position of the little shrub at the bottom-right of the scene.

RIGHT:
Title: *Olea europaea*
Author: Unknown
Language: Italian
Country: Italy
Series/Book: Unknown
Plate: N/A
Publisher: G. B. Paravia (Turin, Rome, Milan, Florence, Naples, Palermo – Italy)
Year: Unknown

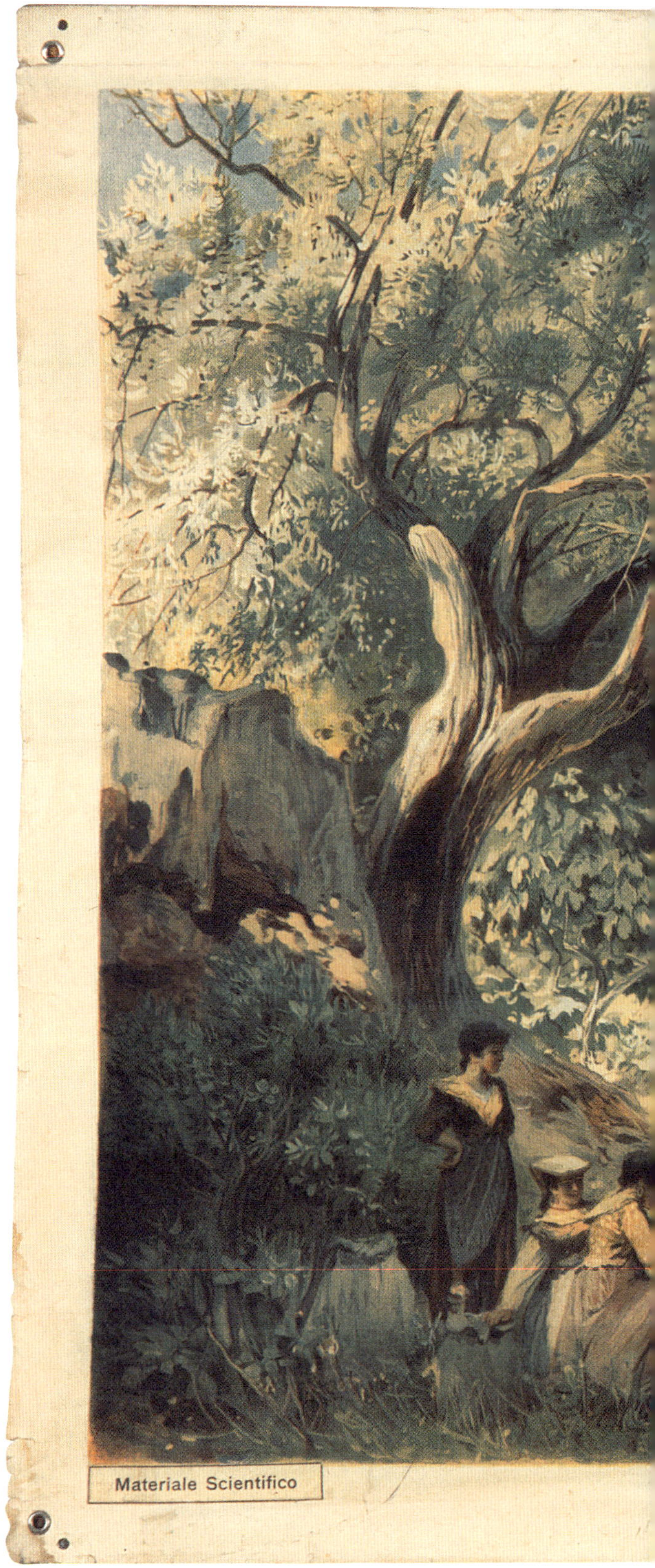

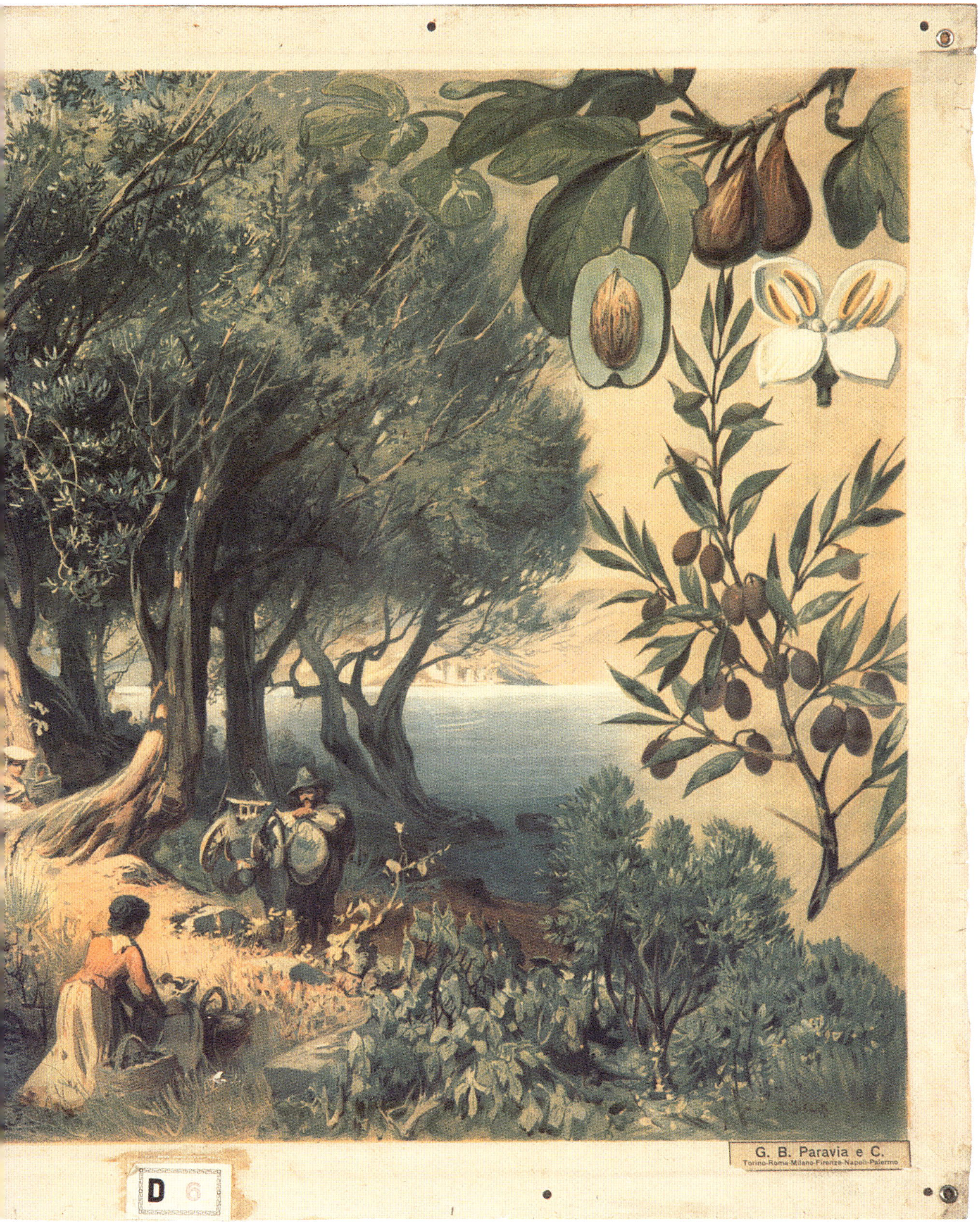
G. B. Paravia e C.
Torino-Roma-Milano-Firenze-Napoli-Palermo
D 6

THIS SPREAD: Von Engleder represented the *Oleaceae* family with two distinctly different species: *Syringa vulgaris* and *Fraxinus excelsior.*

Syringa vulgaris has been widely cultivated in Europe since the beginning of the sixteenth century. Taking its vernacular name from its scent and colour, the common lilac is famous for pale purple panicles and tubular blossoms that amplify a sweet fragrance. Von Engleder's chart includes a flowering sprig (fig. a); a cutting with a full blossom, a bud, and exposed pistil and stamen (fig. b); the cross-section of a flower, revealing anthers (fig. c); an anther (fig. d); a calyx and style (fig. e); dehiscent two-celled, two-seeded fruits (figs. f, g, h); the seed (figs. i, k); and a floral diagram (fig. l).

Fraxinus excelsior (ash tree) is a morphological outlier in the family. Flowers are without petals, and often lacking a calyx. Each tree can bear male flowers, female flowers, bisexual flowers or a mixture – although single-sex trees are more common – further, the same tree can produce all male flowers one year and all female flowers the next (or vice versa). Von Engleder illustrates a sprig of female flowers (fig. a); a sprig of male flowers (figs. b, c); a male flower (fig. d); a bisexual flower (fig. e); the cross-section of an ovary (fig. f); a branch of foliage and young fruits (fig g); an open samara (fig h); a cross-section of a seed within a samara (fig i); a vertical cross-section of a seed (fig. k); a germinating seed (fig. l); and a young plant (fig. m).

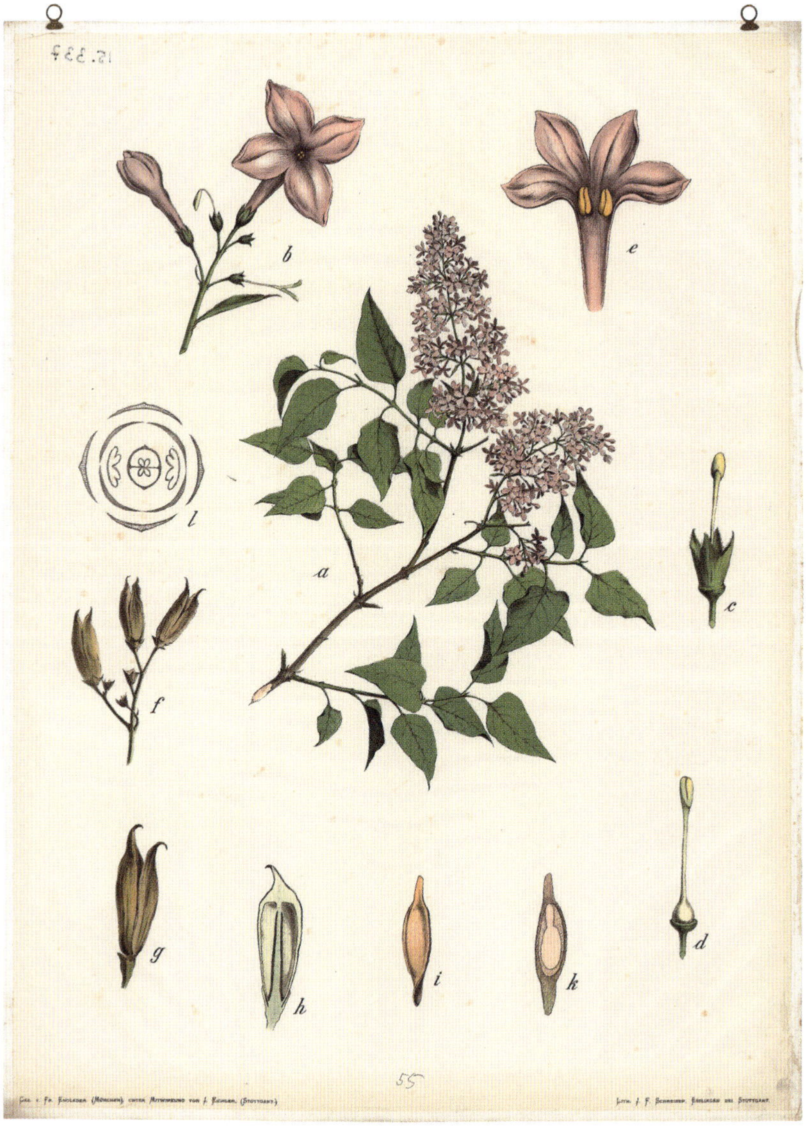

ABOVE & OPPOSITE:
Title, above: *Syringa vulgaris*; Opposite: *Fraxinus excelsior*
Author: Von Engleder; Illustrator: C. Dietrich
Language: German
Country: Germany
Series/Book: *Engleders Wandtafeln für den naturkundlichen Unterricht Pflanzenkunde (Engleder's Wall Charts for Natural History Lessons: Botany)*
Plates: 55; 56
Publisher: J. F. Schreiber (Esslingen, Germany)
Year: 1897

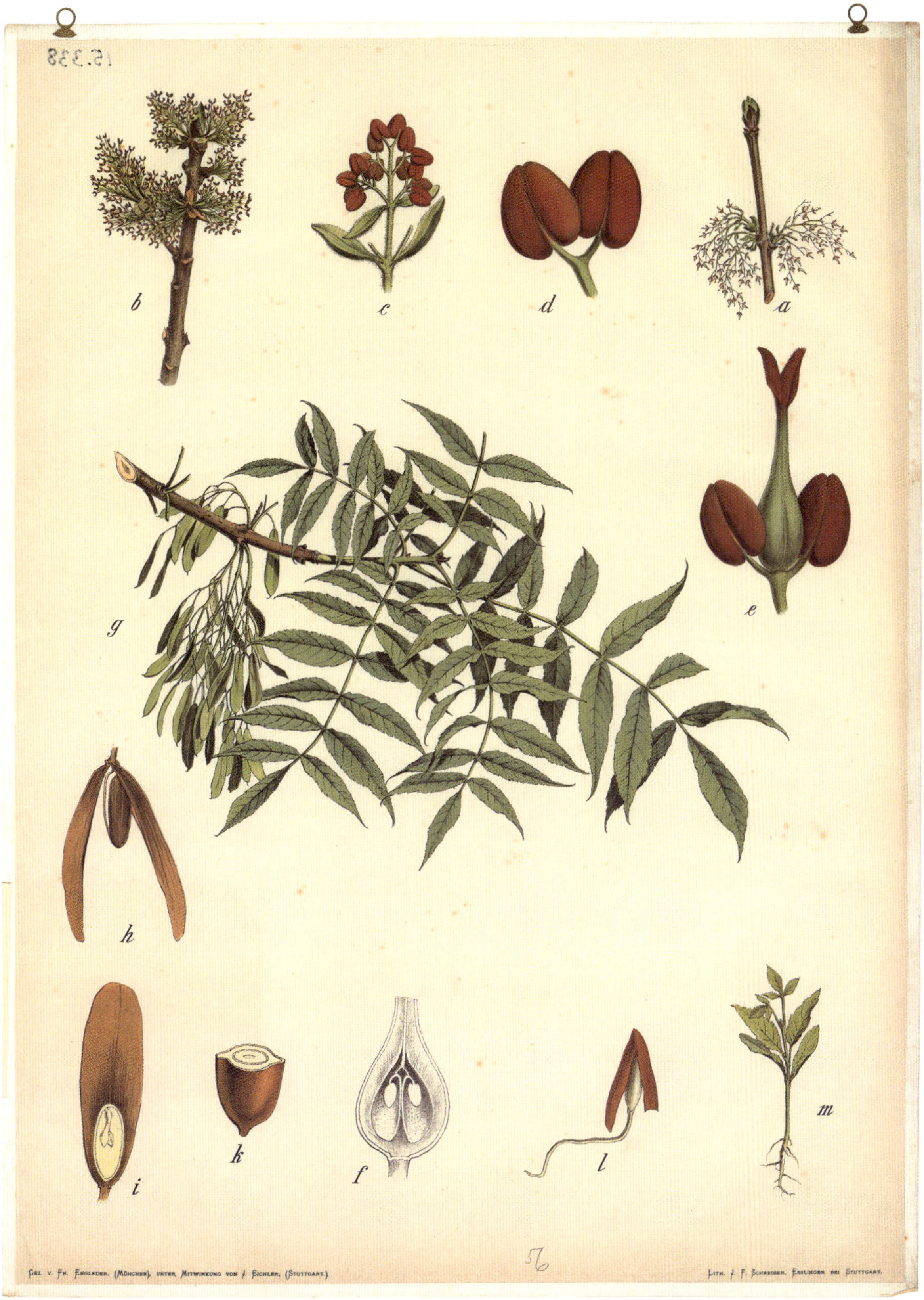
15.338
b
c
d
a
e
g
h
i
k
f
l
m
56
Gez. v. Fr. Engleder. (München), unter Mitwirkung von J. Eichler. (Stuttgart.)
Lith. J. F. Schreiber, Esslingen bei Stuttgart.

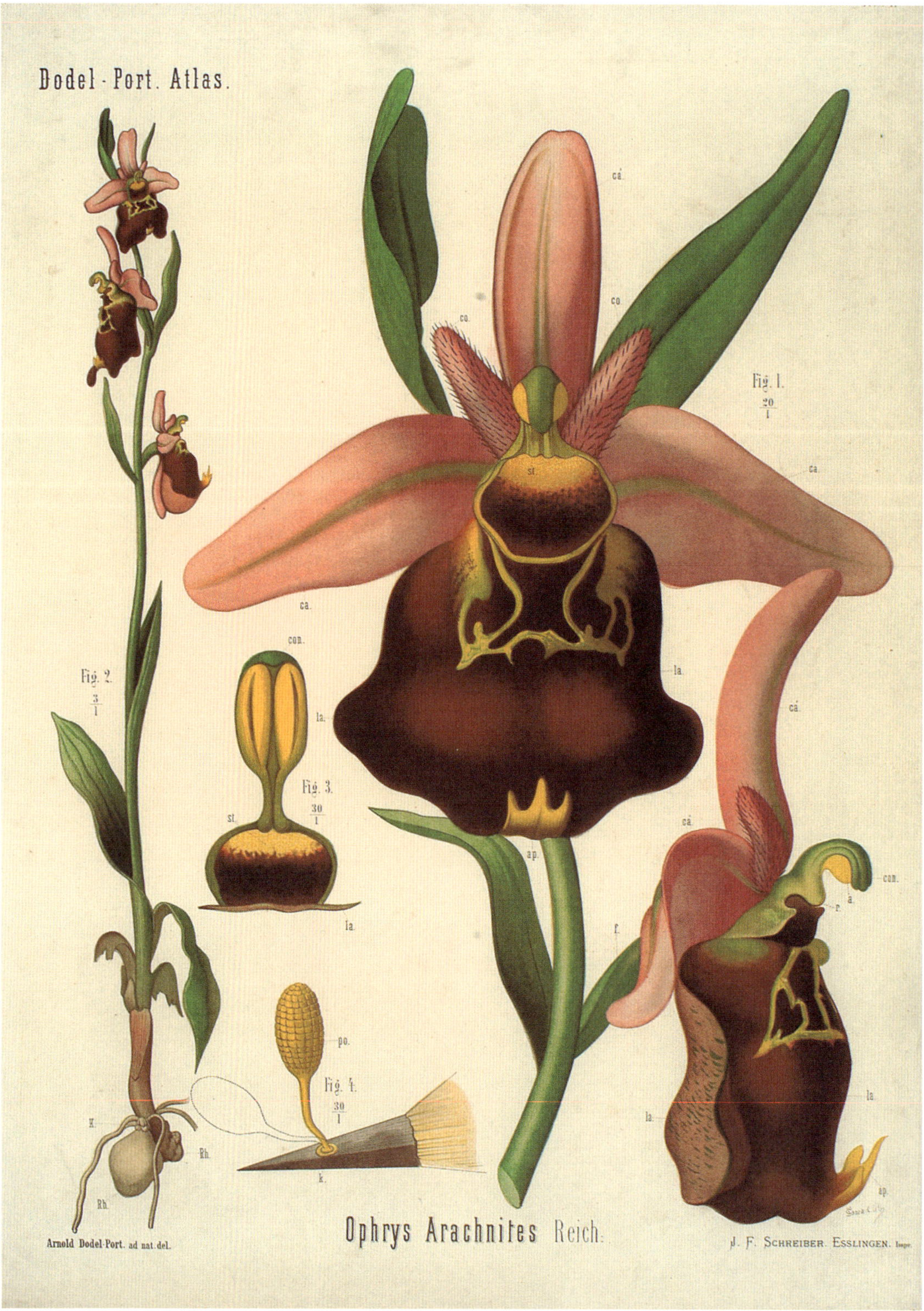

Dodel - Port. Atlas.
Fig. 1.
20/1
Fig. 2.
3/1
Fig. 3.
30/1
Fig. 4.
30/1
Ophrys Arachnites Reich.
Arnold Dodel-Port. ad nat. del.
J. F. Schreiber. Esslingen.

XVII ORCHIDACEAE / ORCHID FAMILY

Vying with *Asteraceae* (see page 32) as the largest family of angiosperms, with over 700 genera and about 28,000 species, the orchid family is widespread in tropical and temperate climates. Orchids are highly variable, with a bewildering array of natural hybrids, all perennial herbs of terrestrial, saprophytic or epiphytic habit. *Vanilla planifolia* has been grown since the time of the Aztecs for its flavorsome seedpods, and the roots of several *Orchis* species yield an edible starch called salep. Many *Orchidaceae* are grown as ornamentals for their beautiful, dramatic, long-lasting, and sometimes fragrant blooms – for example *Cattleya*, *Cymbidium*, *Dendrobium*, *Miltonia*, *Oncidium*, *Paphiopedilum*, *Phalaenopsis* and *Vanda*. The orchid family displays several unique characteristics, especially in the flower.

Characteristics: flower is zygomorphic with six tepals, one modified to labellum or 'lip' – flower is resupine, rotating 180 degrees so that the labellum, at first an upper petal, ends up at the base and forms a landing platform for insect pollinators. Typically, stamens, style and stigma fuse into a 'column,' and pollen often occurs in waxy masses on viscidia (sticky pads). Roots may be rhizomatous or epiphytic; leaves entire with sheathed bases, parallel veins, sometimes plicate; many have pseudobulbs, swollen stem bases used as water storage organs. The fruit, a capsule, produces millions of the smallest seeds in the world.

OPPOSITE:
Title: *Ophrys Arachnites, Reich.*
Authors: Arnold and Carolina Dodel-Port
Language: German
Country: Switzerland
Series/Book: *Anatomisch physiologische Atlas der Botanik* (*The Anatomical & Physiological Atlas of Botany*)
Plate: 36
Publisher: J. F. Schreiber (Esslingen, Germany)
Year: 1878–93

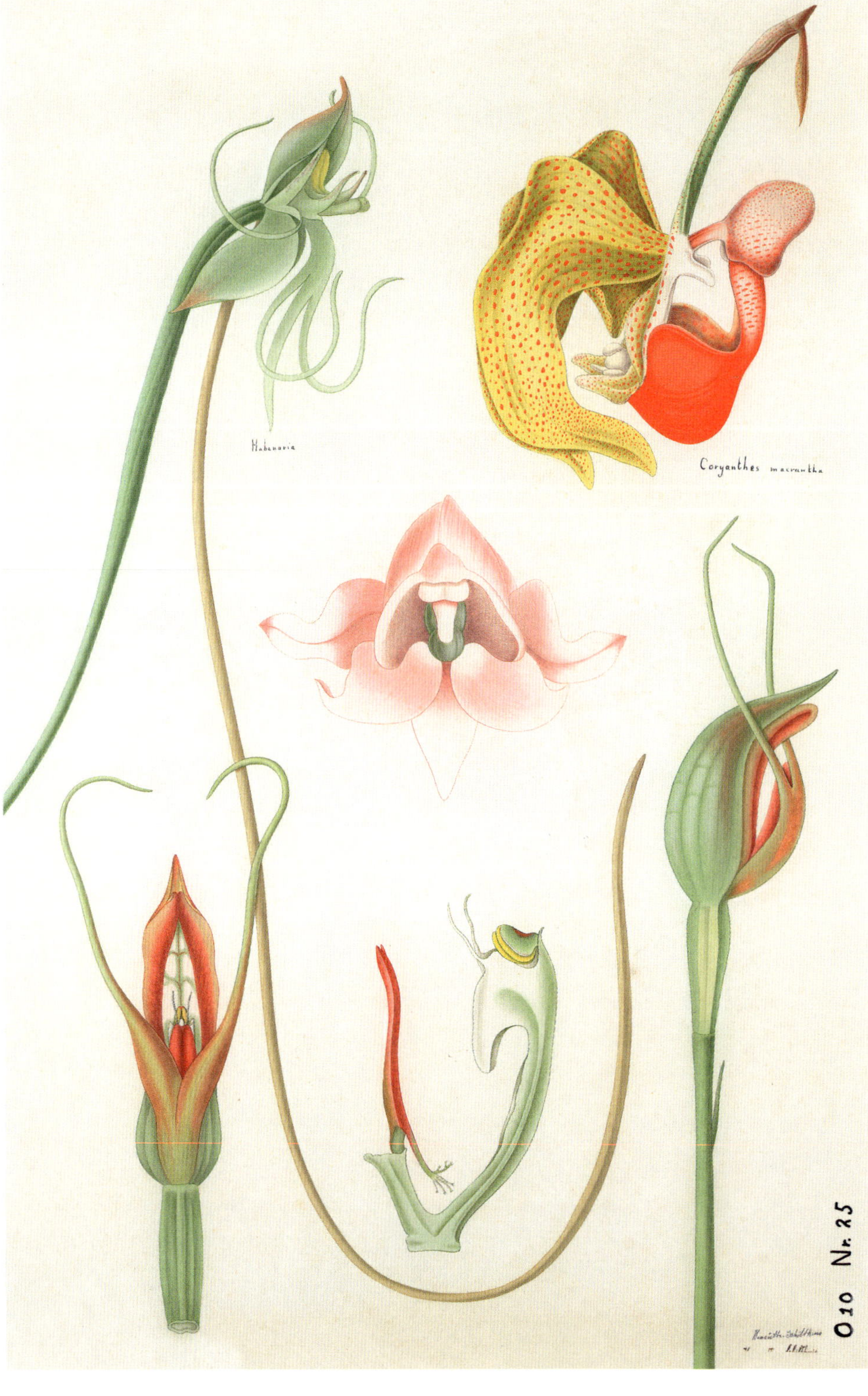
Habenaria
Coryanthes macrantha
O10 Nr. 25

OPENING PAGE: Of the *Orchidaceae* family, Arnold and Carolina Dodel-Port write in their atlas, 'Since this species-rich family is spread almost all over the earth in tropical and temperate zones, and by their striking forms of flowers, [the orchid] immediately assumes the interest of nature lovers; also because the construction of the flower is one of the most interesting things offered by botanical morphology.'

Although the orchid is most commonly seen as a symbol for female sexuality, the name 'orchid' actually derives from 'orchis' – the Greek word for testicle – for the tuberous roots seen at the bottom of figure. 2. Of these two tubers, one carries the above-ground stem and 'appears already limp and empty, while the other bulb bursts with reserve materials.' As the Dodel-Ports explain, 'this year's young tuber is the permanent bud for next year; it will survive the winter to become the root and stem.'

Orchids are also notorious for their sexual trickery, with myriad deceptions for attracting pollinators. The *Ophrys* genus is particularly devious. Also known as the 'bee orchid,' members of the genus imitate the appearance and pheromone scent of a female bee in order to lure male insects into 'pseudocopulation,' which is exactly what it sounds like – the unwitting pollinator attempts to mate with the flower, covering itself in pollen in the process, which it will deposit at the next flower it is fooled by. Here, the Dodel-Ports have depicted *O. arachnites* (late spider orchid), now known as *O. holosericea* subsp. *holosericea*. The earlier and common names no doubt reflect the species' eerily arachnid appearance, but the updated Latin classification, literally meaning bee-flower, may be more accurate as it is a type of bee it is trying to attract, not a spider.

OPPOSITE: Henriette Schilthuis drew at least six wall charts featuring orchids. Here she drew *Coryanthes macrantha*, an epiphytic species that is highly fragrant and pollinated by male euglossine bees, which use their scent to attract females. The so-called 'bucket' (for which the genus receives its common name, 'bucket orchid'), seen on the right -hand side, is a highly modified petal. Attracted by the scent of the blossom, a male bee brushes the lip and falls into the bucket. Immersed in liquid, the bee must force an exit through a small opening at the rear of the flower, where a pollen bundle is deposited onto his back.

Schilthuis was especially deliberate about the angles she portrayed. While she didn't focus on the fruit or cross-section of a species, she would turn a blossom around to explore its total morphology, as we see with her depiction of *C. macrantha*. At the left, meanwhile, *Habenaria bractescens*, a South American bog orchid, sweeps across the chart. *Habenaria bractescens* may be modest compared to its more flamboyant cousins (despite the impressive floral bract that we see sweeping down in a u-shape in the centre of the chart), but Schilthuis demonstrates her artistry. The gentle bog orchid mirrors *Coryanthes macrantha*'s extravagant shape, adding a beautiful symmetry to the chart, and is alloted as much elegance and dignity as the alluring bucket orchid.

OPPOSITE:

Title: *Bloem orchideeën (Orchid Flowers)*
Author: Henriette Schilthuis
Language: N/A
Country: The Netherlands
Series/Book: N/A
Plate: 25
Publisher: Industrial School for Female Youth (Amsterdam, The Netherlands)
Year: c.1880

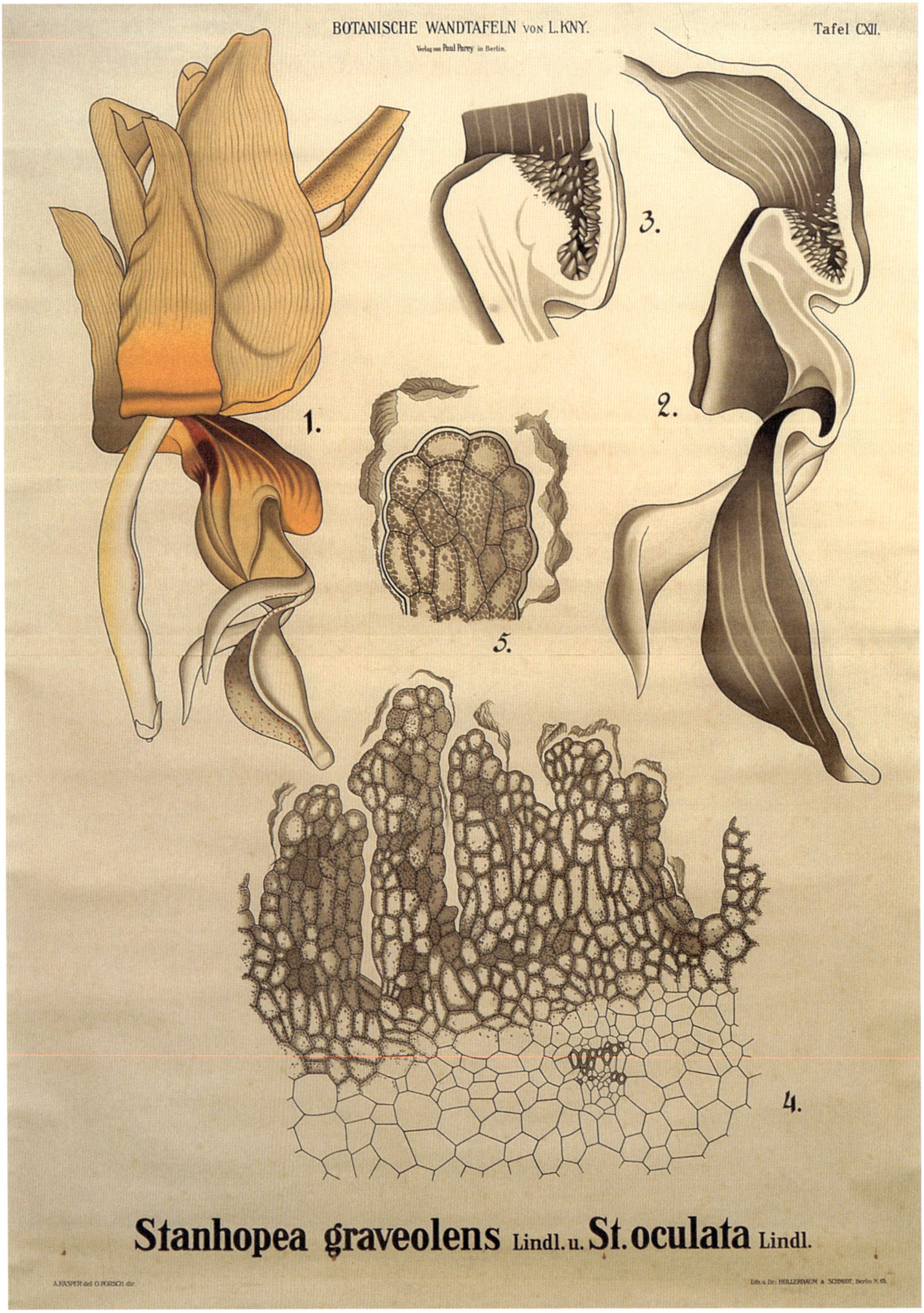
BOTANISCHE WANDTAFELN VON L. KNY.
Verlag von Paul Parey in Berlin.
Tafel CXII.
1.
2.
3.
4.
5.
Stanhopea graveolens Lindl. u. St. oculata Lindl.
A. KASPER del. O. PORSCH dir.
Lith. u. Dr. HOLLERBAUM & SCHMIDT, Berlin N. 65.

THIS SPREAD: Sometimes an accent of colour is more effective than a full-colour chart, as Kny and Kohl demomstrate with their *Orchidaceae* charts.

Leopold Kny uncharacteristically illustrates a full flower alongside his usual cellular inquiries, allowing a palette of yellows and oranges for a vibrant *Stanhopea graveolens*. The *Stanhopea* genus is composed of epiphytic species, all pollinated exclusively by male euglossine bees. These long-tongued bees are attracted by a fragrance produced in the osmophore, a secretory tissue located within the labellum. Apparently, olfactory qualities may be the primary identifying feature among the genus, and visual cues of only secondary importance. Kny reinforces this principle by withholding colour in his illustration of *S. oculata*, a species characterized by papillate osmophores. Four details (figs. 2, 3, 4, 5) show the textured scent tissues. Kny's chart is successful because it portrays two features of the *Stanhopea* genus: first, its morphology (fig. 1), and second, by illustrating a species with contoured fragrance tissue and therefore rendering visible an invisible quality, students are able to visualize *Stanhopea*'s compelling fragrance, a primary method of attracting pollinators.

F. G. Kohl, professor of botany at Germany's University of Marburg, illustrates *Orchis militaris* and *O. purpurea*, two species whose similar inflorescences can lead to their misidentification as each other. Both floral spikes are composed of purple, helmeted blossoms, so at first glance it's curious that Kohl chose to diagram tubers and root systems, rather than focus on the differences between the species, most obviously the shape of the labellum. As it happens, however, distinguishing between the species may be impossible; as the orchid family tends to do, *O. militaris* and *O. purpurea* cross-pollinated with particular frequency in Europe in the nineteenth century. A confounding variety of hybrids can flourish in ecologies with sufficient densities and appropriate habitat conditions. Kohl's chart reassures the student that identification can baffle botanists as well.

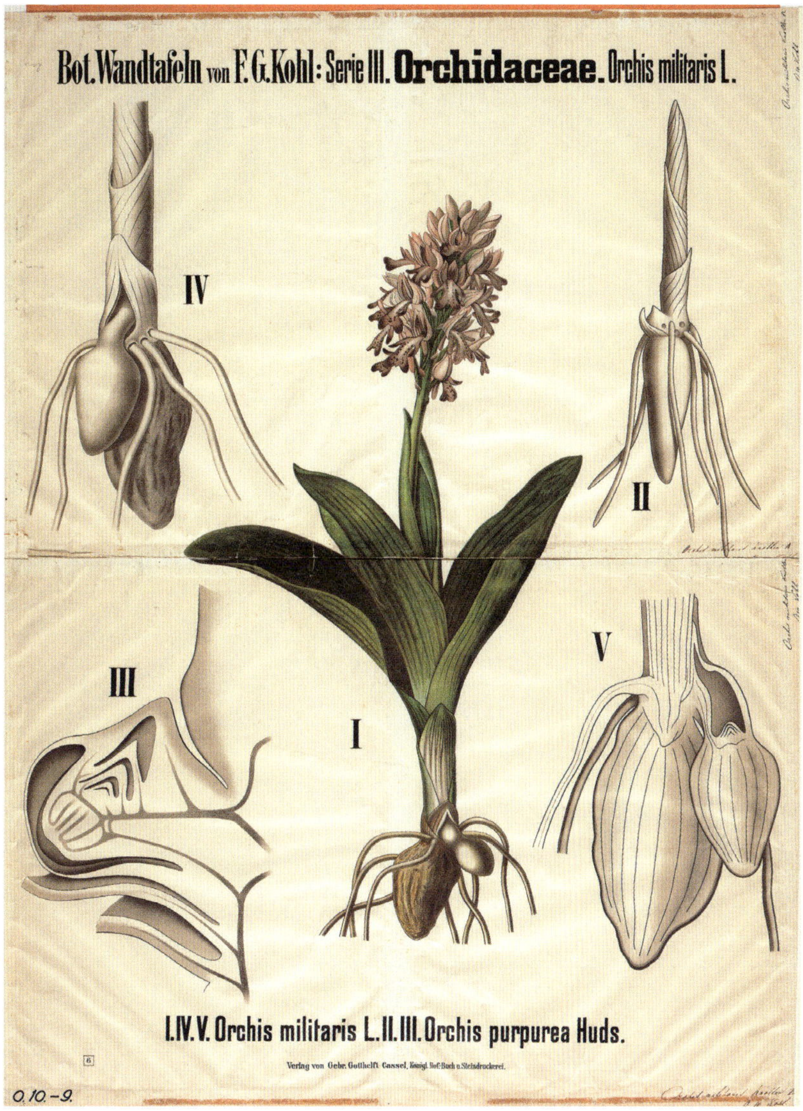

OPPOSITE:
Title: *Stanhopea graveolens Lindl. u., St. Oculata Lindl.*
Author: L. Kny
Language: German
Country: Germany
Series/Book: *Botanische Wandtafeln*
Plate: 112
Publisher: Paul Parey (Berlin, Germany)
Year: 1874

ABOVE:
Title: *Orchis militaris L.*
Author: F. G. Kohl
Language: German
Country: Germany
Series/Book: *Botanische Wandtafeln von F. G. Kohl: Serie III. Orchidaceae*
Plate: O.10–9
Publisher: Gebr. Gotthelft (Cassel, Germany)
Year: 1898

XVIII PAPAVERACEAE / POPPY FAMILY

Plants in the poppy family are principally distributed over temperate regions in the northern hemisphere and number just less than 800 species in 46 genera. They are fairly uniform, being mostly annual to perennial herbaceous plants, with a few subshrubs and shrubs such as *Romneya*. Many, such as *Argemone*, *Corydalis*, *Eschscholzia*, *Glaucium*, *Macleaya*, *Meconopsis* and *Papaver*, are old garden favourites for their pretty, colourful flowers. Some are considered weeds, for example *Chelidonium majus* (greater celandine).

Papaver somniferum has been cultivated for opium, derived from its sap, as far back as 5,000 BCE and is still an important pharmaceutical as well as illegal narcotic. It has been associated with sleep and death in literature and art for centuries, and inspired Thomas De Quincey to write *Confessions of an English Opium-Eater* in 1821. *Papaver rhoeas* (Flanders poppy) became a symbol of remembrance of fallen soldiers after it bloomed in vivid red swathes over the European battlefields of the First World War.

Characteristics: leaves alternate or in basal rosettes, extipulate, often dissected; sap a milky, clear or coloured, sometimes toxic latex. Flower relatively simple; two to three sepals are caduceus (fall early on); petals variable, but usually four to six, often crinkled in bud and silky; numerous stamens, often in a prominent boss; unilocular ovary. Fruit is a dehiscent capsule.

RIGHT:
Title: *Papaver rhoeas L., Papaver argemone L., Chelidonium majus L.*
Author: Otakar Zejbrlík
Language: Czech
Country: Czech Republic
Series/Book: *Lécivé Rostliny* (*Medicinal Plants*)
Plate: Unknown
Publisher: Kropác & Kucharský
Year: 1943

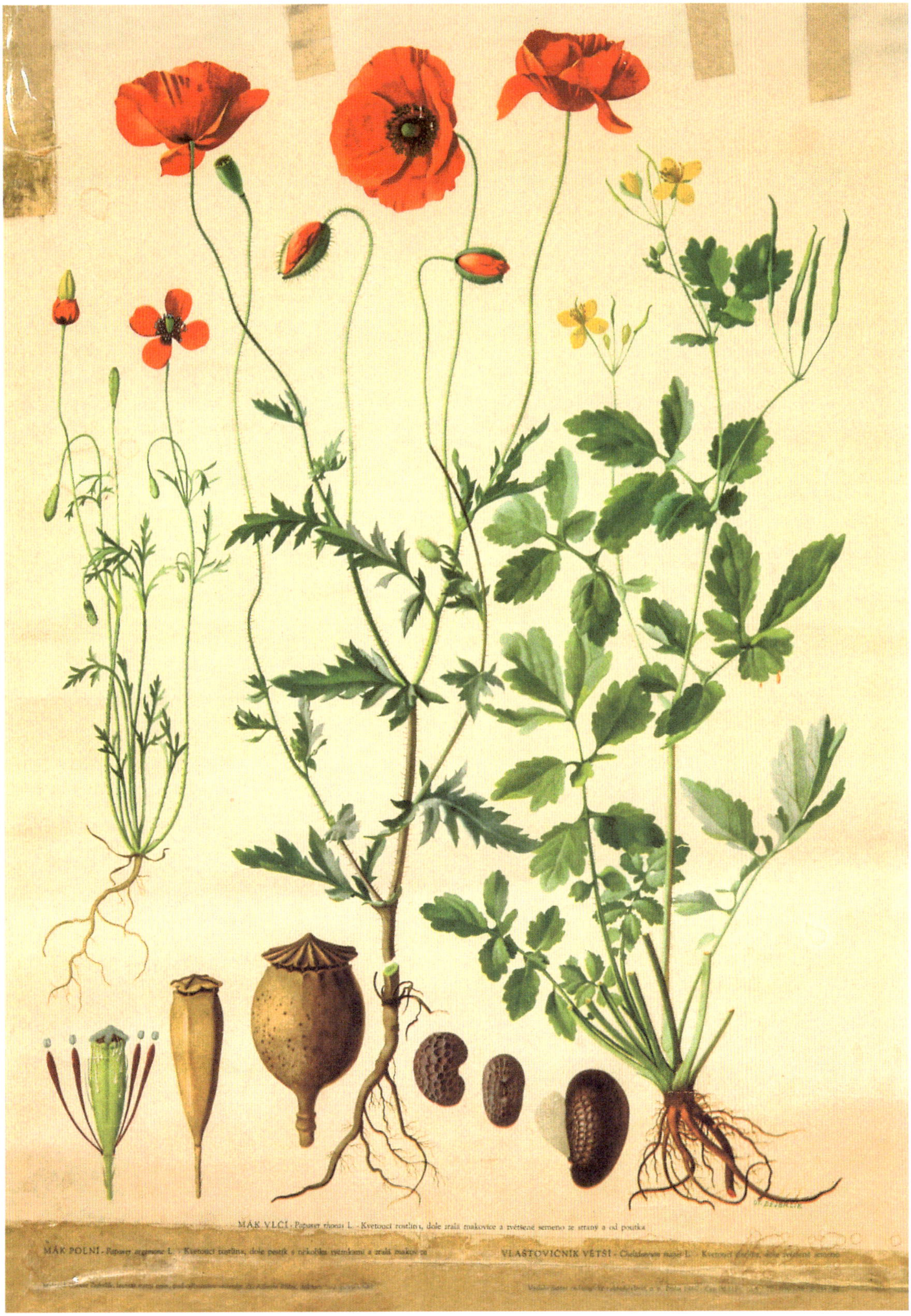
MÁK VLČÍ - Papaver rhoeas L. - Kvetoucí rostlina, dole zralá makovice a zvětšené semeno ze strany a od poutka
MÁK POLNÍ - Papaver argemone L. - Kvetoucí rostlina, dole pestík s několika tyčinkami a zralá makovice
VLAŠTOVIČNÍK VĚTŠÍ - Chelidonium majus L. - Kvetoucí rostlina, dole zvětšené semeno

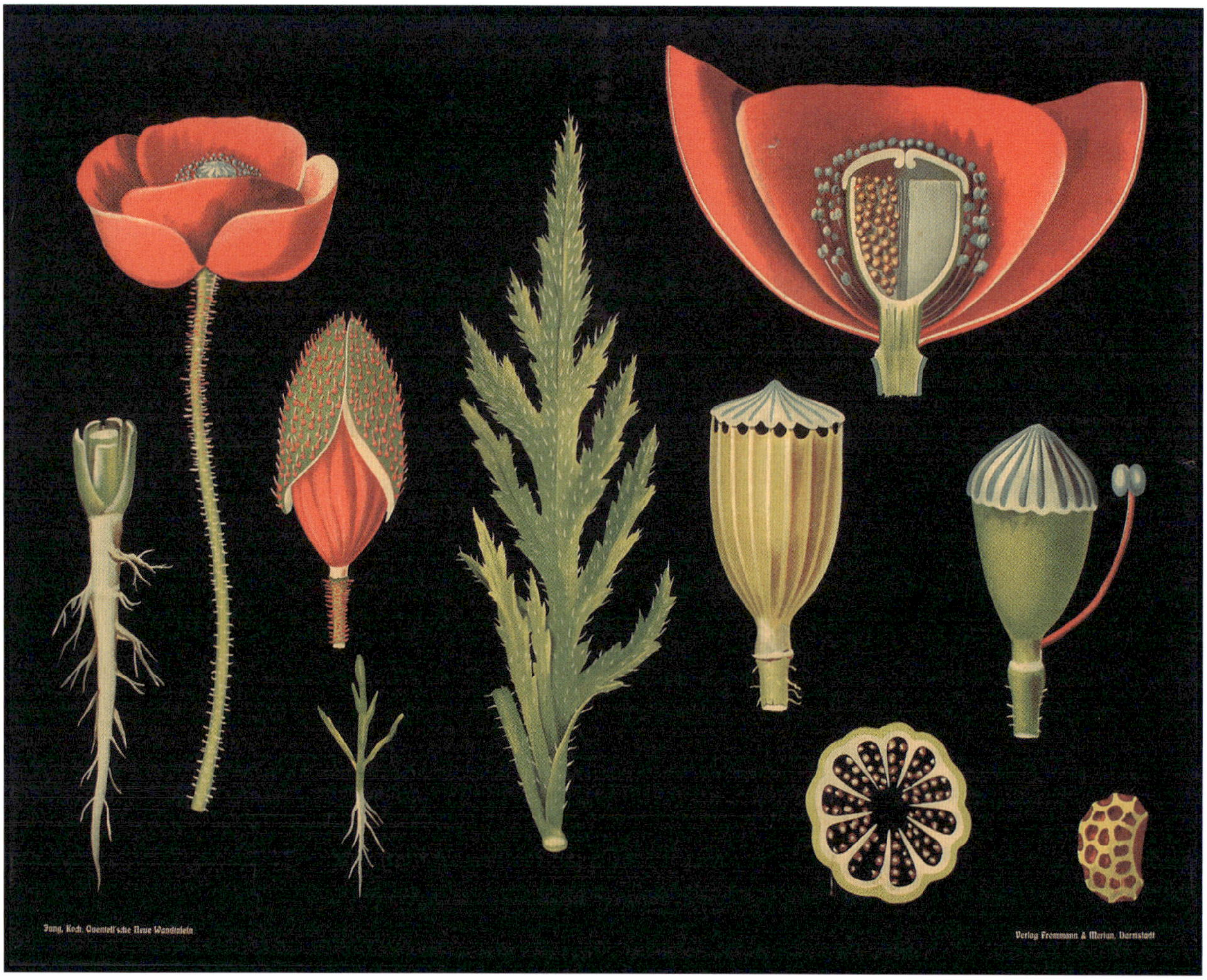

ABOVE:
Title: *Papaver rhoeas*
Authors: Heinrich Jung, Dr. Friedrich Quentell;
Illustrator: Dr. Gottlieb von Koch
Language: N/A
Country: Germany
Series/Book: *Neuen Wandtafeln*
Plate: 7
Publisher: Fromann & Morian (Darmstadt, Germany)
Year: 1902–3

OPENING PAGE: When one thinks of poppies, a dense ruffle of papery red petals and a pepper-shaker seed pod are what usually come to mind. Czech illustrator Otakar Zejbrlík acknowledges a student's expectations by placing *Papaver rhoeas* (Flanders poppy) in the middle of the chart. A central four-petalled blossom exposes her pointillist inner halo of blue-black anthers, while her sisters avert their eyes. At her feet, a stout dehiscent fruit has already lifted its lid and begun to disperse seeds. *Papaver argemone* (now called *Roemeria argemone*, long pricklyhead poppy) is drifting slightly away as though shy about her comparably anemic petal endowment and slender fruit.

Having thus dispelled any doubt that this is a chart of poppies, Zejbrlík slips in an outlier within the *Papaveraceae*. Defiantly waving her leguminous green fruits, buttercup blossoms, and softly serrated leaves, *Chelidonium majus* (greater celandine) grows tall as the black (more precisely, yellow) sheep in the family. The lonely *Chelidonium* genus contains only two species; all lesser celandines belong to the *Ranunculaceae* (buttercup) family. But here, *Chelidonium majus* announces herself as a *Papaveraceae*: after all, she shares her sister's stamen count and pistil structure. And isn't her body also covered with tiny hairs? A detail Zejbrlík's brush has recognized thousands of times.

OPPOSITE: *Papaver rhoeas*'s scarlet and greens are vibrant against Jung, Koch, Quentell's black backdrop. At the left, they show us a mature flower, rendered as a bristly stem and its four petals that make an upturned skirt of a bloom. To its right, a new bud has just begun to shed its spotted turban, with a young plant beginning to grow just beneath it. Dividing the chart is an erect leaf – stiff and pinnately lobed, familiar to anyone who has wandered into a poppy field that has gone to seed. At the top right, Jung, Koch, Quentell illustrate the cross-section of a flower to show the common poppy's signature blue anthers and abundance of seeds. A dehiscent poricidal capsule sits below, exposing the small holes from which it will shake its tiny seeds. The cross-section of a fruit (bottom right) emphasizes its numerous seeds, which account for the scarlet poppy's ubiquity.

While poricidal capsules don't generally disperse far and wide, they will fall generously around the parent plant, creating seed banks that may remain dormant until the soil is disturbed. Hence a second common name – field poppy, for its tendency to easily populate an open field. One plant may produce 60,000 seeds, and given that there is never just one; hundreds of millions of seeds in a field is not improbable. While its ubiquity formerly caused the common poppy to be branded an agricultural weed, it has since come to be a symbol of remembrance for those who fought and died in the First World War, during which the disturbance of the fields where battles were fought and the dead were buried gave the flower opportunity to flourish:

In Flanders fields the poppies blow
Between the crosses, row on row
That mark our place; and in the sky
The larks, still bravely singing, fly
Scarce heard amid the guns below.

(From *In Flanders Fields* by John McCrae, 1915.)

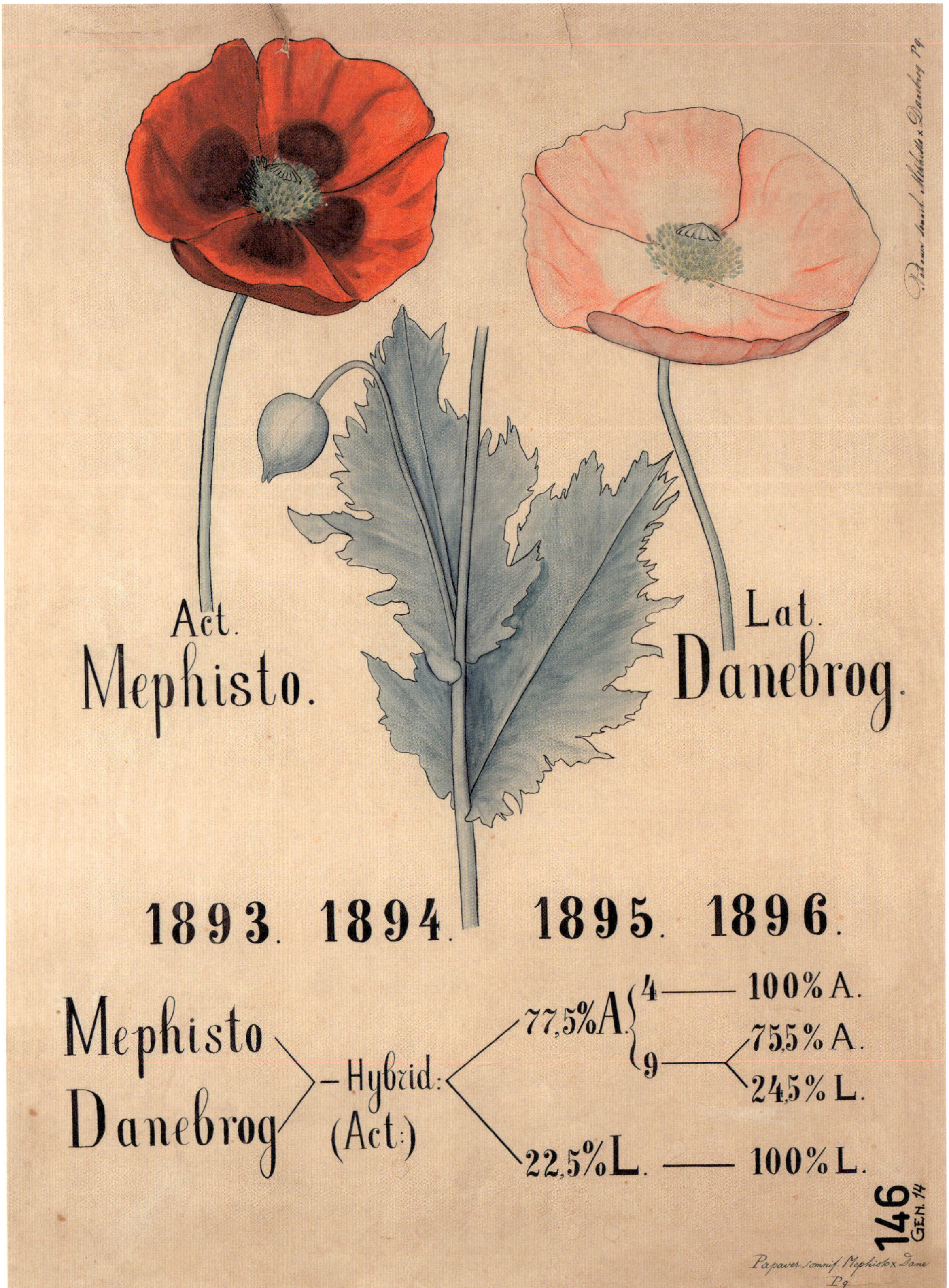
Act.
Mephisto.
Lat.
Danebrog.
1893. 1894. 1895. 1896.
Mephisto
Danebrog
– Hybrid:
(Act:)
77,5% A.
4 — 100% A.
9
75,5% A.
24,5% L.
22,5% L. — 100% L.
146
GEN. 14
Papaver somnif. Mephisto x Dane
Pg.

LEFT: Two poppies, one man, a genetic revolution, and a botanic chart that reads like a love letter. Hugo de Vries was appointed the first professor of botany at the University of Amsterdam in 1878, after a fervent study of plants in his youth (gathering a complete collection of all the plants in the Netherlands), and years corresponding with Charles Darwin as a protégé on principles of heredity. At Darwin's death, de Vries passionately pursued collecting and crossing plants to refine his mentor's flawed theory of hereditary traits based on what he called 'pangenesis.' In particular, he tasked himself with proving that a 'monstrosity' (or mutation) could be transferred from one variety to another. His dazzling (and beautiful) proof arrived in the progeny of two *Papaver somniferum*: the polycephalous black-spotted Mephisto, which, with its abnormal ring of fruit (mutated from stamens to pistils) around the capsule, was his 'monstrous' variety; and the 'normal' pale pink and white Danebrog variety.

Over the course of four years, de Vries produced a near-perfect 3:1 ratio of black- to white-accented polycephalous hybrid poppy blooms – a garden of evidence that grew into his independent discovery of Mendelian segregation. De Vries would soon regale the world with his research, but perhaps never as lovingly as with this chart, rendered with delicate calligraphy and a soft palette.

ABOVE RIGHT: Included in a series titled *Poisonous Plants of Germany*, P. Esser's portrayal of *Papaver somniferum* is almost too enticing. Or perhaps that was his intent – a wall chart as intoxicating as its subject. After all, the species has long been cultivated for its hypnotic and narcotic potential, most notably opium, derived from the milky latex of an unripe seed pod (fig. 5) (the seeds contain only trace amounts).

Esser's chart includes a new bud (fig. 2), cross-section of a flower with its anthers removed (fig. 4), and a central stalk (fig. 1) with beckoning leaves, glabrous and greyish-green, and two stems, one bowing its somniferous head.

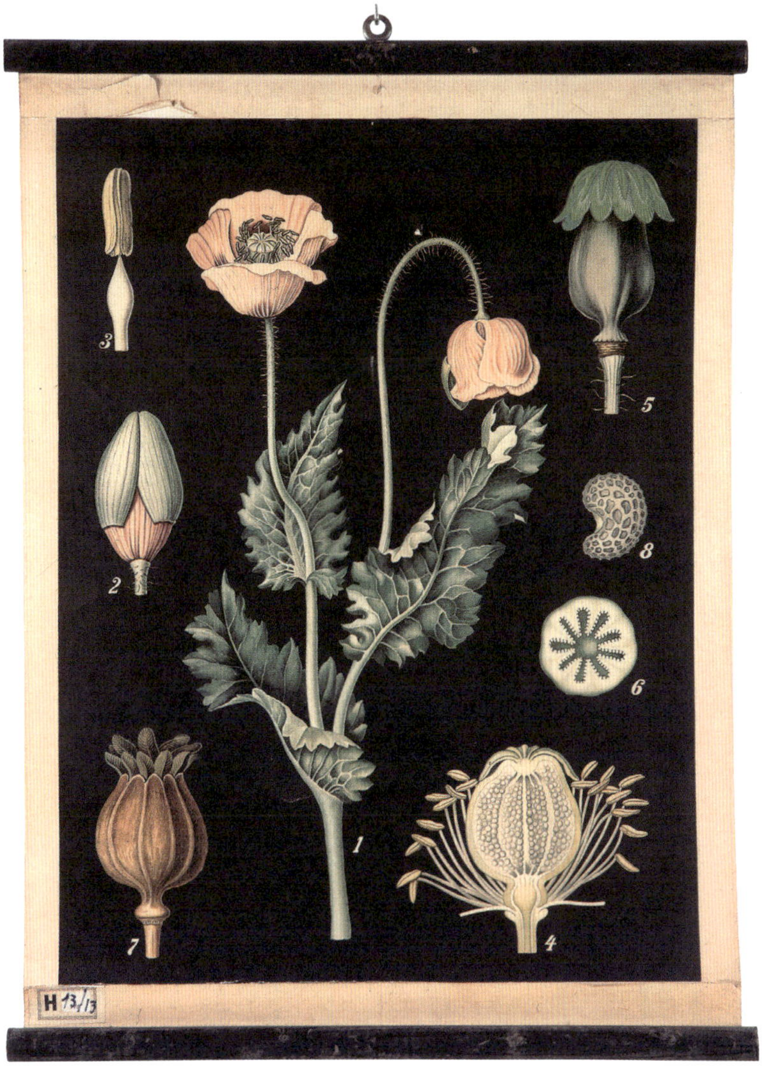

OPPOSITE:
Title: *Papaver somnifera mephisto x danebrog*
Author: Hugo de Vries
Language: Dutch
Country: The Netherlands
Series/Book: N/A
Plate: N/A
Publisher: N/A
Year: 1896

ABOVE:
Title: *Papaver somniferum*
Author: Dr. P. Esser; Illustrator: Carl Bollmann
Language: German
Country: Germany
Series/Book: *Die Giftpflanzen Deutschlands* (*Poisonous Plants of Germany*)
Plate: 8
Publisher: Friedrich Vieweg & Sohn (Braunschweig, Germany)
Year: 1910

RIGHT: Another example of Zippel and Bollmann's mixed-family charts, here they compare the blooms and fruits of two *Papaveraceae* and one *Nymphaeaceae*.

Dicentra spectabilis (now called *Lamprocapnos spectabilis*), commonly known as bleeding heart for its unique morphology resembling a heart shedding a single droplet. *Lamprocapnos spectabilis* grows in arching racemes of up to twenty pendent flowers. Zippel and Bollmann understandably focus on the flower's structure to show us how its petals arch to form a heart, and a pendulous white stigma becomes the drop where later the fruit will form (fig. 1). A pair of inner stamens (figs. 2, 9) are an armature within the heart. The peculiar flower's unusual morphology has lent itself over time to other, equally curious, common names, including Venus's car, lady-in-a-bath, and Dutchman's trousers.

Alongside *Lamprocapnos spectabilis*, a second *Papaveraceae* demonstrates the family's more familiar forms. *Papaver rhoeas* flourishes at the centre of the chart, waving a tiny red bud and a welcoming open blossom.

Finally, the *Nymphaeaceae*, *Nuphar luteum* (now *N. Lutea*) included in the chart perhaps for its gourd-shaped fruit that is not entirely dissimilar to the structured fruit of a poppy.

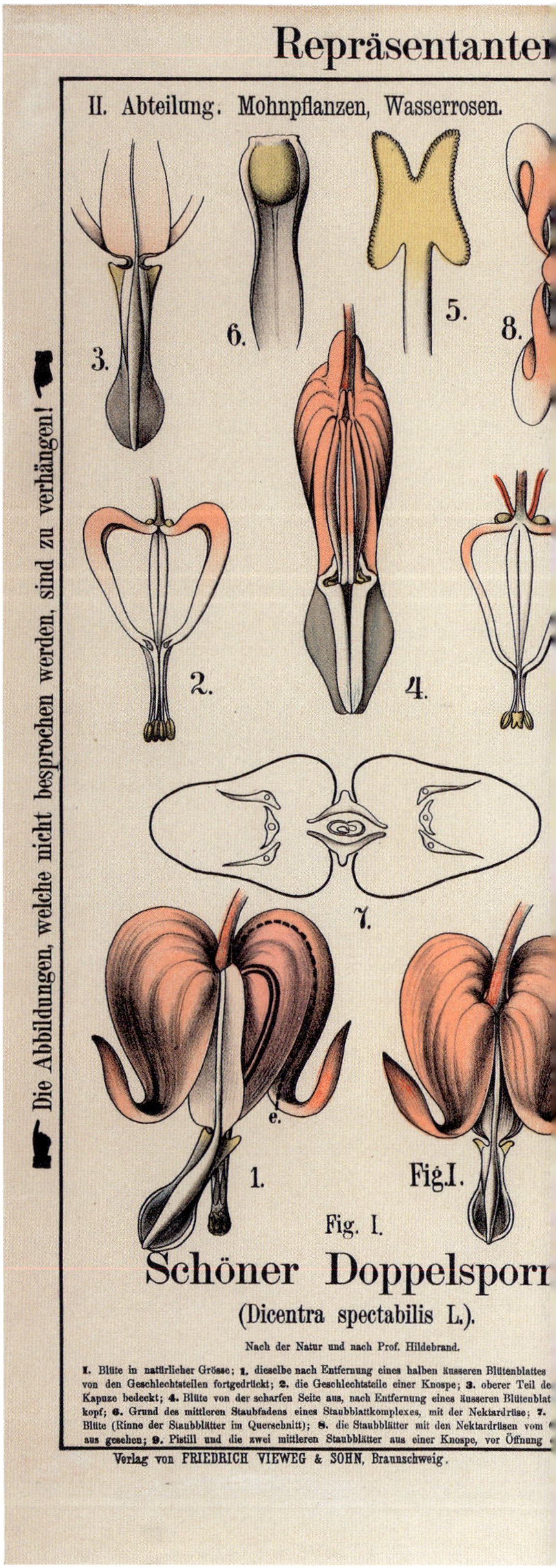

RIGHT:
Title: *Mohnpflanzen, Wasserosen (Poppy Plants, Water Lilies)*
Author: Hermann Zippel; Illustrator: Carl Bollmann
Language: German
Country: Germany
Series/Book: *Repräsentanten einheimischer Pflanzenfamilien (Representatives of Indigenous Plant Families)*
Plate: II Abteilung, 47
Publisher: Friedrich Vieweg & Sohn (Braunschweig, Germany)
Year: 1879–82

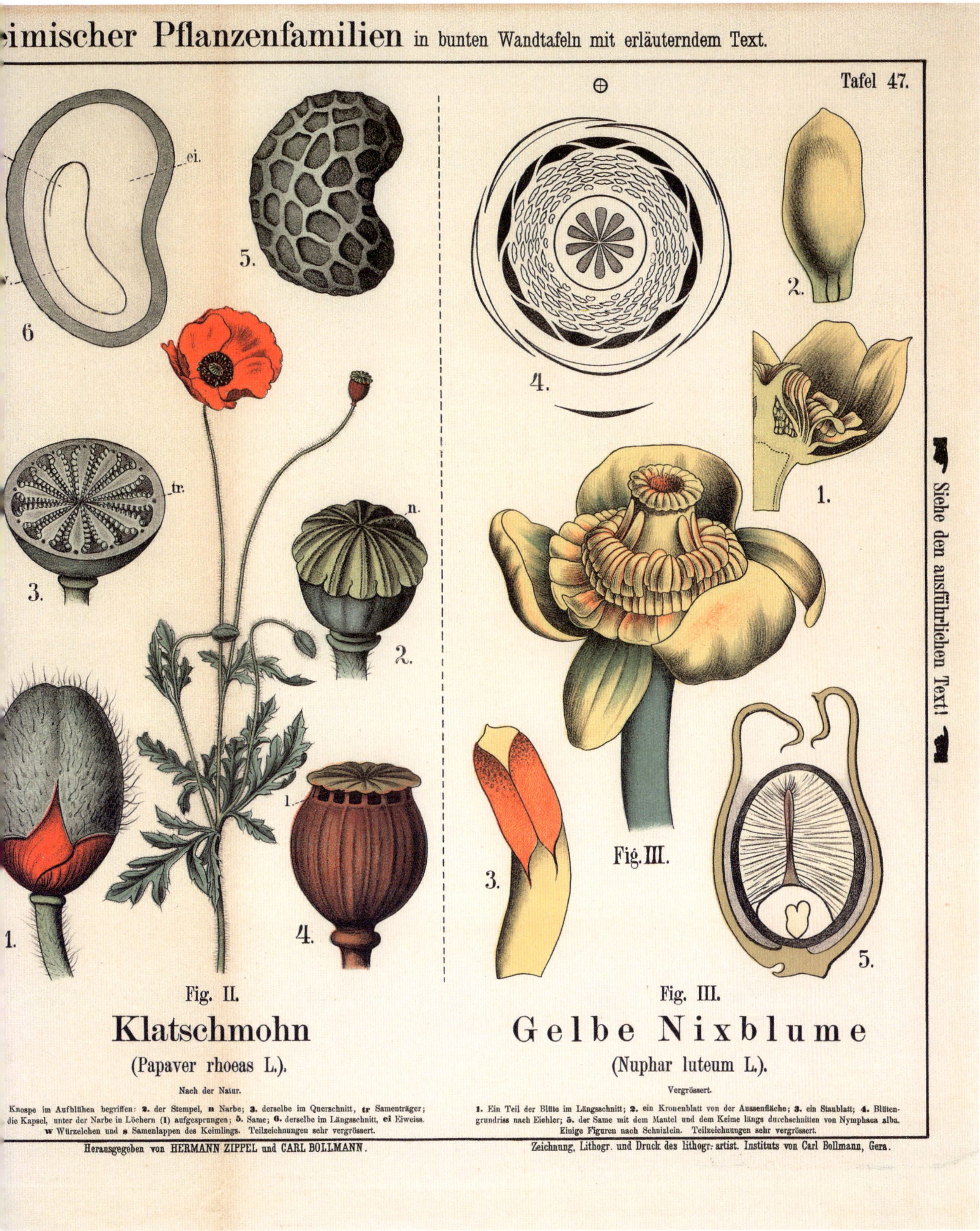

imischer Pflanzenfamilien in bunten Wandtafeln mit erläuterndem Text.
Tafel 47.
ei.
5.
6
3.
tr.
n.
2.
1.
4.
1.
Fig. II.
Klatschmohn
(Papaver rhoeas L.).
Nach der Natur.
Knospe im Aufblühen begriffen: 2. der Stempel, n Narbe; 3. derselbe im Querschnitt, tr Samenträger;
die Kapsel, unter der Narbe in Löchern (1) aufgesprungen; 5. Same; 6. derselbe im Längsschnitt, ei Eiweiss.
w Würzelchen und s Samenlappen des Keimlings. Teilzeichnungen sehr vergrössert.
⊕
2.
4.
1.
Fig. III.
3.
5.
Fig. III.
Gelbe Nixblume
(Nuphar luteum L.).
Vergrössert.
1. Ein Teil der Blüte im Längsschnitt; 2. ein Kronenblatt von der Aussenfläche; 3. ein Staublatt; 4. Blütengrundriss nach Eichler; 5. der Same mit dem Mantel und dem Keime längs durchschnitten von Nymphaea alba.
Einige Figuren nach Schnizlein. Teilzeichnungen sehr vergrössert.
Siehe den ausführlichen Text!
Herausgegeben von HERMANN ZIPPEL und CARL BOLLMANN.
Zeichnung, Lithogr. und Druck des lithogr.-artist. Instituts von Carl Bollmann, Gera.

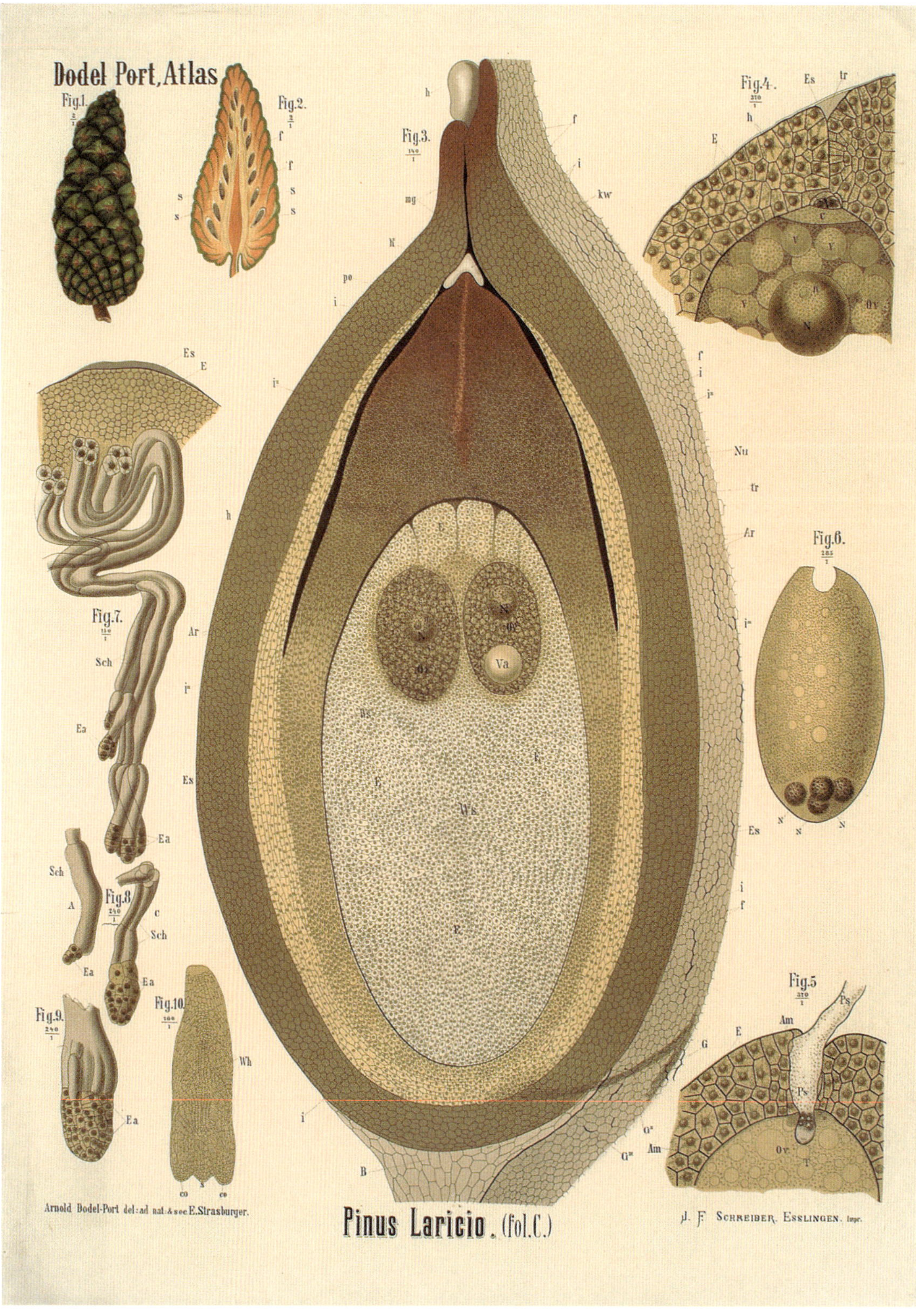
Dodel Port, Atlas
Fig.1.
Fig.2.
Fig.3.
Fig.4.
Fig.5
Fig.6.
Fig.7.
Fig.8
Fig.9.
Fig.10.
Arnold Dodel-Port del: ad nat. & sec E. Strasburger.
Pinus Laricio. (fol.C.)
J. F. Schreiber. Esslingen. impr.

XIX PINACEAE / PINE FAMILY

This is the largest family of conifers, comprising 11 genera and approximately 250 species, and the most substantial in the gymnosperms. They are usually monoecious, evergreen (except for *Larix* and *Pseudolarix*), resinous trees, and occur widely through the northern hemisphere, mostly in temperate areas. Nearly all the genera are popular as garden and landscaping trees and many, such as *Abies*, *Cedrus*, *Picea*, *Pinus* and *Tsuga*, are major sources of softwood timber, wood pulp, resins and essential oils.

Cedrus libani, the stately cedar of Lebanon, has been prized for its timber by many civilizations since ancient times – for example by the Babylonians, Phoenicians and the Egyptians – and into the twentieth century. There are now only ten – all protected – remnants of the once extensive cedar forests in Lebanon. *Pinus longaeva* (bristlecone pine) grows in extreme, subalpine habitats from Nevada to California, but can live for up to 5,000 years, so core samples from surviving trees provide invaluable information for paleoclimatologists.

Characteristics: leaves are needle-like or linear, singly or spirally arranged, sometimes whorled or in short spurs, with resin canals; female cones are large and woody, with numerous, spirally arranged scales, each with two ovules and usually distinct from the bracts; male cones are catkin-like with scales arranged in a spiral, each with two pollen sacs. Bark and buds are scaly; seeds are typically winged (samara).

OPPOSITE:
Title: *Pinus Laricio (fol. C)*
Authors: Arnold and Carolina Dodel-Port
Language: German
Country: Switzerland
Series/Book: *Anatomisch physiologische Atlas der Botanik (The Anatomical & Physiological Atlas of Botany)*
Plate: 27
Publisher: J. F. Schreiber (Esslingen, Germany)
Year: 1878–93

OPENING PAGE: Arnold and Carolina Dodel-Port included three charts of *Pinus laricio* (now *P. nigra* subsp. *laricio*, Corsican pine) in their *Botanic Atlas*. The first two illustrated the morphologies of male and female flowers. The third, here, was requested by teachers using the atlas, which promised to explain how plants reproduce, especially on a cellular level. Nonetheless, they believed such fine details 'should be, for pedagogical reasons, unnecessary here.' Eventually they conceded to the requests of their colleagues with the following rather wordy explanation: 'Thus, I follow their decisive council and therefore address, in the present panel, the intimate sexual activities inside the pollinated ovule of *P. laricio*, to fill the last major gap in the "physiology of reproduction".'

THIS SPREAD: From the intimate close-up to two full-length portraits; Kautsky and Beck's formats were different than any other. Their sole subjects were trees, and all charts centrally featured a beautiful full-grown specimen in a picturesque, species-characteristic landscape. For *Abies alba* (European silver fir, opposite), a mountainous horizon, and sparkling streams; here, a coastal garden and Mediterranean breezes for *Pinus pinaster* – appropriate enough for the 'maritime pine.'

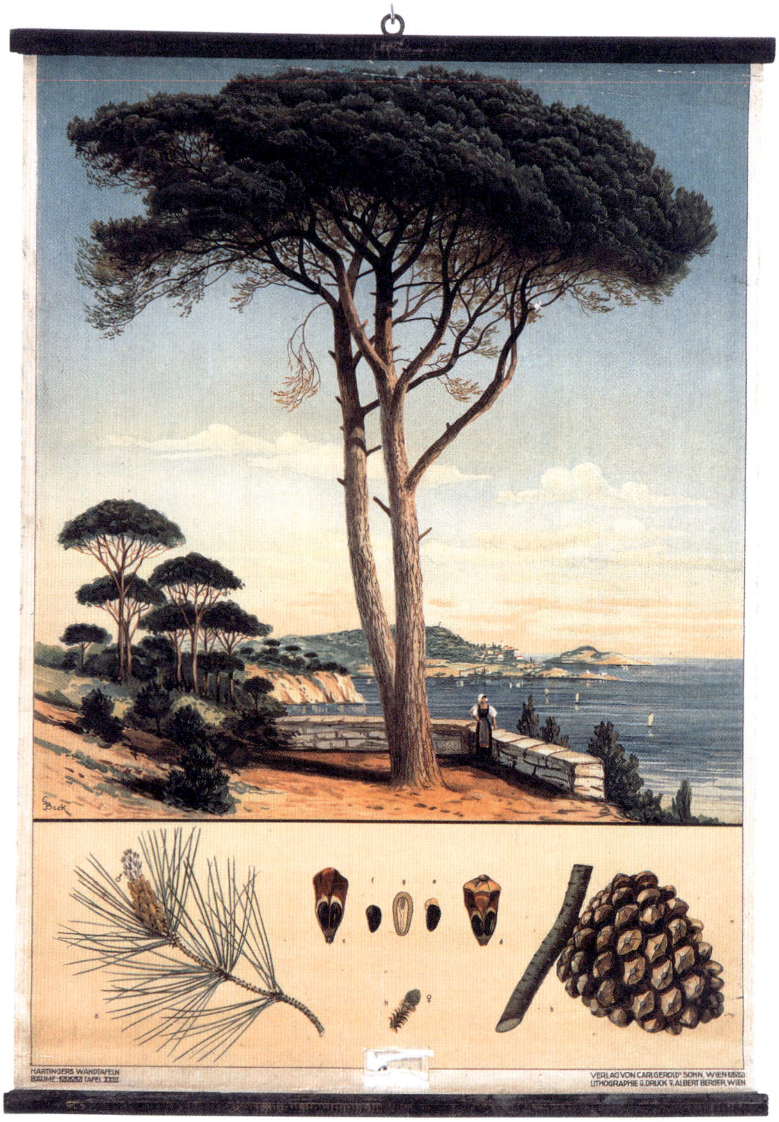

Pinus pinaster is a popular ornamental tree that prefers wet winters and dry summers and grows quickly. Particularly hard wood qualifies it as an important source of timber. The lower portion of the chart features details from the tree, including a spring of needles, several large-winged seeds for which the species is known, and a female cone.

On the opposite page, *Abies alba* towers above a grove of smaller cousins. Reaching a height of about 130–160 feet (40–50m), the European silver fir is not the tallest tree in the family; *Pseudotsuga menziesii* var. *menziesii* (coast Douglas-fir), *Picea sitchensis* (Sitka spruce) and *Abies procera* (noble fir) are but a few that can reach double its normal height. These towering species, however, are all native of North America, whereas their European cousins rarely exceed 200 feet (60m) – and of the European species, *A. alba* is the tallest, recorded to have reached 223 feet (68m), which explains its precedence in Kautsky and Beck's chart.

The inset details feature a cross-section of bark, a spring of needles, a cone and several seeds.

ABOVE & RIGHT:
Title, above: *Pinus pinaster*; Right: *Abies alba*
Authors: Johann Kautsky, G. v. Beck
Language: German
Country: Czech Republic
Series/Book: *Hartinger's Wandtafeln*
Plates: 23; 9
Publisher: Carl Gerold's Sohn (Vienna, Austria)
Year: c.1880

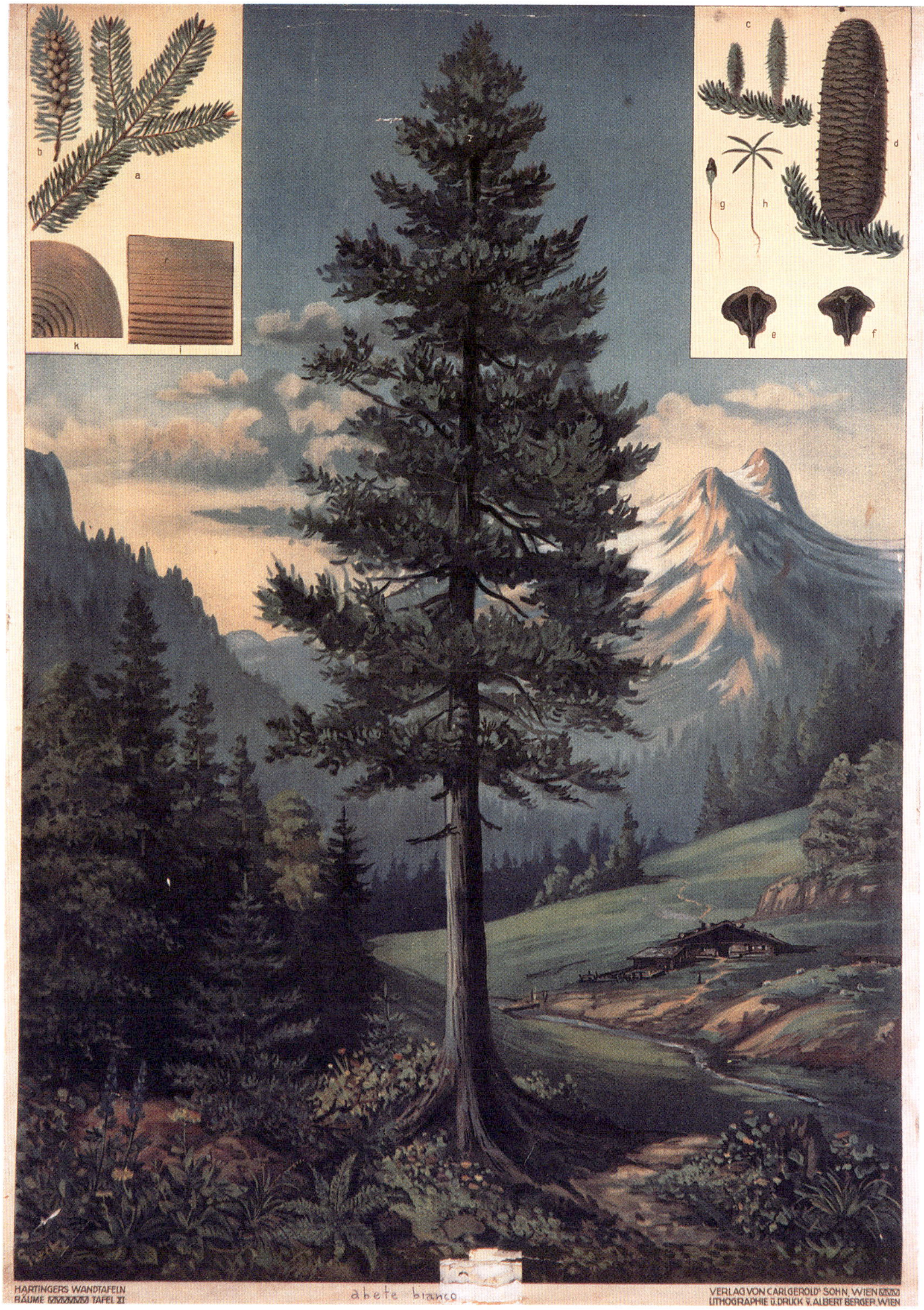
HARTINGERS WANDTAFELN
BÄUME TAFEL XI
abete bianco
VERLAG VON CARL GEROLD'S SOHN, WIEN
LITHOGRAPHIE U. DRUCK V. ALBERT BERGER, WIEN

OPPOSITE: While perhaps ersatz to the tactile experience of handling a specimen directly – that is to say walking away with pollen-dusted fingers and a stubborn smear of sap – wall charts could offer students a comprehensive picture of a plant's life cycle in one glance. Thus there is little disconnect between cone and ovary, pollen and seed.

Here, Jung, Koch, Quentell illustrate a *Pinus sylvestris* (Scots pine) as a double branch, on which are borne many needle pairs, two pollen-bearing staminate cones, and two female seed cones – one young (green) and one mature (brown). In the upper-right corner, a pair of exposed needles, and a younger pair, still sheathed in a supporting scale.

A mature pollen cone is brown (bottom right) and composed of rows of double pollen-bearing stamens. Above the full cone is a stamen pair yet to split. Above that, a single grain of immature pollen. The immature seed cone has green carpels tipped with red (lower left), each of which bears two tender ovules (above). These ovules are a remarkably small detail in the chart, but a significant feature of the pine – were it not for exposed seeds, the gymnosperm (cone-bearing) could be an angiosperm (flowering).

Below the female cone and its ovules are a pollinated carpel and its two protected samaras. Each wing bears a single seed which germinates and discards the seed coat; here, unlike with angiosperms, there are no fruits, just seeds and a dispersal wing.

At the bottom of the chart a few young roots crawl through the soil, under the moon of a stem's pale cross-section.

OVERLEAF: This chart is not so different from the one opposite, but it's worth including for its exquisite details, gorgeous composition and lush gradient colouring, all of which create a vibrancy that Jung, Koch, Quentell attempt with their characteristic black canvas. It's a bit more difficult for the illustrator to achieve a dramatic gestalt with a white backdrop; considerably more attention must be afforded to contrast, arrangement and details – all of which Otto Schmeil achieves.

Especially lovely is his treatment of pollen, both in words and image. His book describes the wind blowing through the branches of a pollinating tree, 'kidnapping him in considerable clouds, [so that] after a thunderstorm, the forest ... is covered by a yellow layer. "It rained brimstone," say people who cannot explain the origin of the yellow paths.' Schmeil goes on to describe a pollen grain (fig. 5), carried on each side with an air-filled cavity: 'The long distance this hot air balloon can travel is apparent from where we find his pollen, often miles away from any pine tree.'

In the upper-left corner, a young pollen-bearing scale (fig. 3) has just ruptured to disperse its pollen. To its left, a pair of vertically adjacent scales shed their pollen; as Schmeil reminds his readers, pines are wind-pollinated. So, in the absence thereof, pollen merely trickles onto the stamen below. And in the absence of that, pollen would fall to a lower cone; this is why the tree's female cones are generally produced on upper branches – to prevent self-pollination (an adaptive feature quietly alluded to in Jung, Koch, Quentell's chart).

One also wonders why neither includes a representation of the grand tree in its entirety, especially considering Schmeil's attention to detail and his devotion to *Pinus sylvestris* as an organism to which Germany's great forests – and its people – owe a debt of gratitude: 'Within the pine tree, the welfare of so many people is intimately linked.'

OPPOSITE:
Title: *Pinus silvestris*
Authors: Heinrich Jung, Dr. Friedrich Quentell
Illustrator: Dr. Gottlieb von Koch
Language: N/A
Country: Germany
Series/Book: *Neue Botanische Wandtafeln*
Plate: 41
Publisher: Fromann & Morian (Darmstadt, Germany); Hagemann (Düsseldorf, Germany)
Year: 1928; 1951–63

OVERLEAF:
Title: *Pinus silvestris*
Author: Otto Schmeil
Language: German
Country: Germany
Series/Book: *Botanische Wandtafeln*
Plate: 3
Publisher: Quelle & Meyer (Leipzig, Germany)
Year: 1907

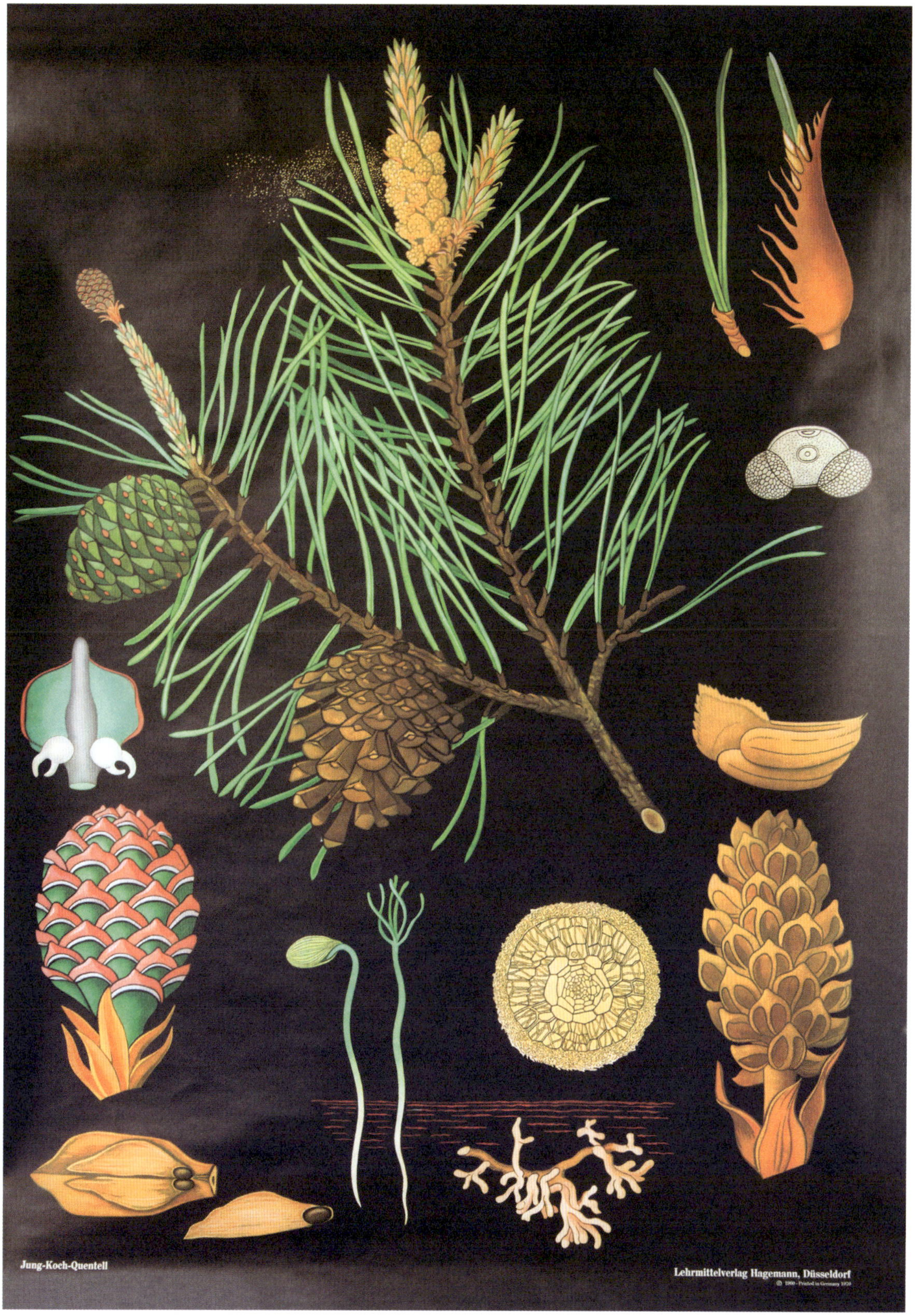
Jung-Koch-Quentell
Lehrmittelverlag Hagemann, Düsseldorf

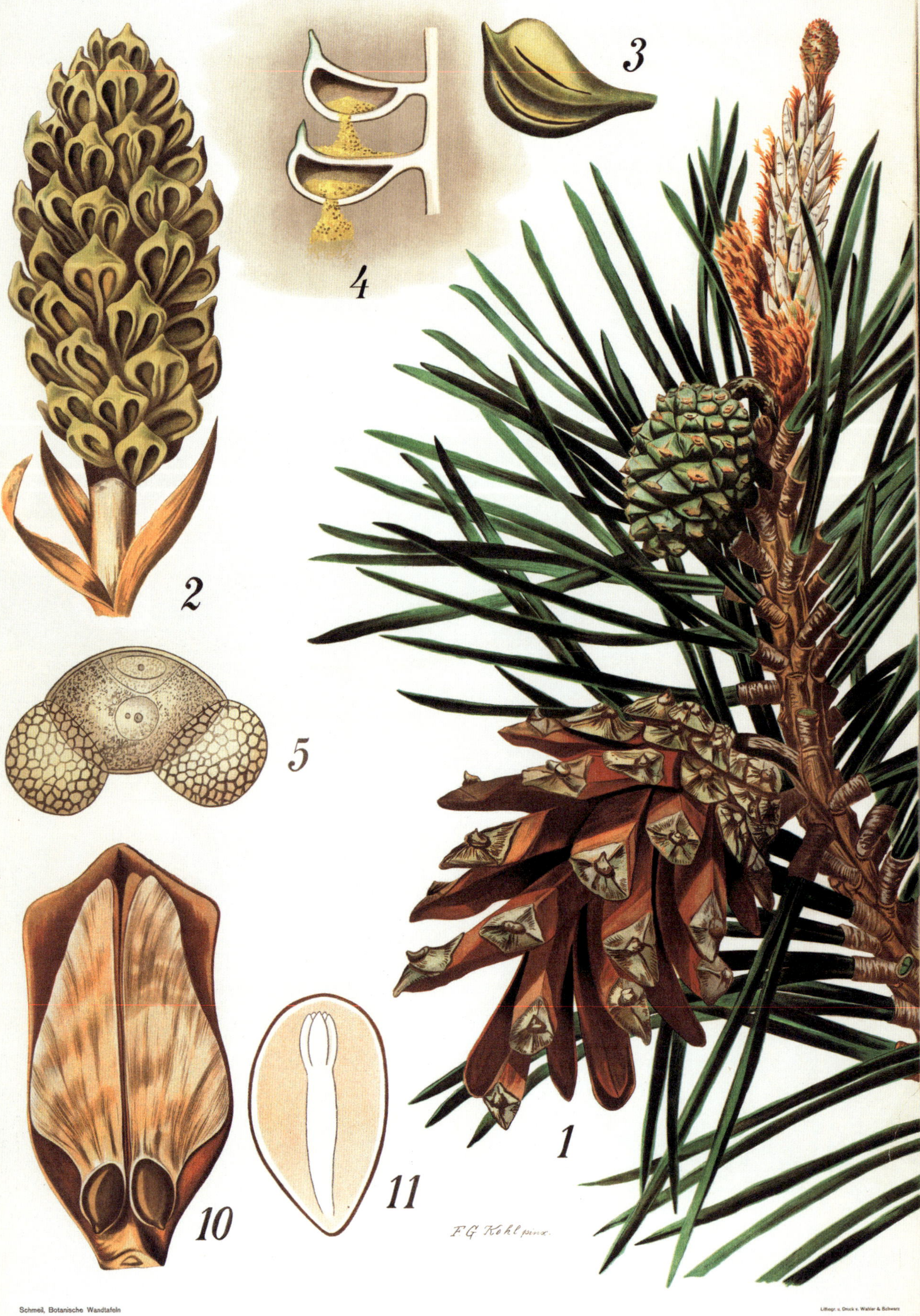

Schmeil, Botanische Wandtafeln

Lithogr. u. Druck v. Wahler & Schwarz

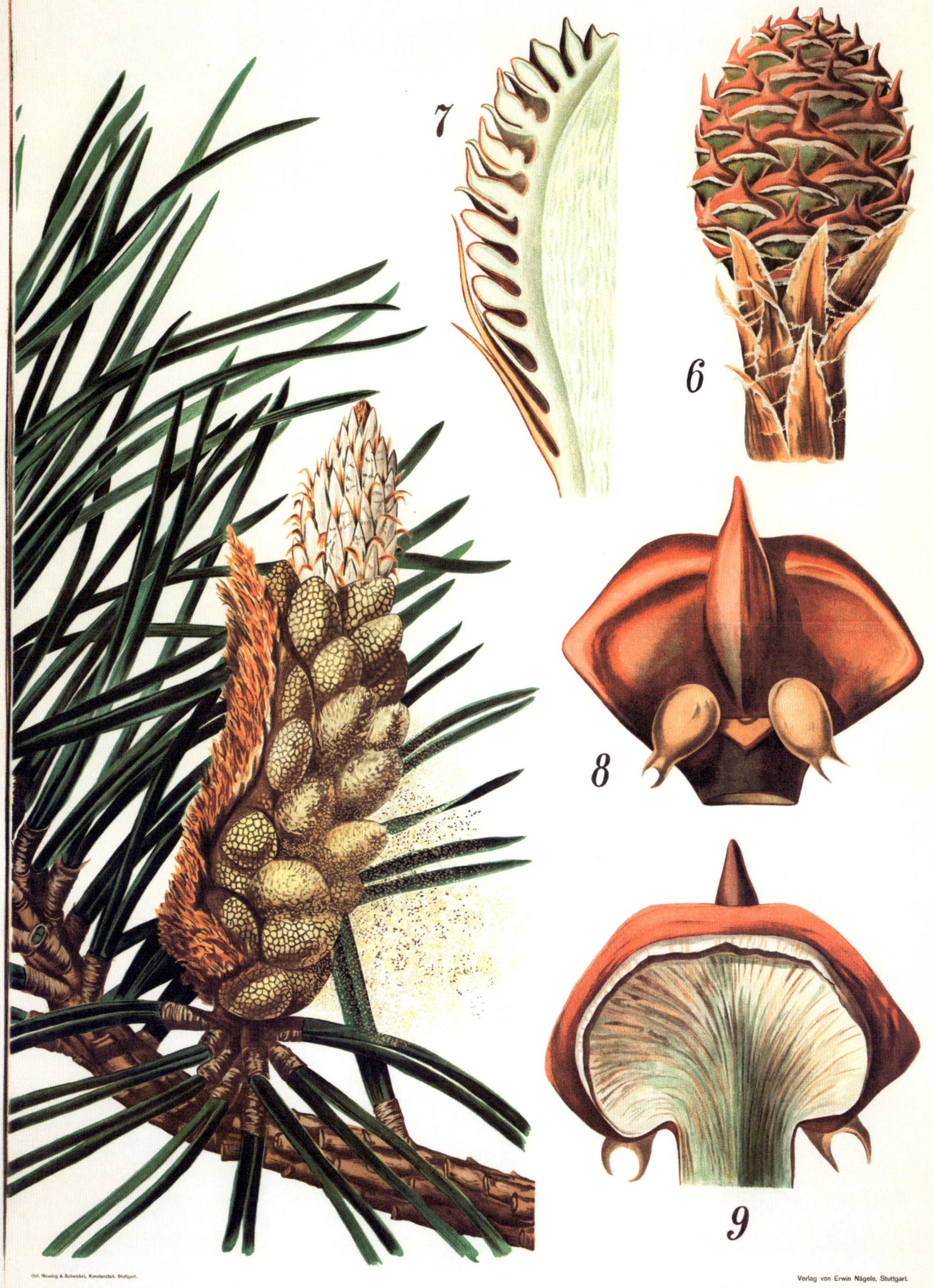

(Inh. Messing & Schwabe), Kunstanstalt, Stuttgart.
Verlag von Erwin Nägele, Stuttgart.

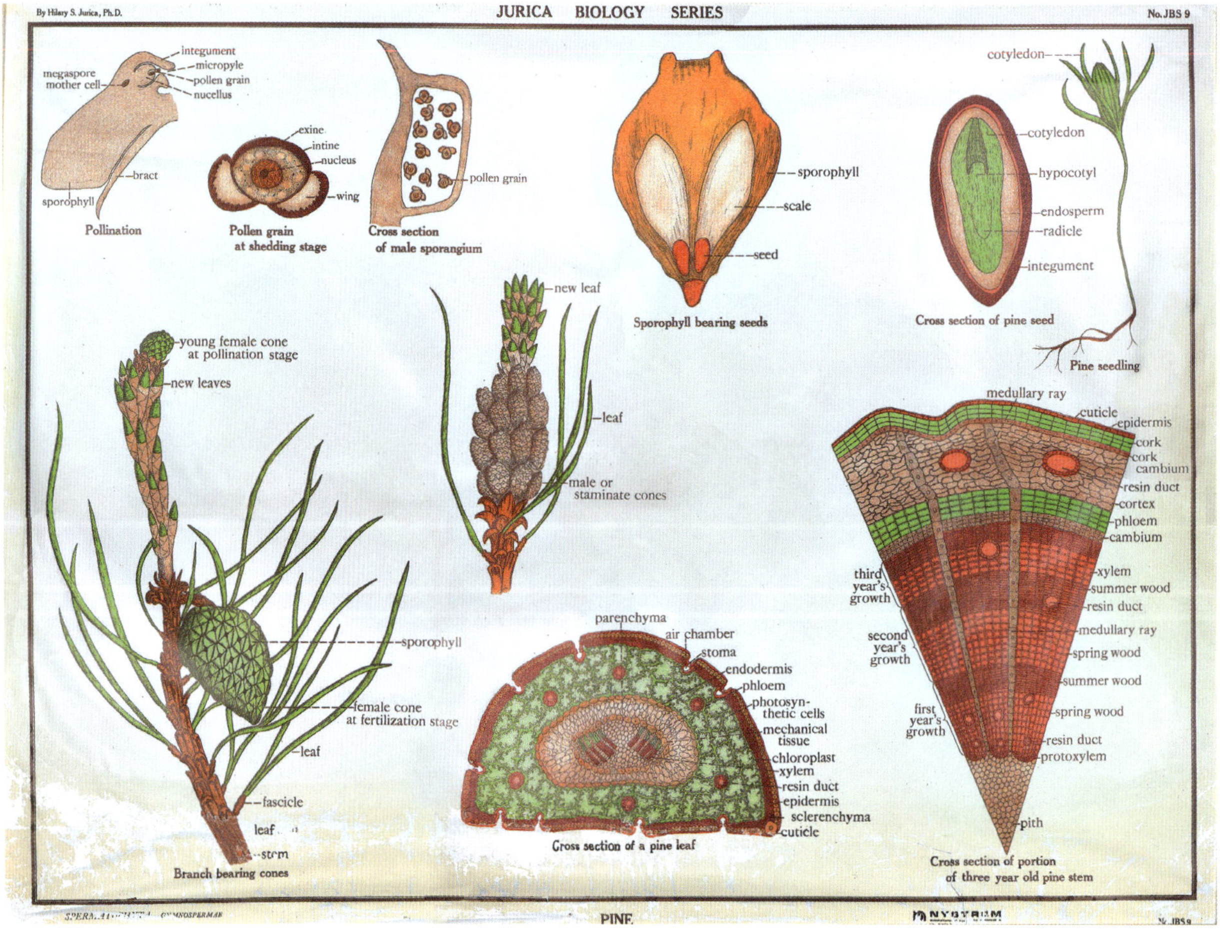

ABOVE: Another pine, a cross-section of wood and needle, on a charming board authored by a friar of Slovakian descent who grew up in Cloverdale, Illinois in the US After earning his Ph.D. in botany, Father Hilary Jurica went on to teach college courses and collect botanic specimens throughout the country, which he employed to supplement his curricula, and to illustrate his teaching boards. Never intended to be used abroad, Jurica's charts include straightforward labels for the morphology of each plant part.

At the left, two young female cones perch at the top of a stem (to prevent self-fertilization). To their right, a cross-section of the stem includes stoma, pores in the epidermis of a leaf or stem that allow for photosynthesis and respiration. The stoma of most plants open during the day, as they are here. These air ducts also allow water vapour to evaporate, however, so the size of the aperture can change in response to environmental conditions.

A cross-section of a three-year-old pine trunk at the bottom right shows growth over two seasons of each year, as well as outer epidermis, phloem, and cambium. Medullary rays ribbon perpendicularly through the growth rings, allowing a radial transmission of compounds from the outer cambium to the inner, older, infection-prone wood. Both the stem and trunk are spotted with resin ducts, or intercellular chambers that secrete resin. Common to conifers, these ducts are a tree's primary defense system against insect attacks.

ABOVE:
Title: *Pine*
Author: Hilary Jurica
Language: English
Country: US
Series/Book: *Jurica Biology Series*
Plate: 9
Publisher: A. J. Nystrom & Co.
Year: c.1920s

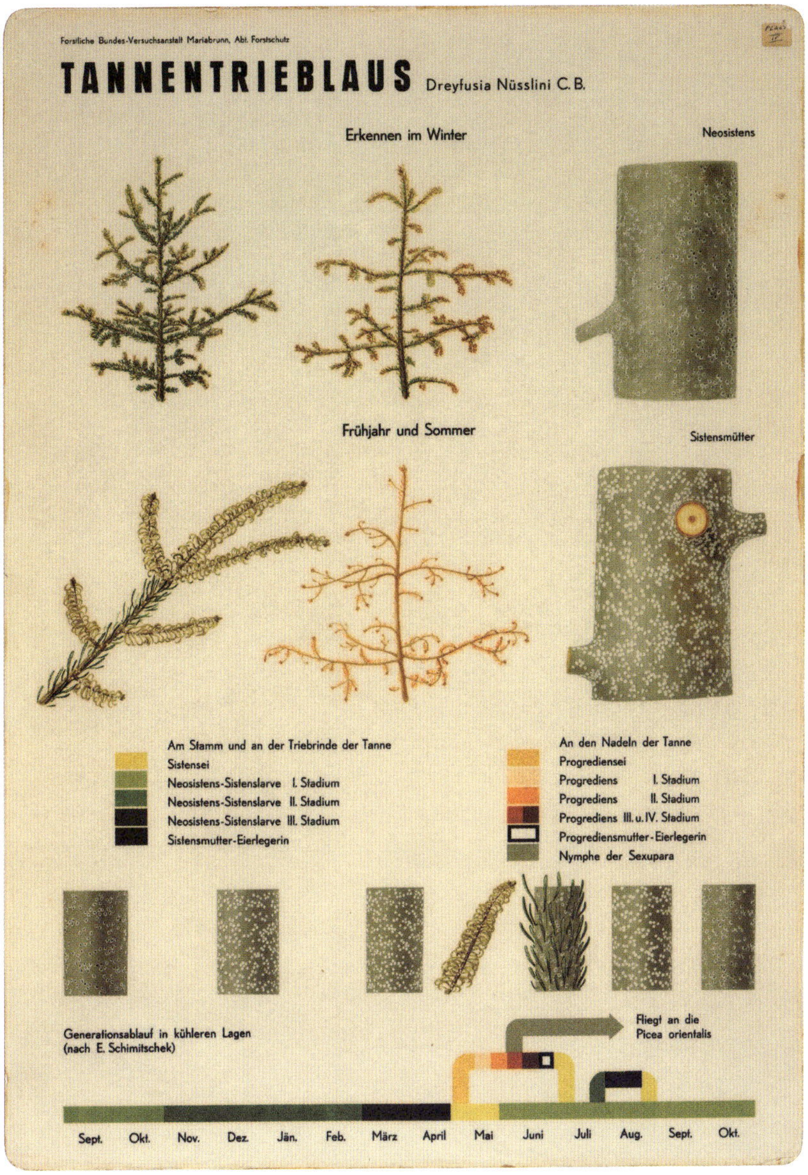

ABOVE: One such problem insect is the aphid, whose cycle is outlined in this German chart. The aphid in question, *Dreyfusia nusslini* (now more commonly known as *D. nordmannianae*, or the woolly fir aphid), was introduced to central Europe from the Caucasus in 1840, and had arrived in Germany by 1880. This aggressive insect has the ability to literally suck a young plant dry. One of the most damaging pests for pines and conifers, this chart may help students understand the pest's complex life cycle and stages of attack, but unfortunately there is little to be done about infestations except to cull and burn severely affected trees.

ABOVE:
Title: *Tannentrieblaus* (*Woolly Fir Aphid*)
Author: Unknown
Language: German
Country: Germany
Series/Book: Unknown
Plate: IV
Publisher: Forstliche Bundes-Versochsanstalt Mariabrunn, Abt. Forstchutz (Federal Forest Research Centre; Forest Protection)
Year: Unknown

XX ROSACEAE / ROSE FAMILY

This is a family with almost 5,000 species in roughy 110 genera, found mainly in north temperate regions of the world. Members of the rose family are woody and herbaceous plants, mainly perennial, but including a few annuals. Featuring some of the best-known garden trees and shrubs, the family includes *Cotoneaster*, *Exochorda*, *Kerria*, *Pyracantha* and *Rosa*, and wild-growing trees such as *Crataegus monogyna* (hawthorn) and *Sorbus* (mountain-ash or rowan). Other garden plants include the *Alchemilla* and *Geum*. The family also encompasses many fruits of temperate regions, among them the apple, cherry, peach, pear, plum, raspberry and strawberry. *Rosa × damascena* is commercially cultivated for attar of roses, used in the perfume industry.

Characteristics: stipules at base of each leaf; five sepals; five free petals; flower always hypanthium (cup-like structure composed from the fused petal, sepal and stamen bases); stamens usually numerous; fruit variable, often pome (apple-like), drupelets (blackberry-like), drupe (stone fruit), dry capsules or dry nutlets.

OPPOSITE:
Title: *Rosa canina*
Authors: Heinrich Jung, Dr. Friedrich Quentell
Illustrator: Dr. Gottlieb von Koch
Language: N/A
Country: Germany
Series/Book: *Neue Botanische Wandtafeln*
Plate: 9
Publisher: Fromann & Morian (Darmstadt, Germany); Hagemann (Düsseldorf, Germany)
Year: 1928; 1951–63

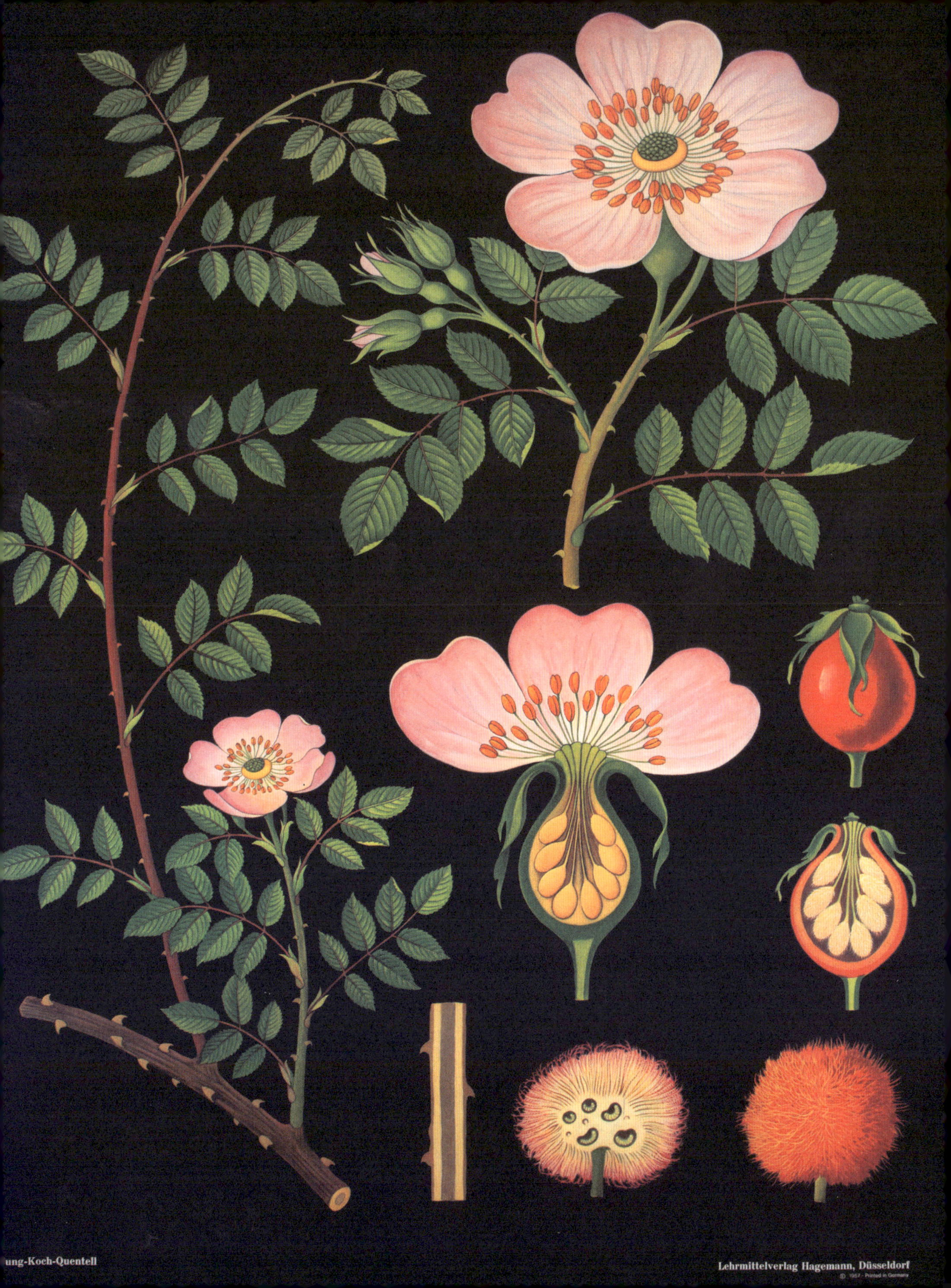
ung-Koch-Quentell
Lehrmittelverlag Hagemann, Düsseldorf
© 1957 · Printed in Germany

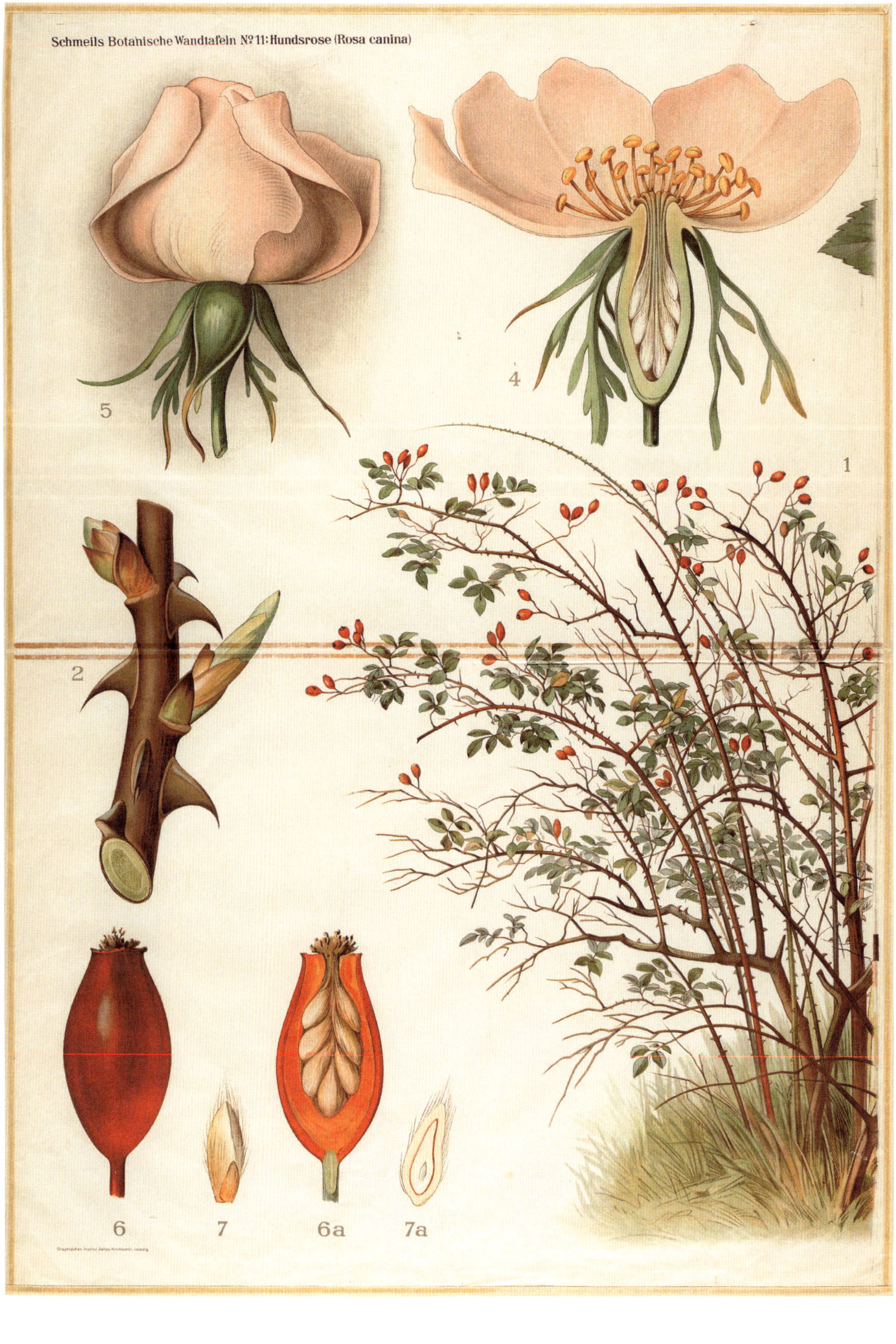
Schmeils Botanische Wandtafeln №11: Hundsrose (Rosa canina)
5
4
1
2
6
7
6a
7a

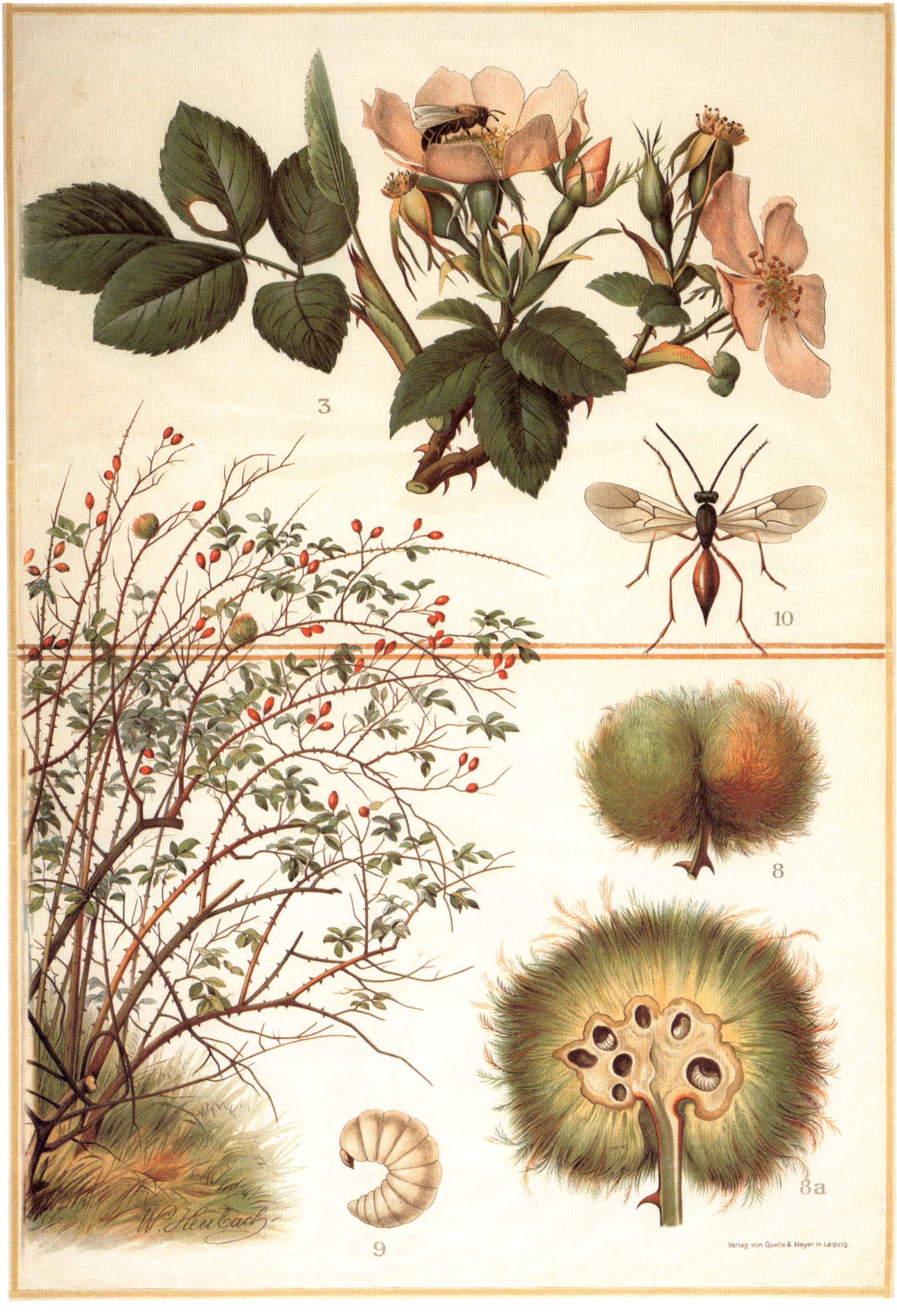

3
10
8
8a
9
W. Heubach
Verlag von Quelle & Meyer in Leipzig.

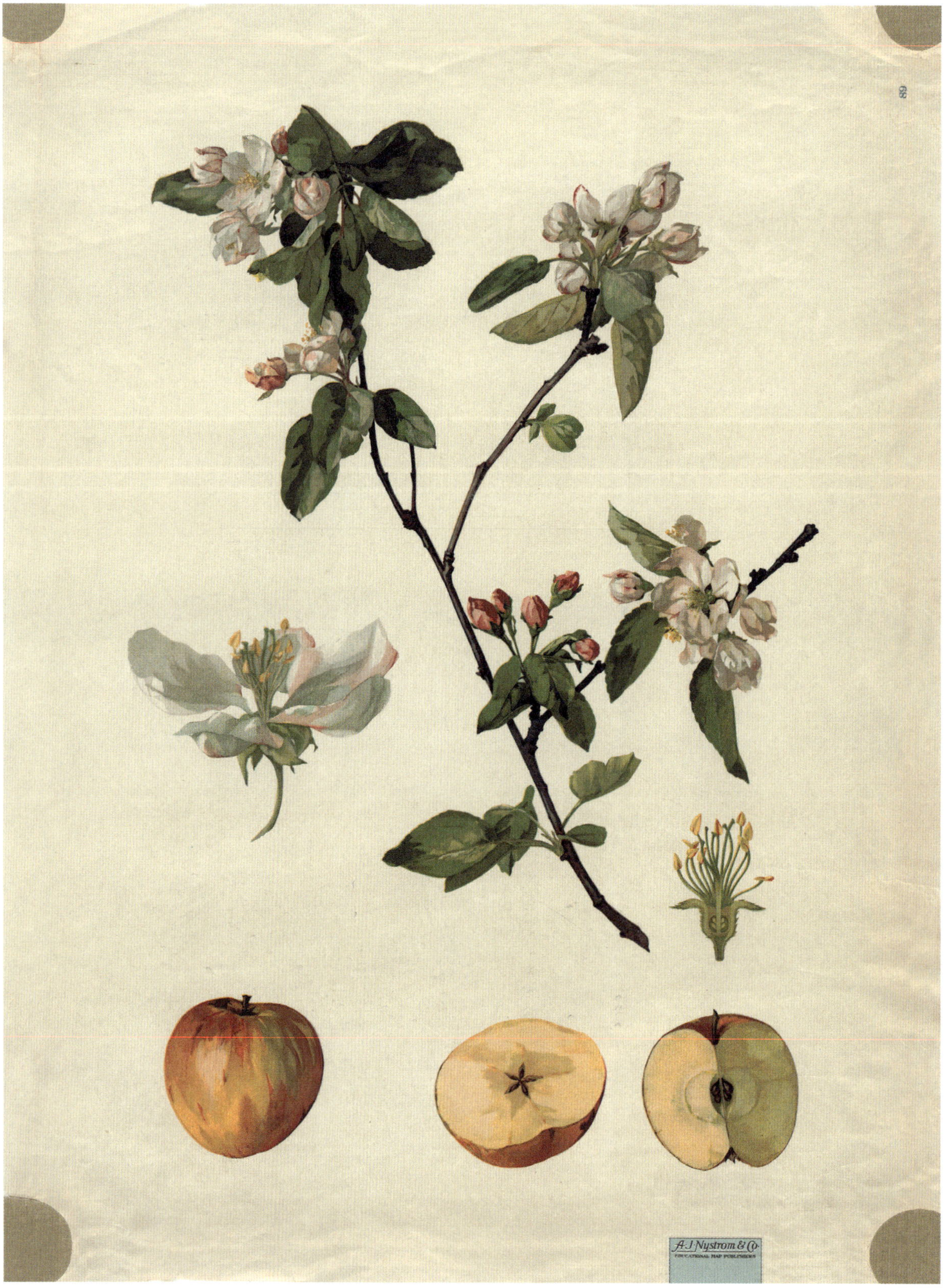
A. J. Nystrom & Co.
EDUCATIONAL MAP PUBLISHERS

PREVIOUS PAGE:
Title: *Hundsrose (Rosa canina)*
Author: Otto Schmeil
Language: German
Country: Germany
Series/Book: *Botanische Wandtafeln*
Plate: 11
Publisher: Quelle & Meyer (Leipzig, Germany)
Year: 1907

OPPOSITE:
Title: *Malus domestica*
Author: Von Engleder; Illustrator: C. Dietrich
Language: N/A
Country: Germany
Series/Book: *Engleders Wandtafeln für den naturkundlichen Unterricht Pflanzenkunde (Engleder's Wall Charts for Natural History Lessons: Herblore)*
Plate: 31
Publisher: J. F. Schreiber (Esslingen, Germany)
Year: 1897

OPENING & PREVIOUS PAGE: Two different authors, and two quite different interpretations of *Rosa canina* – although the charts cover many of the same elements. Jung, Koch, Quentell open this chapter with *R. canina*, presented as elegantly as though she were a cultivated lady in the rose court. In fact, the dog rose is a ragamuffin, growing wild in country hedgerows and scrambling up through treetops. Yet these pretty peasant roses aren't without claims to aristocracy: the heraldic rose of the age of chivalry was based on *R. canina*.

Otto Schmeil, on the previous page, gives us the wild rose in its countryside habitat, and the plant looks quite different. Schmeil's *R. canina* is a spiny thicket, and a close up of the inflorescences reveals a scruffy clump of flowers at all different stages of life, and imperfect leaves – just as we might see in life, although this presentation is of course also useful for didactic purposes.

Besides the flowers and details of their thorny stems, both authors show us in cross-section the ripening seeds within the fruit that will become a bright red hip beloved by wildlife and herbalists for its high nutritional content. And at the bottom right of both charts, a curiosity with a quaint name – Robin's pincushion. The name 'Robin' refers to the puckish character of English folklore, Robin Goodfellow, who must have been supposed to have planted them there. In fact this is a common gall, caused by the larvae of *Diplolepis rosae*, that can be found on the stems of *Rosa canina* from late summer. Within these fibrous nests, between 40 and 60 gall wasp larvae are growing, and will hatch out in the fall when the pincushion turns brown and dry. Schmeil goes a step further than Jung, Koch, Quentell, illustrating the culprit gall wasp and its larvae offspring.

OPPOSITE: Sparsely drawn and absent of text, Von Engleder's charts were internationally versatile and seen around the world. Unlike many charts, these were authored and illustrated by the same hand. Engleder was a teacher in Munich, and designed his series for elementary classroom instruction. His illustrations were neither too idealized, nor too specific. The fruit of *Malus domestica*, the common apple, for example, is rendered as a simple fleshy circular pome – not particular to any one cultivar, and easily identifiable regardless of geography. At the same time, the fruit is depicted gently bruised thanks to Engleder's commitment to verisimilitude. As with many charts, this realism gives the illustrations authority and accessibility.

As a general rule, Von Engleder approached each element as an individual, equidistantly distributed, and padded with a comfortable orbit of white space. The chart is comfortable to look at, while directing the eye between objects. For example, two pairs that are not adjacent yet demand comparison: the vertical (figs. c, f) and horizontal (figs. d, g) cross-sections of the fruit and ovary, which provide the student with a portrait of an apple's changing anatomy.

OPPOSITE: Alois Pokorny's beautiful charts would be at home in an herbarium drawer or a high-end florilegium, never mind the classroom wall. His work with live plants influenced his aesthetic – his illustrations resemble real plants splayed on a specimen sheet, roots and all.

Fragaria vesca (woodland strawberry) is native to Europe and Asia. It grows wild and bears small edible fruits with a remarkable fragrance, inspiring the Latin name of the genus. As strawberry plants do, Pokorny's grows vertically and horizontally, with a winding runner from which new growth has begun.

Despite an adherence to verisimilitude, Pokorny employs magnification – arguably a requisite of any botanic illustration intending to fulfill a didactic role. In the lower-right corner, the cross-section of a fruit has been unexpectedly inverted, a gesture Pokorny presumably intended to mirror the flower at centre-left, also in cross-section to show the numerous ovules and eager anthers.

Among the leaves Pokorny has added two details that underscore the chart's commitment to realism: at left, the red freckles of common spot, a widespread disease caused by the fungus *Mycosphaerella fragariae*; to its right, a leaf's evenly serrated contour has been interrupted by a hungry insect.

OPPOSITE:
Title: *Fragaria vesca*
Author: Alois Pokorny
Language: N/A
Country: Germany
Series/Book: *Botanische Wandtafeln*
Plate: N/A
Publisher: Smichow (Neubert, Germany)
Year: 1894

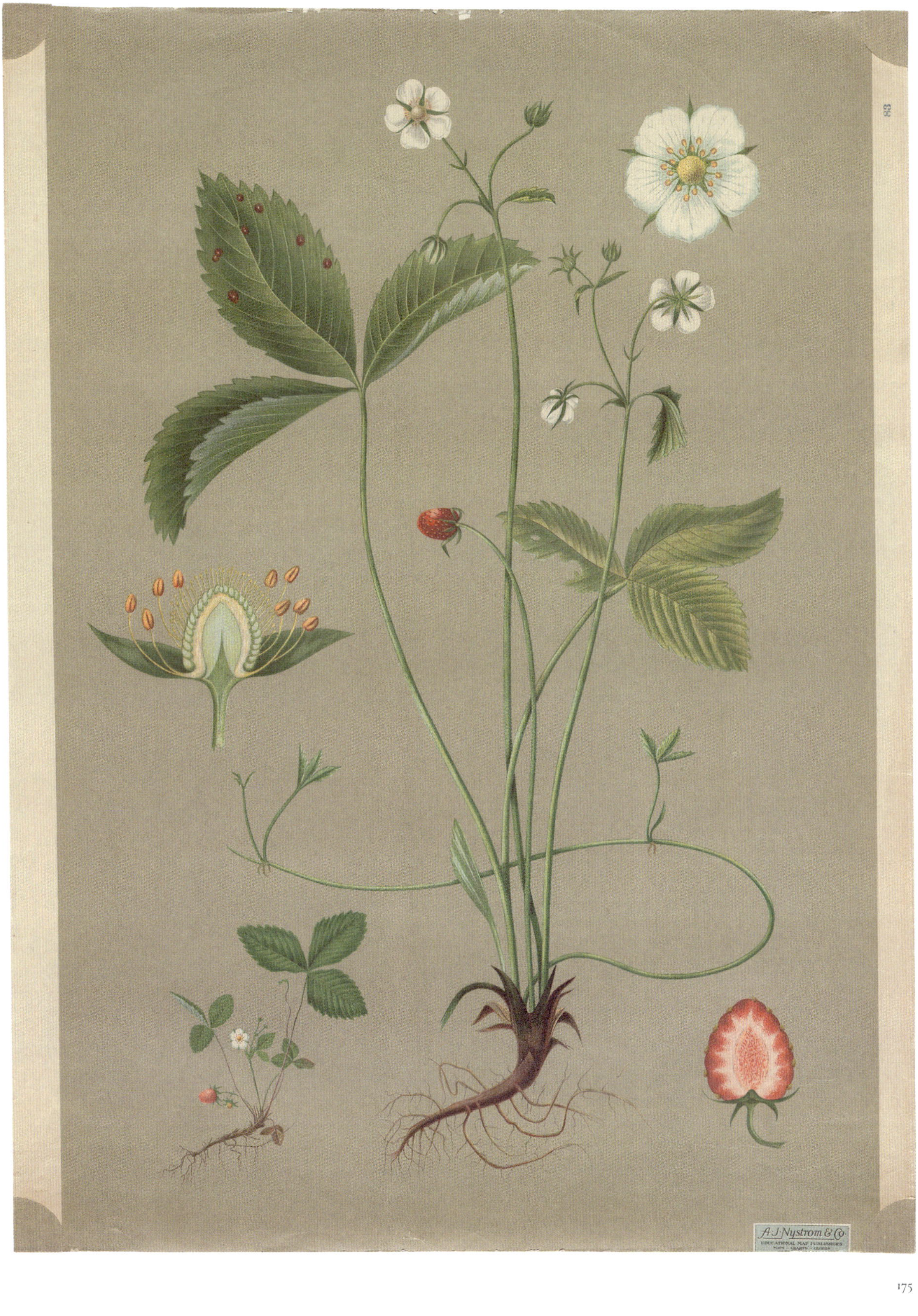
A·J·Nystrom & Co·
EDUCATIONAL MAP PUBLISHERS
MAPS - CHARTS - GLOBES

RIGHT: Like Von Engleder, André and Madeleine Rossignol also managed the authorship and illustration of their charts, but their audience was specifically Francophone. Their boards would become ubiquitous throughout France, their success attributed to their effective execution, as well as the social landscape in post-Liberation France. Poised to re-engage education, the country embraced wall charts as a tool to aid new teachers.

The Rossignols' illustrations were approachable and universal, rendered in a style that would be familiar to young students, in a format not unlike a storybook. A hierarchy of straitlaced sans-serif labels were added within the chart, allowing students to make a direct association between an object and its name. Emphasis was on object, not detail. One could easily count the ovary's nine ovules and the fruit's nine seeds.

Flat colour fields and simple lines were the rule, while the full plant illustrated at left with shadow and detail is an exception adhered to throughout their flower boards, providing a nice contrast and a context for the stylized illustrations of the individual parts.

Here, the Rossignols illustrate *Rosa canina*. The name 'Eglantier' may sound elegant to non-Francophones; however the name in fact derives from the Old French 'aiglantin,' which in turn comes from the Latin 'acus,' simply meaning 'needle' – referring, of course, to the rose's thorny stems. In modern French, 'Églantine' refers to wild roses and in particular the *R. canina*.

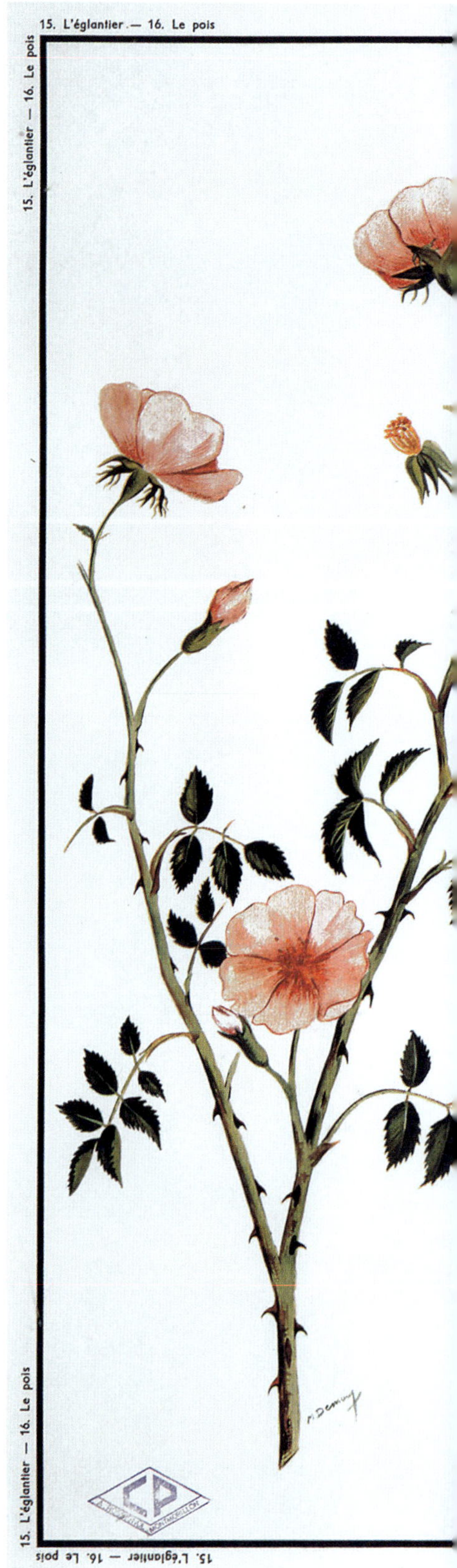

RIGHT:
Title: *L'Eglantier*
Authors: Madeleine and Andre Rossignol
Language: French
Country: France
Series/Book: N/A
Plate: 15
Publisher: Éditions Rossignol
Year: c.1950s

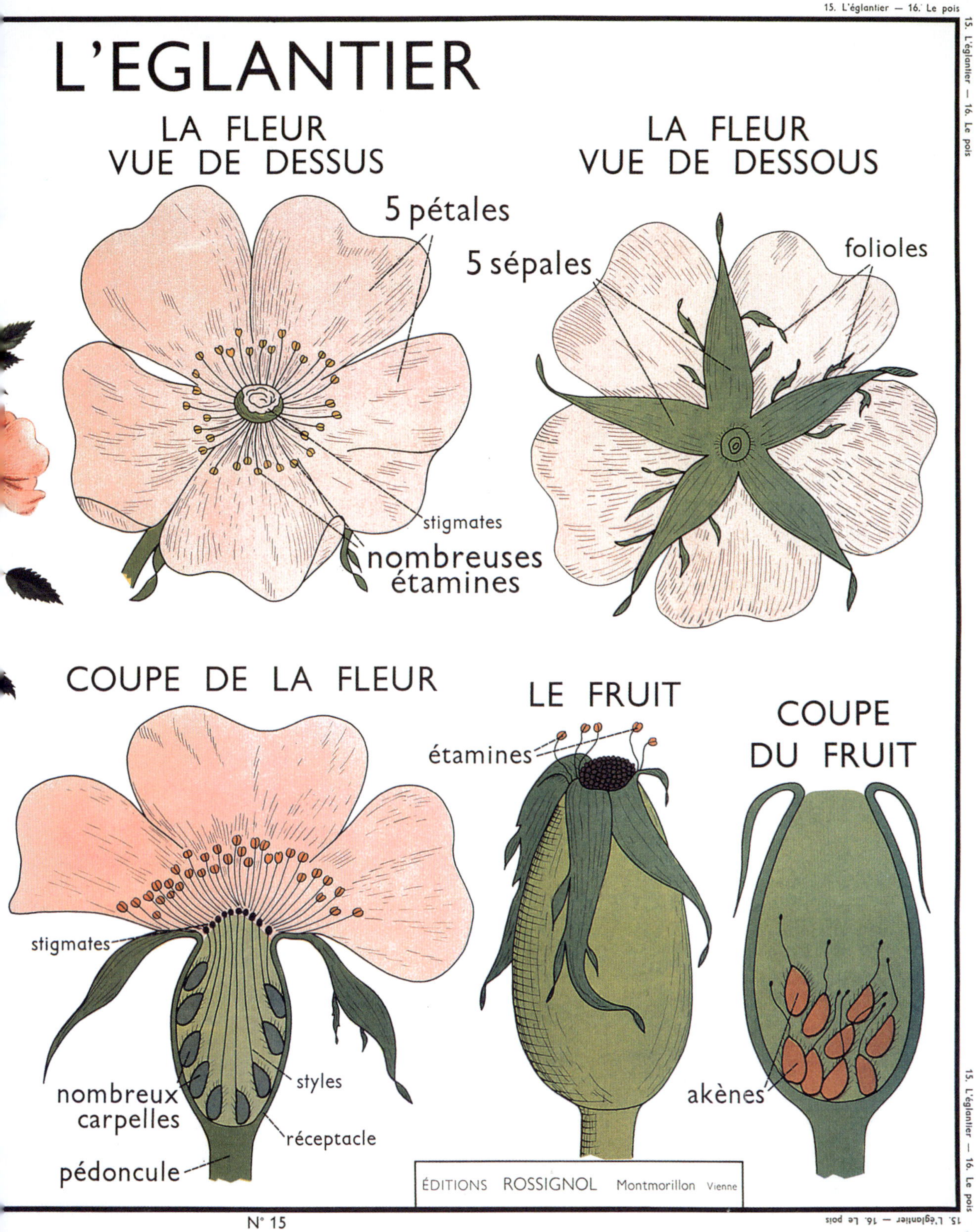
15. L'églantier — 16. Le pois
L'EGLANTIER
LA FLEUR VUE DE DESSUS
5 pétales
stigmates
nombreuses étamines
LA FLEUR VUE DE DESSOUS
5 sépales
folioles
COUPE DE LA FLEUR
stigmates
nombreux carpelles
styles
réceptacle
pédoncule
LE FRUIT
étamines
COUPE DU FRUIT
akènes
ÉDITIONS ROSSIGNOL Montmorillon Vienne
N° 15

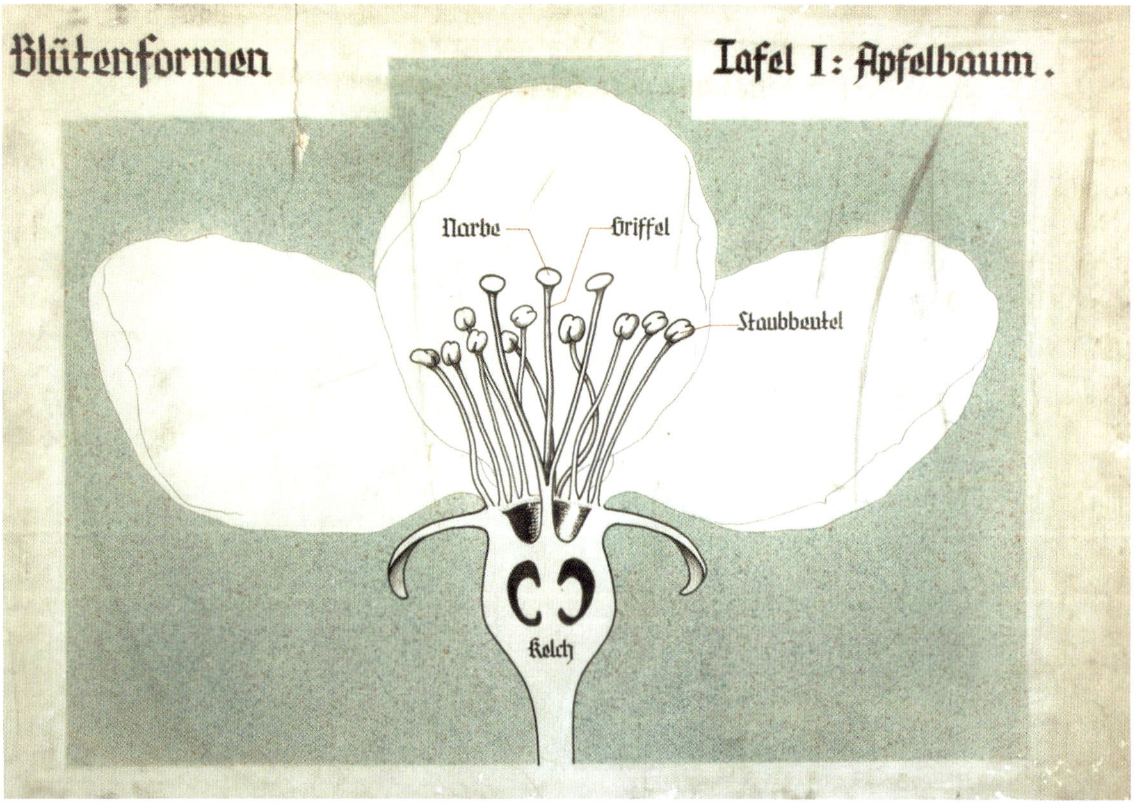
Blütenformen
Tafel I: Apfelbaum.
Narbe
Griffel
Staubbeutel
Kelch

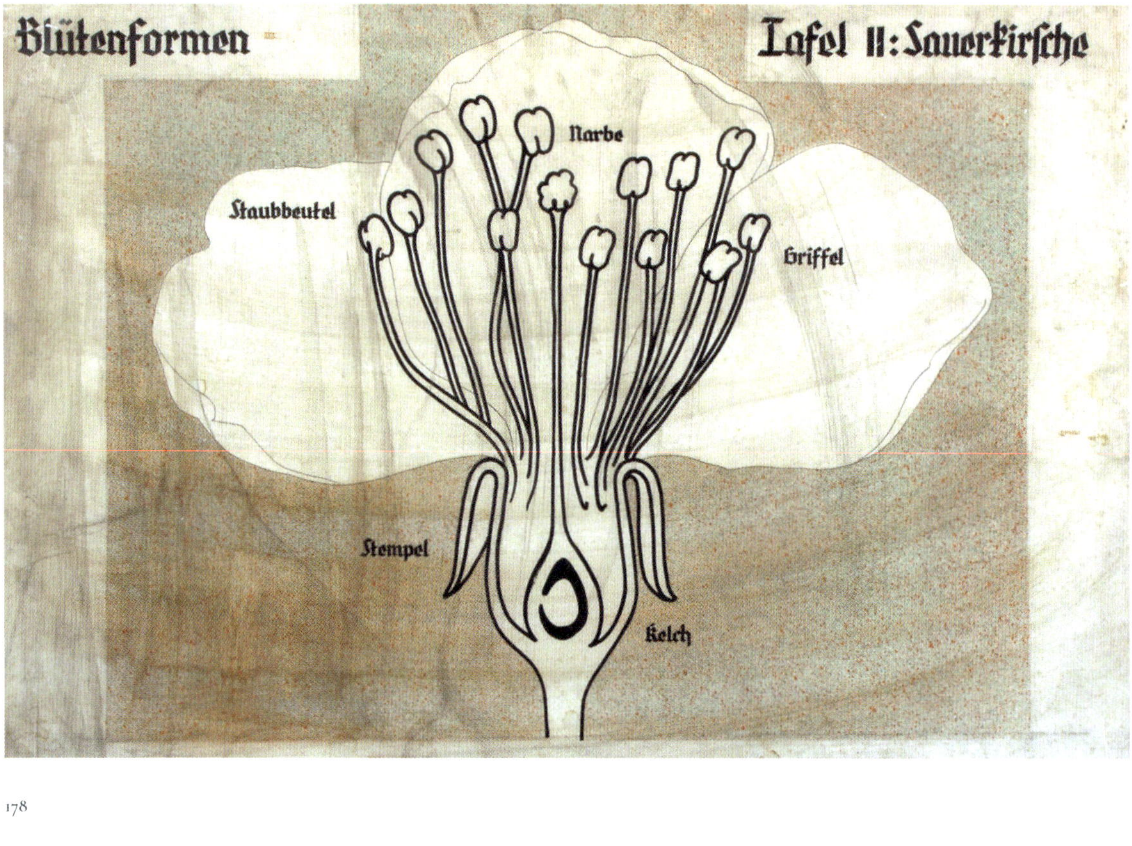
Blütenformen
Tafel II: Sauerkirsche
Narbe
Staubbeutel
Griffel
Stempel
Kelch

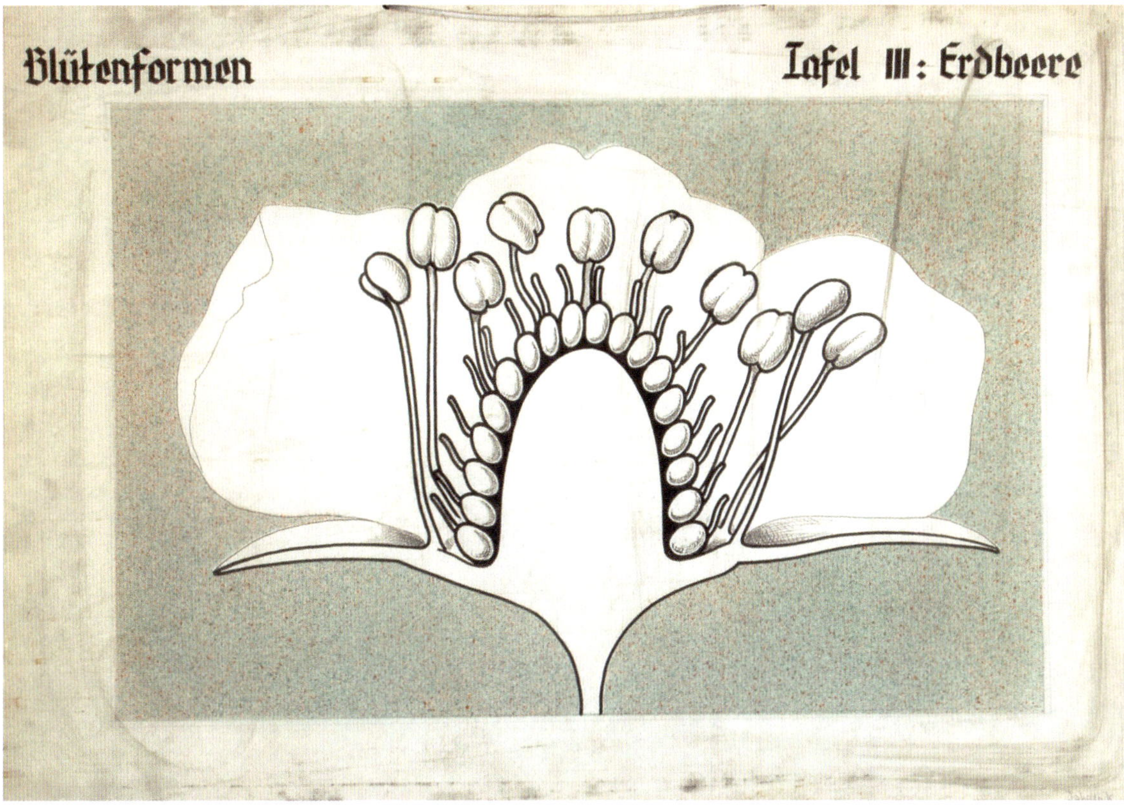

THIS SPREAD: Three German charts provide an easy comparison of *Rosaceae* reproductive organs. *Tafel I, Apfelbaum*; *Tafel II, Sauerkirsche*; *Tafel III, Erdbeere* compare, respectively, apple, sour cherry, and strawberry.

In all three panels, the flowers' three petals are rendered as fields of white (articulated by hairline contours that appear to have been added by a student's pencil), thereby relinquishing emphasis to the anthers, filaments, stigmas, styles, ovaries and ovules. This approach effectively shows three major fruits of the *Rosaceae* family: the apple's multi-seeded pome, the cherry's single-seeded drupe, and the strawberry's enlarged fleshy receptacle covered with achenes.

OPPOSITE & ABOVE:
Title, opposite top: *Apfelbaum* (*Apple*);
Opposite bottom: *Sauerkirsche* (*Sour Cherry*);
Above: *Erdbeere* (*Strawberry*)
Author: Unknown
Language: German
Country: Germany
Series/Book: *Blütenformen* (*Flower Types*)
Plates: 1; 2; 3
Publisher: Unknown
Year: Unknown

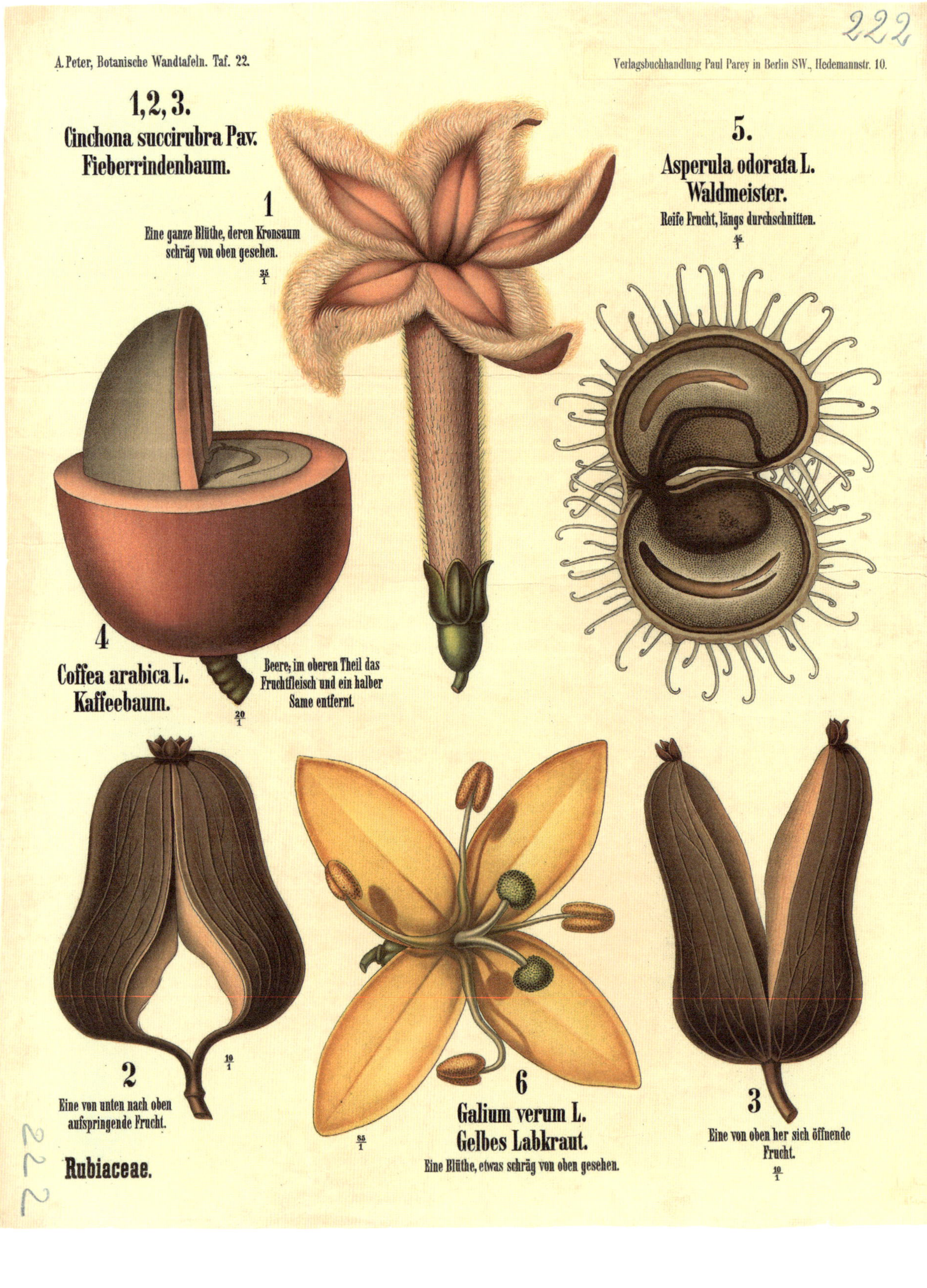
A. Peter, Botanische Wandtafeln. Taf. 22.
Verlagsbuchhandlung Paul Parey in Berlin SW., Hedemannstr. 10.
1, 2, 3.
Cinchona succirubra Pav.
Fieberrindenbaum.
1
Eine ganze Blüthe, deren Kronsaum schräg von oben gesehen.
35/1
5.
Asperula odorata L.
Waldmeister.
Reife Frucht, längs durchschnitten.
45/1
4
Coffea arabica L.
Kaffeebaum.
Beere; im oberen Theil das Fruchtfleisch und ein halber Same entfernt.
20/1
2
10/1
Eine von unten nach oben aufspringende Frucht.
Rubiaceae.
6
Galium verum L.
Gelbes Labkraut.
Eine Blüthe, etwas schräg von oben gesehen.
85/1
3
Eine von oben her sich öffnende Frucht.
10/1

XXI RUBIACEAE / COFFEE FAMILY

This large family has about 14,00 species in 615 genera, but few genera are widely grown. The *Rubiaceae* are widely distributed, however most occur in the tropics. Although principally woody shrubs, they are variable in habit, having adapted to a range of habitats, from arid desert to rainforest, and also include herbaceous species, vines and some trees. Two genera provide important crops: coffee from *Coffea* and quinine from *Cinchona*. Popular ornamentals include *Asperula*, *Bouvardia*, *Coprosma*, *Galium*, *Houstonia* and *Ixora*; the flowers are often fragrant, as with *Gardenia*.

The family also includes *Rubia tinctorum* (madder), which was long used as a source of red dye, and the invasive weed *Galium aparine* (cleavers or goosegrass). *Balmea stormiae* (ayuque) is the only species in its genus, with vivid red blooms, and is becoming a rare plant in Mexico, where it is regularly cut for use as a Christmas tree. A few *Rubiaceae* such as *Hydnophytum* and *Myrmecodia* are caudiciform succulents with epiphytic habits; they are known as 'ant plants' because their swollen stems have cavities that provide perfect habitats for ants.

Characteristics: leaves simple and entire, opposite or whorled, stipules sometimes modified to interpetiolar false leaves; inflorescence typically a cluster, cyme, corymb, panicle or spike; four to five sepals and petals, occasionally twelve fused of each, calyx sometimes reduced or absent; four to five stamens sympetalous (alternating with corolla lobes); the diverse fruit can come in the form of berry, capsule, schizocarp, or drupe.

OPPOSITE:
Title: *Rubiaceae*
Author: A. Peter
Language: German
Country: Germany
Series/Book: *Botanische Wandtafeln*
Plate: 10
Publisher: Paul Parey (Berlin, Germany)
Year: 1901

OPENING PAGE: A. Peter's chart offers a diverse selection of the flowers and fruit of various members of the *Rubiaceae* family, including those indigenous to Germany as well as the more exotic. The native curiosities he examples include the opened seed of *Asperula odorata* (now *Galium odoratum*, woodruff), whose tiny hooked bristles allow it to cling to passing animals for dispersal. The yellow blossom of the other German natural, *G. verum* (yellow bedstraw), is recognizable for its pointedly symmetrical four-petalled form, but here it is artfully portrayed, with shadows and swirling stamen, separate from the dense clusters in which these blossoms grow.

For the more exotic varieties, Peter gives us a cutaway section of a berry of *Coffea arabica* – yes the coffee bean, within its ripe red fruit. *Cinchona succirubra* (now called *C. pubescens*, quina, or fever tree) may be best known for the curative powers of its bark, but Peter focuses on its reproduction – its long, tubular and rather furry blossom, and two views of a seed opening, respectively, from the bottom up, and from the top down.

OPPOSITE: When European conquistadors brought malaria to South America, the disease, which was already widespread throughout much of Europe, found its antidote in *Cinchona*, a native plant known locally as the 'fever tree' for its curative properties. As the story goes, in the mid-seventeenth century, the Countess of Chinchón, wife of the viceroy or Peru, fell ill with malaria while visiting the country, and was nursed with a tea of cinchona bark. Quickly ferried to European hospitals the remedy was ambiguously referred to as 'Peruvian bark' in various anti-malarial concoctions. King Louis XIV's physician prescribed 'seven grams of rose leaves, two ounces of lemon juice, and a strong decoction of the cinchona bark served with wine.'

Yet the tree was unknown, and doubtlessly apothecaries offered ersatz barks. Finally, the species was identified. A 1797 description of the genus reads, 'The Peruvian Bark was made use of during a whole century, without its being known from what tree it was taken ... almost inaccessible, its native country not being easily visited by naturalists ... And this ignorance would have still continued, had not some botanists obtained an opportunity of seeing it in its native country.' Thus Zippel and Bollmann may well have decided that the *Cinchona* was a very important genus to represent in an educational chart. It's peculiar, however, that the authors omitted any reference to the ameliorative bark; perhaps a seed pod (fig. 5) and flower (figs. 1, 2) were deemed sufficiently distinctive for identification.

OPPOSITE:
Title: *Fieberrindenbaum (Cinchona Calisaya, var. Josephina Weddell)*
Author: Hermann Zippel; Illustrator: Carl Bollmann
Language: German
Country: Germany
Series/Book: *Ausländische Kulturpflanzen in farbigen Wandtafeln (Foreign Crops in Coloured Wall Panels)*
Plate: I Abteilung, 17
Publisher: Friedrich Vieweg & Sohn (Braunschweig, Germany)
Year: 1879–82

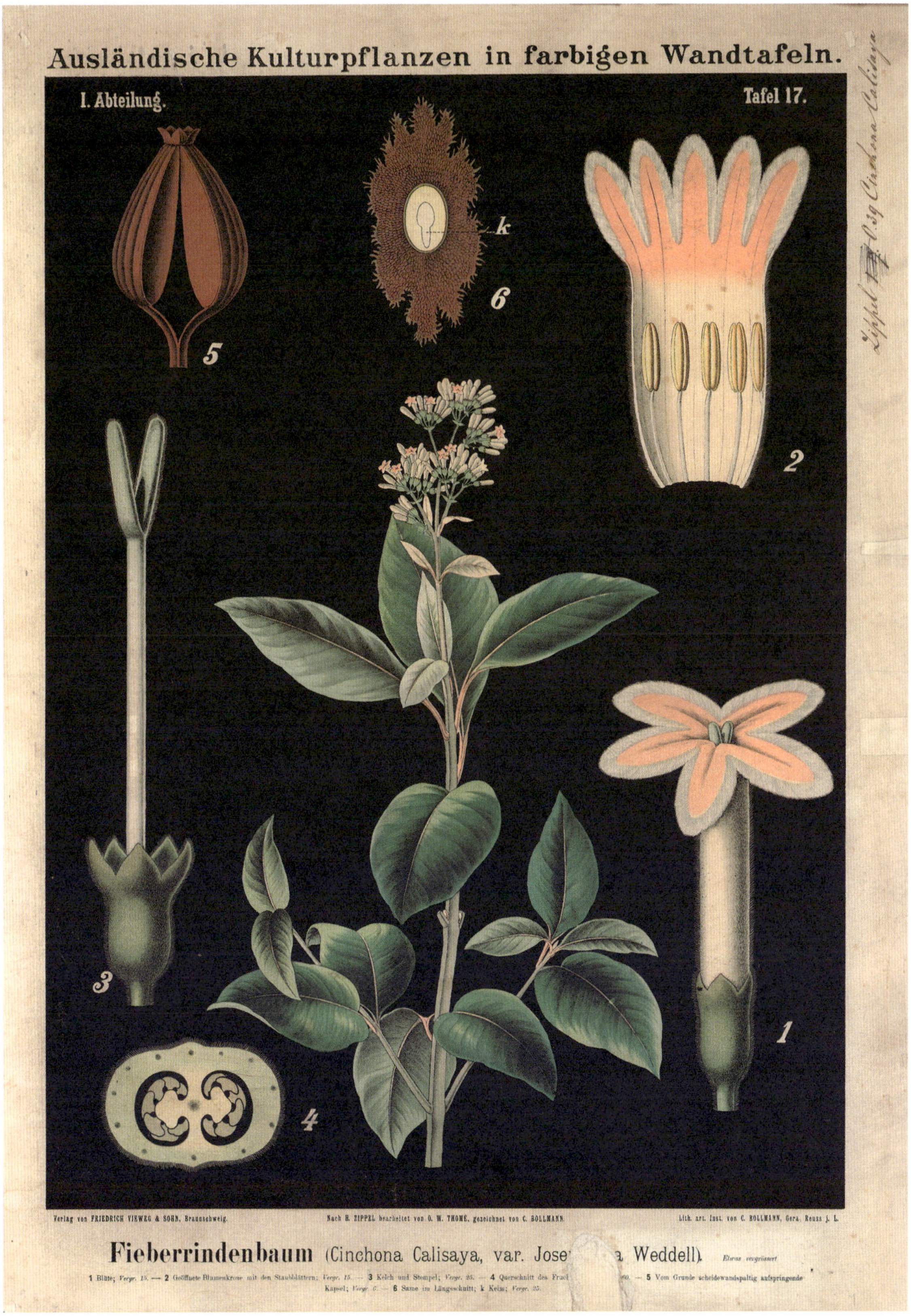
Ausländische Kulturpflanzen in farbigen Wandtafeln.
I. Abteilung.
Tafel 17.
k
6
5
2
3
1
4
Verlag von FRIEDRICH VIEWEG & SOHN, Braunschweig.
Nach H. ZIPPEL bearbeitet von O. W. THOMÉ, gezeichnet von C. BOLLMANN.
Lith. art. Inst. von C. BOLLMANN, Gera, Reuss j. L.
Fieberrindenbaum (Cinchona Calisaya, var. Jose a Weddell).
Etwas vergrössert
1 Blüte; Vergr. 15. — 2 Geöffnete Blumenkrone mit den Staubblättern; Vergr. 15. — 3 Kelch und Stempel; Vergr. 25. — 4 Querschnitt des Fruc 60. — 5 Vom Grunde scheidewandspaltig aufspringende
Kapsel; Vergr. 6. — 6 Same im Längsschnitt; k Keim; Vergr. 25.

RIGHT: *Coffea arabica* originated in Ethiopia's forest highlands, where it is said that a goatherd observed his flock eating red berries and becoming very excitable. He tried it himself, felt a mild high, and soon the country was chewing on red berries. Apocryphal though the story may be, a fascination with the bean spread across the Red Sea, where pilgrims to Mecca planted coffee seeds en route. By the fifteenth century, coffee culture had grown in Turkey, Persia, Egypt and North Africa.

Around 1890, when this chart was produced in Germany, *C. arabica* crops were thriving in Latin America and the Caribbean. A coffee plantation is harvested, while in the upper left, a branch shows the progression of maturity: jasmine-scented blossoms and berries that mature from pale green to the darkest red. Shaken into baskets or picked by hand, the ripe fruits contain two beans that are ready to percolate.

Similar to Schmeil's depictions of plants in their ecologies, this series portrays economic crops in their habitat of harvest. It is not particularly instructional, merely environmental.

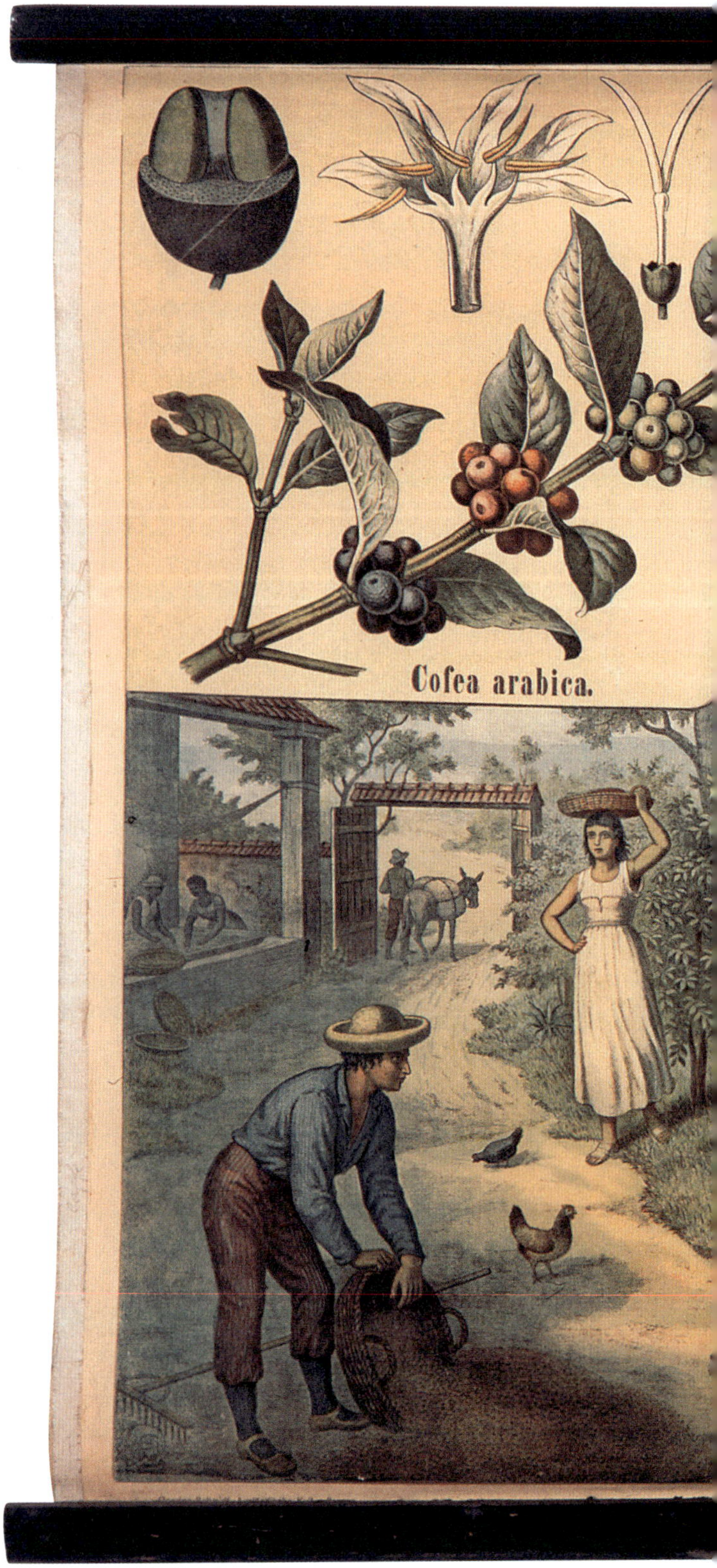

RIGHT:
Title: *Coffea arabica / Kaffee*
Author: Unknown
Language: N/A
Country: Germany
Series/Book: *Goering-Schmidt Ausländische Kulturpflanzen (Goering-Schmidt Foreign Crops)*
Plate: 1
Publisher: F. E. Wachsmuth (Leipzig, Germany)
Year: 1890

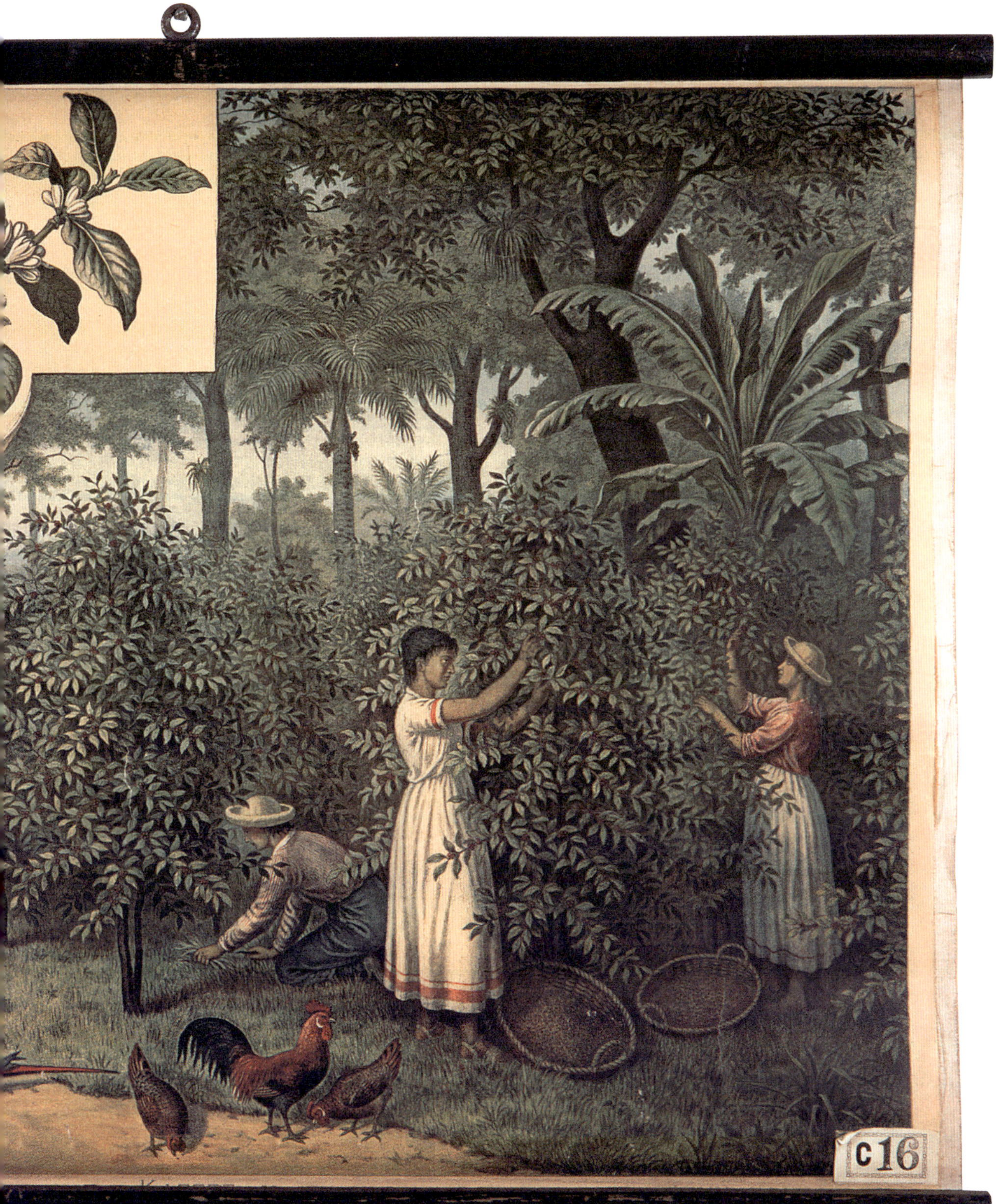
C16

XXII SOLANACEAE / NIGHTSHADE FAMILY

The nightshade family is large, with some 2,700 species in just over 100 genera. The plants are globally distributed, but concentrated in tropical and South America. *Solanaceae* are quite variable, ranging from annual to perennial herbaceous species, some with tuberous or rhizomatous roots, to shrubs, small trees, and climbers. Although there are some important vegetables – eggplant, Cape gooseberry, chilli and sweet peppers, potato, tomatillo, tomato – in this family, it is also known for its poisonous plants. In fact, green parts of most species are poisonous if ingested, but there are some famously deadly plants, such as *Atropa belladonna* (deadly nightshade).

Paradoxically, some poisonous species also provide medically valuable ingredients to the pharmaceutical industry; these useful genera include *Atropa*, *Brugmansia*, *Duboisia* and *Scopolia*. Another commercially important plant is *Nicotiana*, the source of tobacco. In the past, nicotine was used extensively in insecticides and pesticides. The most popular garden *Solanaceae* include *Browallia*, *Brunfelsia*, *Cestrum*, *Datura*, *Petunia*, *Physalis* and *Solanum crispum*.

Characteristics: leaves are spirally arranged, extipulate, simple to pinnatisect, often with a characteristic scent; branching habit due to sympodial shoots (apical buds cease to grow, prompting laterals to grow); often hairy, sometimes with thorns or prickles; most often five sepals, petals, and stamens – sepals are sometimes fused and enclosing fruit; corolla with five lobes; fruit is a capsule or berry.

OPPOSITE:
Title: *Solanum tuberosum*
Author: Von Engleder; Illustrator: C. Dietrich
Language: N/A
Country: Germany
Series/Book: *Engleders Wandtafeln für den naturkundlichen Unterricht Pflanzenkunde (Engleder's Wall Charts for Natural History Lessons: Herblore)*
Plate: 6
Publisher: J. F. Schreiber (Esslingen, Germany)
Year: 1897

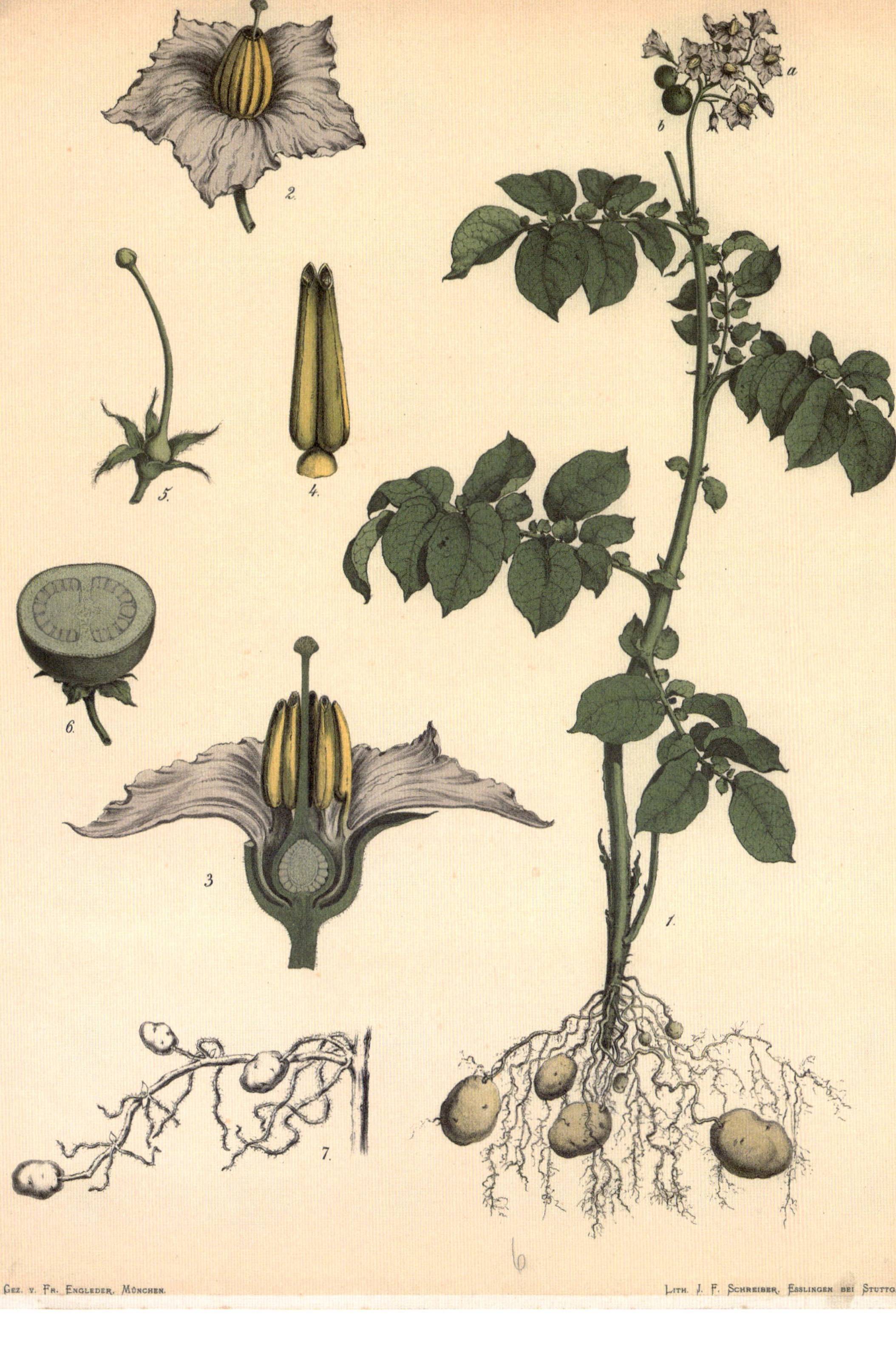

Gez. v. Fr. Engleder, München.

Lith. J. F. Schreiber, Esslingen bei Stuttgart.

OPENING PAGE: Von Engleder represents the full potato plant from subterranean tuber to flowers (fig. 1a) and tomato-like fruit, which contains hundreds of seeds (fig. 1b). Given that the potato is one of the most important crops in the world, it's rather curious that Von Engleder devotes his attention to its sexual organs rather than to its tubers. And while *Solanum tuberosum* can reproduce sexually, asexual reproduction, straight from the tuber, is the norm in agriculture.

If, however, you wanted to create a new variety of potato, the flowers and their fruits would suddenly be needed – asexual reproduction produces only clones of the parent, while sexual reproduction encourages genetic diversity. Potatoes are well suited for this: when coaxed into fruiting you will usually see a huge variance among the offspring. Growers can then select their seedlings of choice, and either refine the plant through further breeding, or allow the plant to clone itself.

A word of warning: like almost every part of *S. tuberosum* (including its tuber, when green) – and like most *Solanaceae* – the tempting fruit of the potato is poisonous.

OPPOSITE: At first glance it's surprising that Zippel and Bollmann's chart of *Capsicum longum* (now called *C. annuum* var. *annuum*) doesn't pay more homage to the scarlet fruit, economically the most valuable part of the plant, infused as it is with capsaicin, a chemical that 'gives the Spanish pepper its fiery-hot taste' and the species its Latin name. The small pepper is diminutive alongside a flower (fig. 1), enlarged to show the curious stamen formation making a ring around the bulbous ovary, and the jewels on a crown of petals (also fig. 2); a young ovary lined with two columns of ovules (fig. 3); and a cross-section of the hollow fruit (fig. 5) with an abundance of seeds. Dominating the chart is a cutting of a flowering plant, just beginning to grow heavy with pendulous, nascent green fruits.

It's helpful to recall, then, that the chart's purpose was educational. Why focus on the already-familiar pepper when students are less likely to recognize the flowers and leaves? Where one may find the chart lacking, the companion text summarily addresses: 'The fruits are odorless in their natural state, but dried and crushed are very sharp, causing violent sneezing, a burning taste, and a lasting heat … irritating to the digestive organs, and, in large doses, causing inflammations and reddening of the skin'.

OPPOSITE:
Title: *Spanischer Pfeffer (Capsicum longum DC)*
Author: Hermann Zippel; Illustrator: Carl Bollmann
Language: German
Country: Germany
Series/Book: *Ausländische Kulturpflanzen in farbigen Wandtafeln (Foreign Crops in Coloured Wall Panels)*
Plate: III Abteilung, 15
Publisher: Friedrich Vieweg & Sohn (Braunschweig, Germany)
Year: 1889

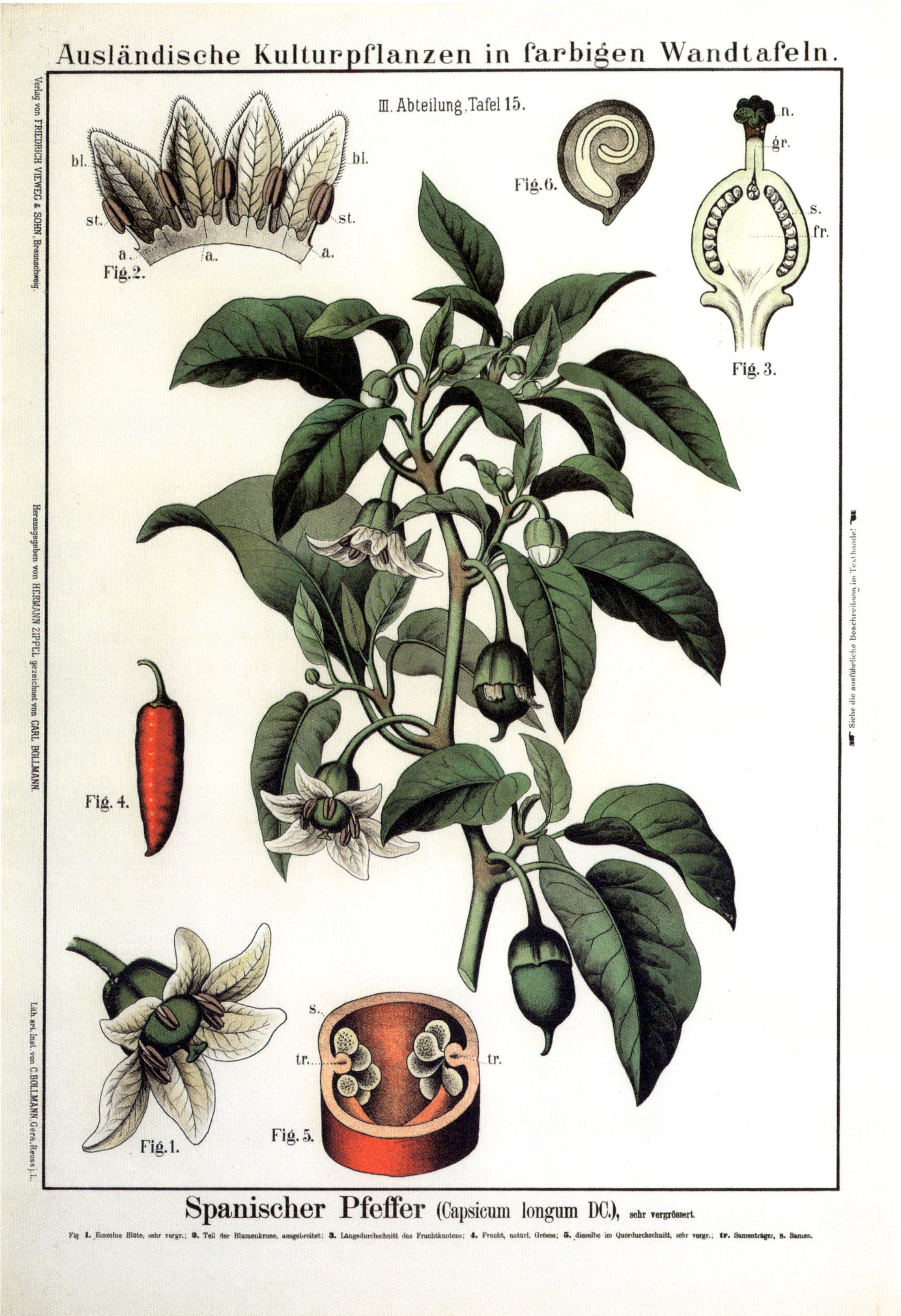
Ausländische Kulturpflanzen in farbigen Wandtafeln.
III. Abteilung, Tafel 15.
Verlag von FRIEDRICH VIEWEG & SOHN, Braunschweig.
Herausgegeben von HERMANN ZIPPEL gezeichnet von CARL BOLLMANN.
Lith. art. Inst. von C. BOLLMANN, Gera, Reuss j. L.
Siehe die ausführliche Beschreibung im Textbande!
bl.
st.
a.
Fig. 2.
Fig. 6.
n.
gr.
s.
fr.
Fig. 3.
Fig. 4.
s.
tr.
Fig. 5.
Fig. 1.
Spanischer Pfeffer (Capsicum longum DC.), sehr vergrössert.
Fig 1. Einzelne Blüte, sehr vergr.; 2. Teil der Blumenkrone, ausgebreitet; 3. Längsdurchschnitt des Fruchtknotens; 4. Frucht, natürl. Grösse; 5. dieselbe im Querdurchschnitt, sehr vergr.; tr. Samenträger, s. Samen.

RIGHT: The darker side of the *Solanaceae* family, *Hyoscyamus niger* (henbane) and *Datura stramonium* (thorn apple or jimson weed) have been written about and illustrated for centuries. An 1831 medical text writes of *D. stramonium*: '*Stramonium* produces intoxication, nausea, delirium, loss of sense, drowsiness, a sort of madness and fury; loss of memory, convulsions, sense of suffocation, paralysis of the limbs, cold sweats, excessive thirst, dilatation of the pupil, tremblings, and death.' Of henbane, it warns: 'The whole plant has a strong foetid narcotic smell, and abounds in a clammy juice of a similar odor. The root has a sweetish taste, which has caused it to be sometimes mistaken for that of the parsnip.' And the two verboten species were a popular subject for wall charts – failure to recognize either species could be perilous. Hence we see in the chart at right a thorough deconstruction of the two species from root to fruit, by Czech illustrator Otakar Zejbrlík.

OPPOSITE: Illustrating *Datura stramonium*, Pokorny gives us no more than what is necessary: a flowering plant to illustrate flowers young and mature, an ample number of leaves, fruits new and dehiscent, a severed stem, and a root system. A vertical cross-section of the flower and a horizontal slice of the seeded fruit complete Pokorny's profile – as always, elegant and precise.

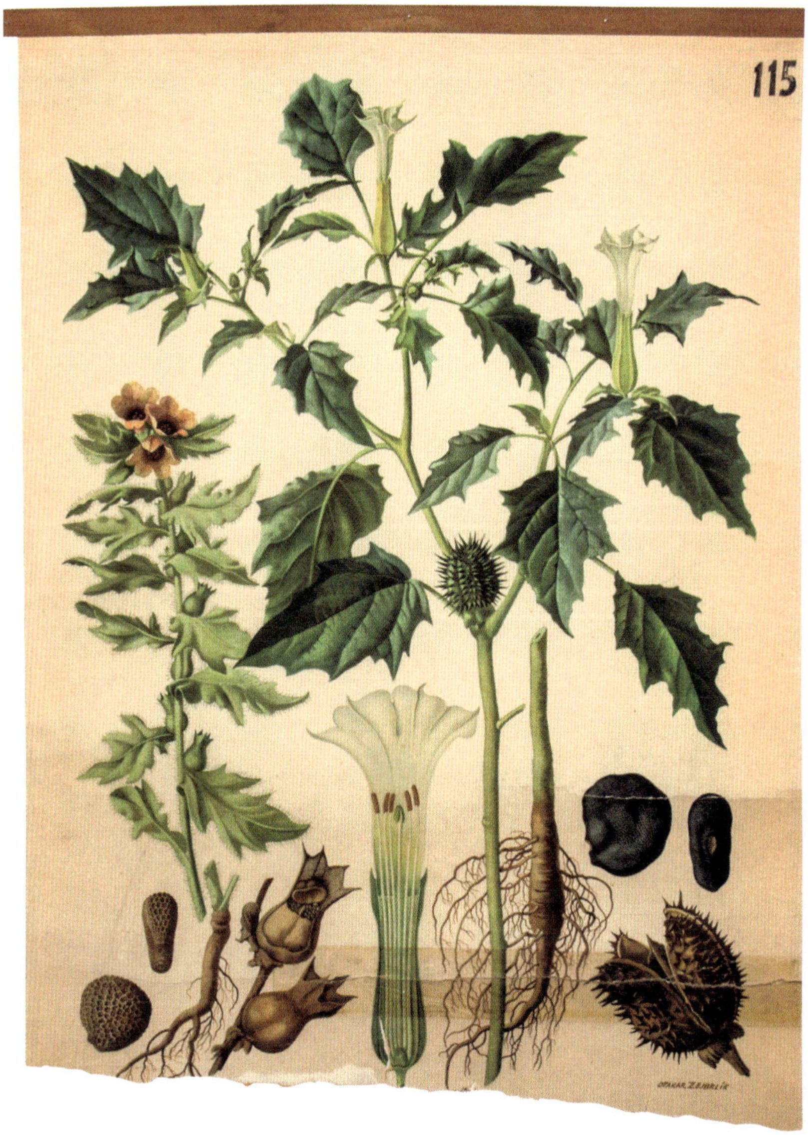

ABOVE RIGHT:
Title: *Hoscyamus niger*
Author: Otakar Zejbrlík
Language: Czech
Country: Czech Republic
Series/Book: *Léčivé Rostliny (Medicinal Plants)*
Plate: Unknown
Publisher: Kropác & Kucharský
Year: 1943

OPPOSITE:
Title: *Datura stramonium*
Author: Alois Pokorny
Language: N/A
Country: Germany
Series/Book: *Botanische Wandtafeln*
Plate: Unknown
Publisher: Smichow (Neubert, Germany)
Year: 1894

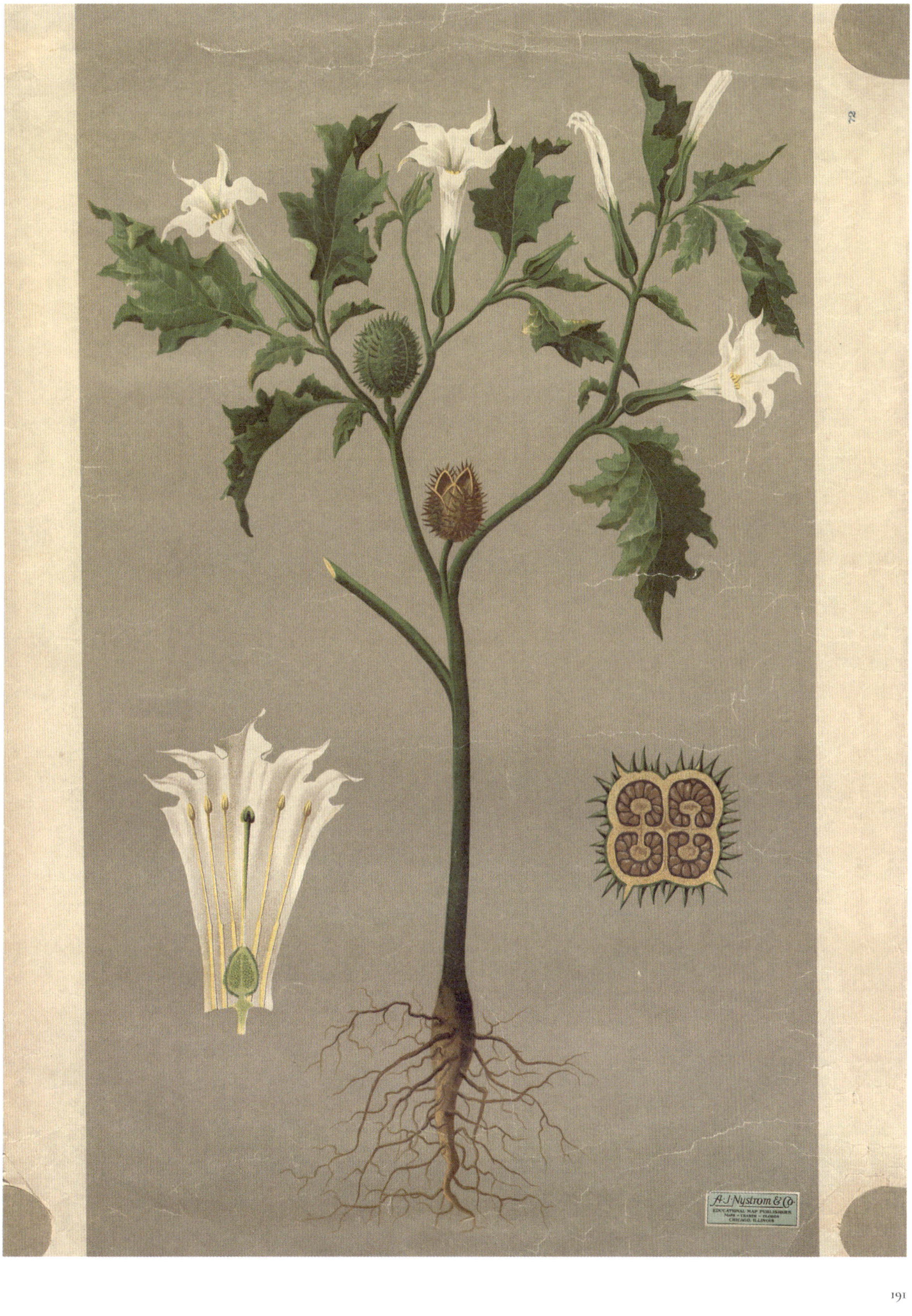
72
A·J·Nystrom & Co·
EDUCATIONAL MAP PUBLISHERS
MAPS – CHARTS – GLOBES
CHICAGO, ILLINOIS

OPPOSITE: Zippel and Bollmann's wall chart of *Nicotiana tabacum* is everything one would expect: illustrations of the flower (fig. 1), corolla and stamens (fig. 2), a cross-section of an ovary and a style (fig. 3), calyx and capsule (fig. 4), seed (figs. 5, 6), and a central specimen with its cyme of funnel-shaped flowers and stalk of leaves, 'deep green on the upper side, paler below, covered with short glandular hairs; somewhat sticky, ribbed, slightly undulating,' prominently displayed as the tobacco plant's feature attraction and appropriately dominating the chart.

More interesting than Zippel and Bollmann's illustrations is their companion text, in which they editorialize the effects of tobacco. As always, the text progresses from basic morphology – 'Capsules are oval, narrower above, longer than the calyx' – to cultivation – 'tobacco thrives best north and south of the Equator between the fifteenth and thirty-fifth degrees' – to consumption, which is where the commentary begins. Under the topic titled 'Constituents of Tobacco,' they write: 'The leaves of all species of tobacco smell, in fresh condition, more or less disgusting and taste bitter; the cause is the nasty poisons they contain.' And finally, within 'Use and Impact of Tobacco,' they remark: 'The use of smoking tobacco, chewing tobacco, and snuff is known and spread throughout the earth. The enjoyment of tobacco has highly narcotic effects, irritating and numbing the nerves. The fresher the tobacco, the stronger the effects. A newcomer will experience the narcotic effects bringing vomiting, diarrhea, headache, numbness, and fear (fearing tobacco); older smokers perceive these effects less ... only smoke when you're fully grown and very healthy!' As to the history of its popularity, they write: 'Chewing tobacco is common with the sailors, especially in North America in the male population.' And finally, how the poisonous plant reached their country, 'Smoking spread from Spain very quickly; it was brought to Germany during the thirty-year war by foreign troops and was only smoked by sailors and marines; soon, however, also by the upper classes.'

OPPOSITE:
Title: *Virginischer Tabak* (Nicotiana Tabacum Linné)
Author: Hermann Zippel; Illustrator: Carl Bollmann
Language: German
Country: Germany
Series/Book: *Ausländische Kulturpflanzen in farbigen Wandtafeln* (*Foreign Crops in Coloured Wall Panels*)
Plate: I Abteilung, 2
Publisher: Friedrich Vieweg & Sohn (Braunschweig, Germany)
Year: 1899

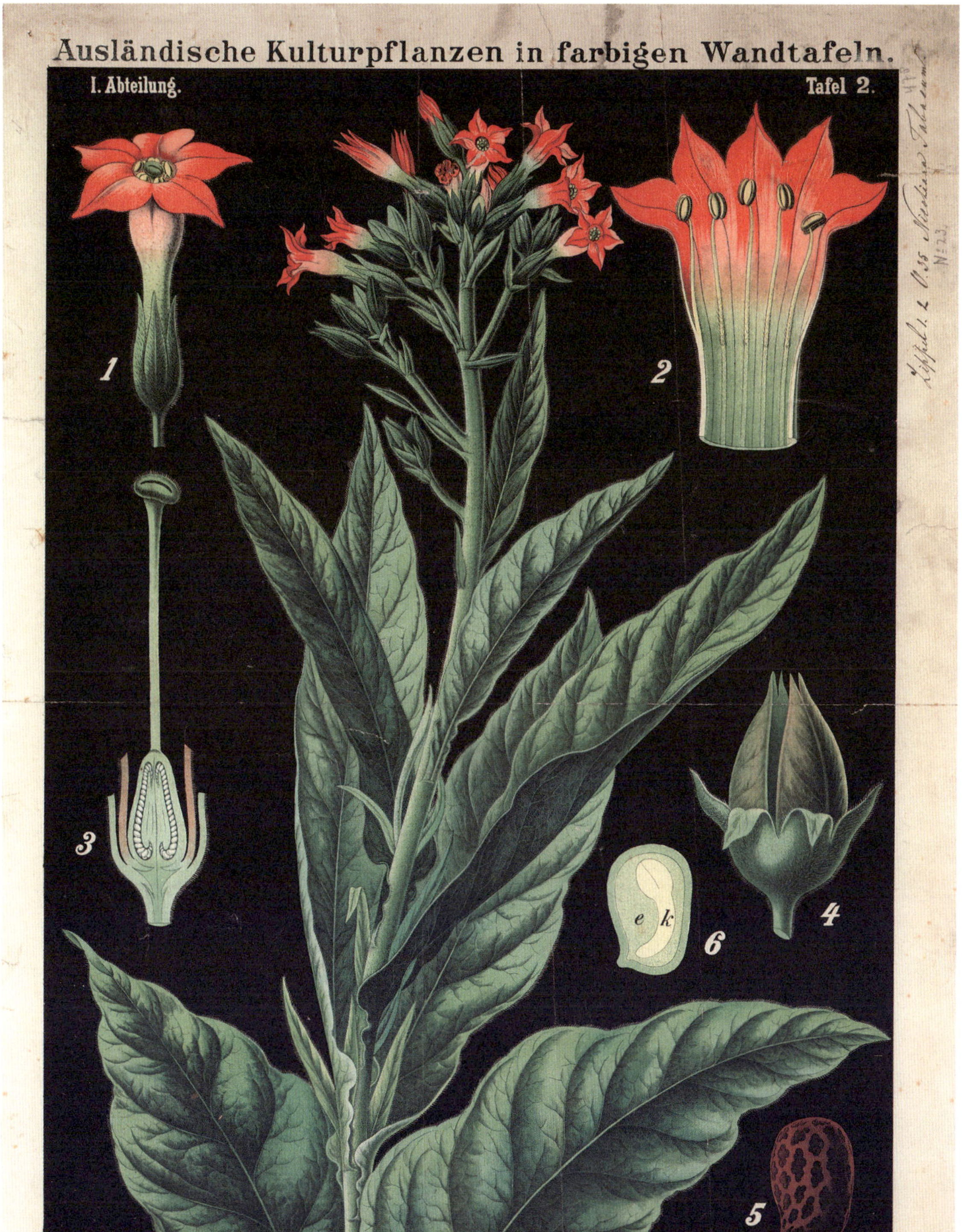
Ausländische Kulturpflanzen in farbigen Wandtafeln.
I. Abteilung.
Tafel 2.
1
2
3
4
5
6
e k
Verlag von FRIEDRICH VIEWEG & SOHN, Braunschweig.
Nach H. ZIPPEL bearbeitet von O. W. THOMÉ, gezeichnet von C. BOLLMANN.
Virginischer Tabak (Nicotiana Tabacum Linné
1 Blüte; Vergr. 2½. — 2 Geöffnete, ausgebreitete Blumenkrone mit den Staubblättern; Vergr. 3. — 3 Stempel und unterer Teil der Blüte; letzterer nebst Fr
4 Im Kelche sitzende, aufgesprungene Kapsel; Vergr. 5. — 5 Same; Vergr. 48. — 6 Same im Längsschnitt: k Keimling, e Samen
Tabak

XXIII VITACEAE / GRAPE FAMILY

The 20 genera comprise around 1,000 species, but this small family includes a plant that has been prized since agriculture began – *Vitis*, the grapevine. Genera in this family occur in the northern hemisphere or pantropic regions. *Vitis* species are found from North America to East Asia, but the species originally cultivated for wine, grapes and raisins is the Mediterranean *V. vinifera*. Ancient Romans and Greeks even gave the grape vine its own deity – the Roman god Bacchus and Greek god Dionysus.

Ornamental plants are mostly grown for their foliage, such as *Cissus alata* (grape ivy), *Parthenocissus quinquefolia* (Virginia creeper) and *P. tricuspidata* (Boston ivy). Although most *Vitaceae* are woody climbers, some *Cissus* and all *Cyphostemma* are succulents, including the endangered *C. juttae* from deserts in Namibia, which has weird, yellowish, trunk-like caudices.

Characteristics: leaf-opposed tendrils; leaves lobed and divided, simple or compound; bark often peeling; typically five sepals, petals and antepetallous stamens; flower in terminal or leaf-opposed cymes; fruit a pulpy berry, in clusters.

OPPOSITE:
Title: *Vitis*
Author: Alois Pokorny
Language: N/A
Country: Germany
Series/Book: *Botanische Wandtafeln*
Plate: Unknown
Publisher: Smichow (Neubert, Germany)
Year: 1894

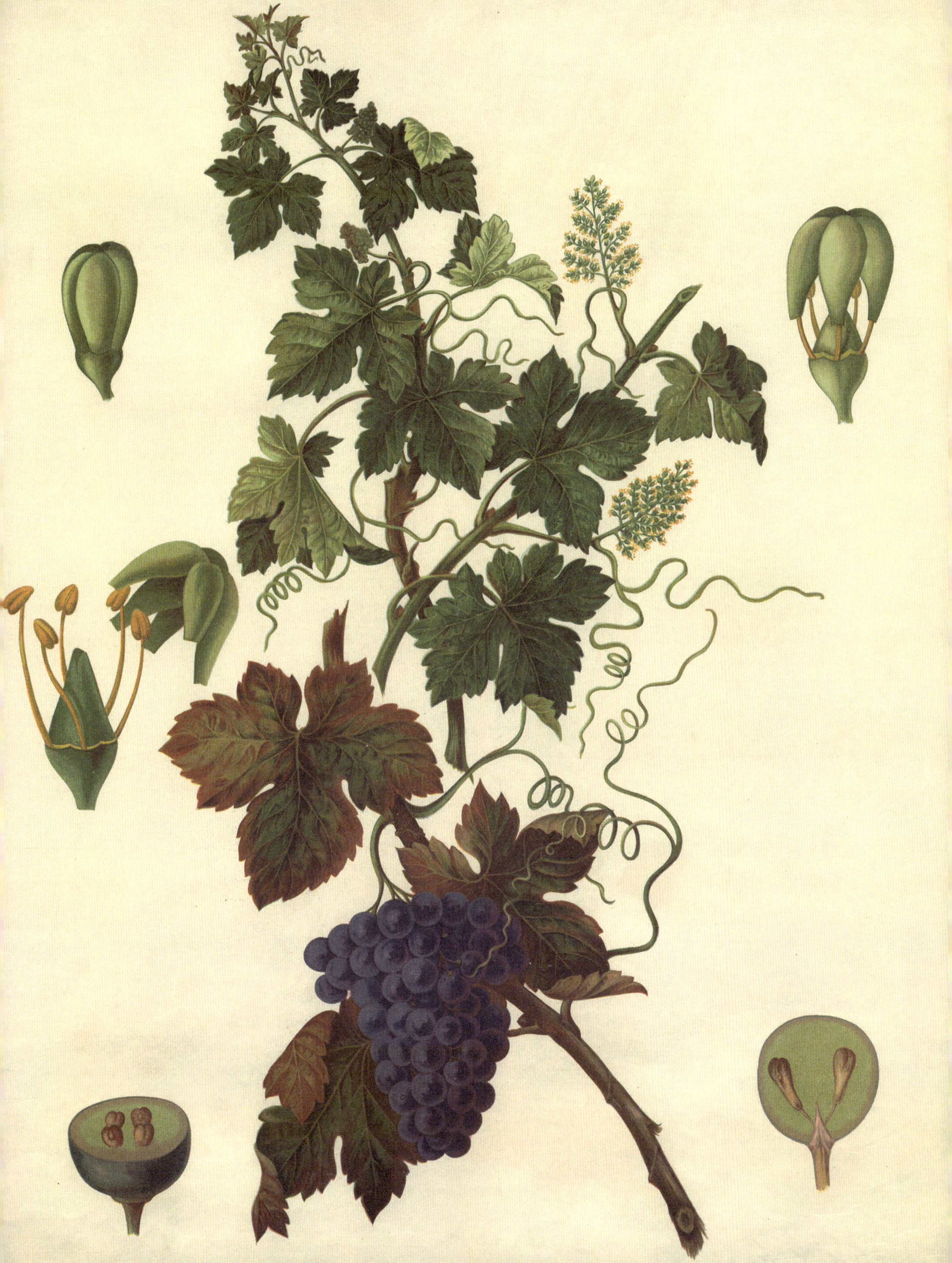

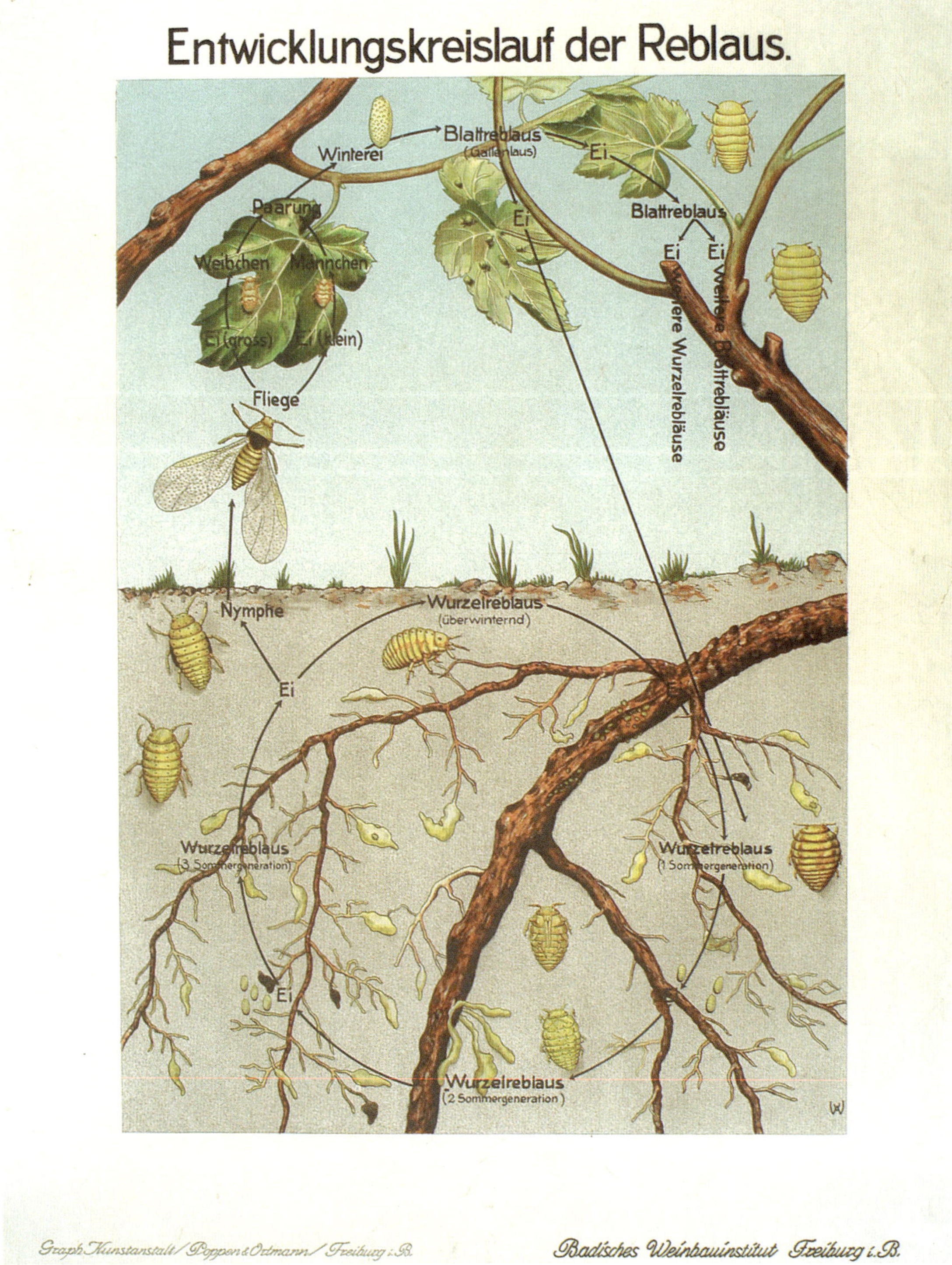
Entwicklungskreislauf der Reblaus.
Winterei
Blattreblaus
(Gallenlaus)
Ei
Paarung
Weibchen
Männchen
Ei (gross)
Ei (klein)
Fliege
Ei
Blattreblaus
Ei
Ei
Weitere Wurzelrebläuse
Weitere Blattrebläuse
Nymphe
Wurzelreblaus
(überwinternd)
Ei
Wurzelreblaus
(3 Sommergeneration)
Wurzelreblaus
(1 Sommergeneration)
Ei
Wurzelreblaus
(2 Sommergeneration)
Graph. Kunstanstalt / Poppen & Ortmann / Freiburg i. B.
Badisches Weinbauinstitut Freiburg i. B.

OPENING PAGE: Exquisitely rendered on a canvas that prized white space, it's easy to forget that Alois Pokorny's illustrations were intended for classroom displays. With no indication of a key, a scale, or a publisher, his prints would be comfortable in a royal florilegium or gallery. But for their lack of explanation, Porkorny's charts are worth a close study.

Here, for example, Pokorny offers more than the classic rendering of the *Vitis* fruit, in the three corollas that hover around the vine. The five petals of the *Vitis* flower are very small, about 0.2 inches (5 mm) long, and pale green, emerging as a familiar bud. As the plant matures, however, the unique structure reveals itself. The petals are united at the apex, like a corolla turned upside down. This cap or 'calyptra' detaches from the flower base and detaches as a whole, uncorking the five stamens within, when the bud is ready for reproduction. The inflorescences can be seen at the upper right-hand side of Pokorny's chart, haloed yellow with the stamens, prefiguring the form of the mature grape clusters, like the one at the bottom of the chart, that they will grow into.

OPPOSITE: Here, and on the following pages, a variety of approaches to studying *phylloxera* reflected the need for a comprehensive understanding of the dreaded plague that destroyed most of Europe's vineyards in the late nineteenth century. *Phylloxera* are tiny sap-sucking insects, which feed on the roots and leaves of grapevines, have a complex life cycle of up to eighteen stages, divided into four forms – sexual, leaf, root, and winged – as illustrated by this chart from Badisches Weinbauinstitut (the Baden State Viticulture Institute).

The first form begins with *phylloxera* eggs on the underside of young grape leaves (upper left of the chart). Upon hatching, male and female immediately mate, and die, but not before the female lays one winter egg in the bark of the vine's trunk. The leaf form begins when the nymph hatches, climbs to a leaf and lays eggs parthenogenetically in a leaf gall. These nymphs will either move to other leaves, or to the roots where they begin new infections. In this third phase, they perforate the root to find nourishment, infecting it with a poisonous secretion that prevents healing, eventually killing the vine. These nymphs will lay eggs for up to seven further generations each summer. In turn, these offspring spread to roots of any adjacent plant. The generation of nymphs that hatch in autumn will hibernate within the roots until next spring, when the sap rises again. Now, the cycle begins again with new eggs laid on the underside of leaves. The fourth form, winged, occurs in humid areas, where the cycle begins the same, only the nymphs have grown wings and can fly to uninfected vines.

OPPOSITE:
Title: *Entwicklungskreislauf Reblaus* (*Development Cycle of Phylloxera*)
Author: Unknown
Language: German
Country: Germany
Series/Book: N/A
Plate: N/A
Publisher: Popper & Ortmann with Badisches Weinbauinstitut (Baden, Germany)
Year: Unknown

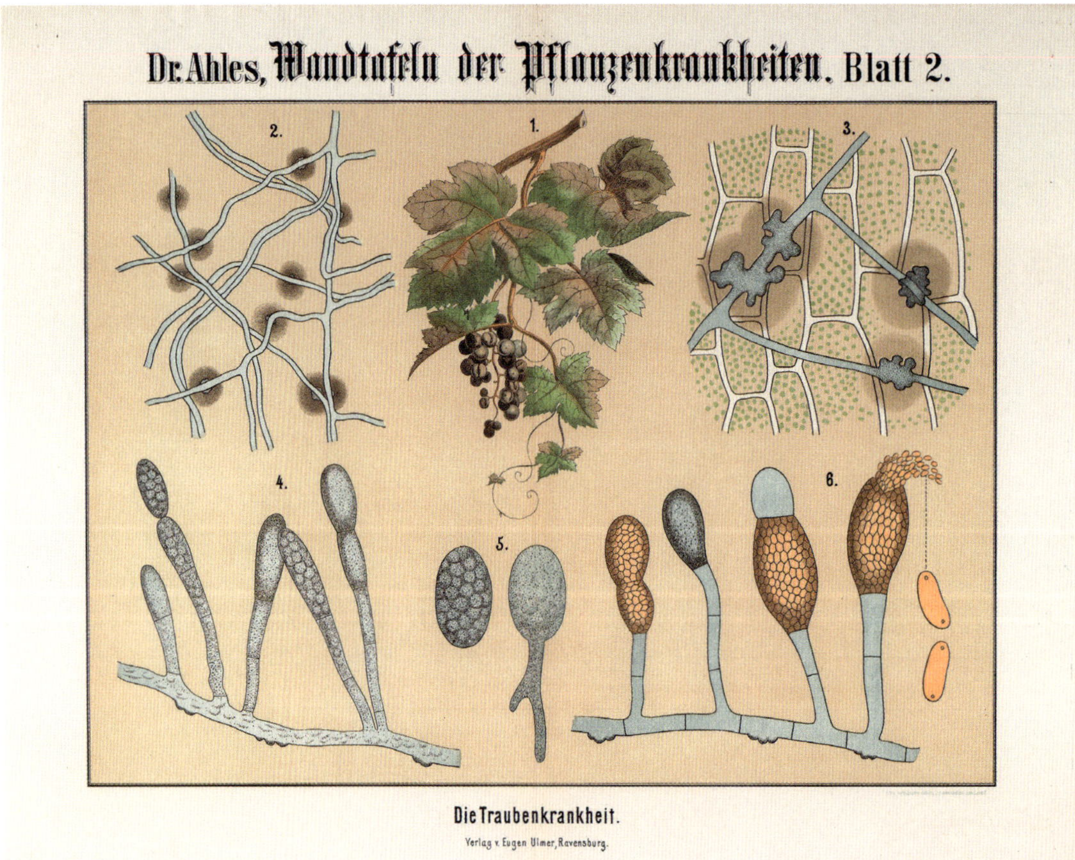

ABOVE:
Title: *Die Traubenkrankheit* (*Grape Diseases*)
Author: Dr. Wilhelm Ahles
Language: German
Country: Germany
Series/Book: *Wandtafeln der Pflanzenkrankheiten: mit text, Vier Feinde der Landwirthschaft: Das Mutterkorn und der Rost der Getreides, Die Kartoffel- und Traubenkrankheit* (*Wall Charts of Plant Diseases: with text, four enemies of agriculture: Ergot and the rust of corn, the potato, and grape disease*)
Plate: 2
Publisher: Eugen Ulmer (Ravensburg, Germany)
Year: 1874

OPPOSITE:
Title: *Weinrebe* (*Vitis vinifera Linné*)
Author: Hermann Zippel; Illustrator: Carl Bollmann
Language: German
Country: Germany
Series/Book: *Ausländische Culturpflanzen in farbigen Wandtafeln* (*Foreign Crops on Coloured Wall Panels*)
Plate: II Abteilung, 21
Publisher: Friedrich Vieweg & Sohn (Braunschweig, Germany)
Year: 1899

ABOVE: Dr. Ahles explores the affliction of *phylloxera* at the cellular level. Figure 1 shows us the appearance of the affected plant, with the discolouration to its leaves; figures 2 to 6 zoom in to show us what is happening under the microscope. Galls on the leaves are produced by the excrescences of the leaf-stage insects. Within these galls is a parent insect, surrounded by her eggs, which begin to hatch just a few days later (fig. 6). By his stage-by-stage magnification, Dr. Ahles impresses upon us the rather terrifying scale of these pests.

OPPOSITE: While their chart is titled *Vitis vinifera*, Zippel and Bollmann give equal attention to *phylloxera*'s forms. They include an illustration of the victim as a fruit (fig. 1), in bloom (fig. 2), a capped corolla (fig. 3), a bloom after throwing off its crown (fig. 4) to expose the gland ring (fig. d), the cross-section of an ovary (fig. 5) where the hint of a grape can already be seen, and its tiny seed (fig. 6). The remainder of the chart is dominated by magnified *phylloxera* forms, beginning with a swollen root which had been infected by *phylloxera* nymphs (fig. 7), and remaining forms of the *phylloxera*'s life cycle.

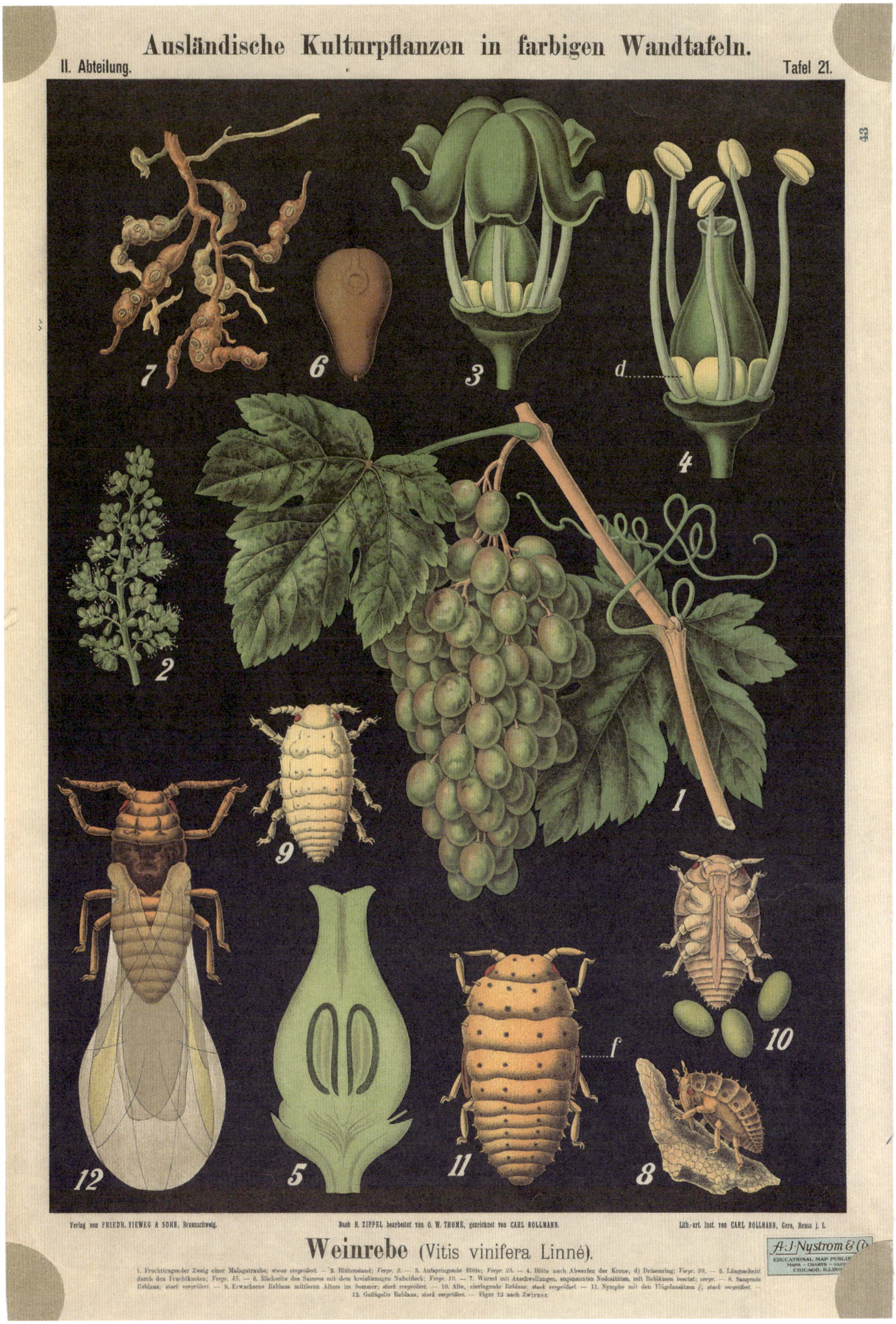

Ausländische Kulturpflanzen in farbigen Wandtafeln.
II. Abteilung.
Tafel 21.
43
7
6
3
d
4
2
1
9
12
5
11
f
10
8
Verlag von FRIEDR. VIEWEG & SOHN, Braunschweig.
Nach H. ZIPPEL bearbeitet von O. W. THOMÉ, gezeichnet von CARL BOLLMANN.
Lith.-art. Inst. von CARL BOLLMANN, Gera, Reuss j. L.
Weinrebe (Vitis vinifera Linné).
1. Fruchttragender Zweig einer Malagatraube; etwas vergrößert. — 2. Blütenstand; Vergr. 2. — 3. Aufspringende Blüte; Vergr. 25. — 4. Blüte nach Abwerfen der Krone; d) Drüsenring; Vergr. 30. — 5. Längsschnitt durch den Fruchtknoten; Vergr. 45. — 6. Rückseite des Samens mit dem kreisförmigen Nabelfleck; Vergr. 10. — 7. Wurzel mit Anschwellungen, sogenannten Nodositäten, mit Rebläusen besetzt; vergr. — 8. Saugende Reblaus; stark vergrößert. — 9. Erwachsene Reblaus mittleren Alters im Sommer; stark vergrößert. — 10. Alte, eierlegende Reblaus; stark vergrößert. — 11. Nymphe mit den Flügelansätzen f; stark vergrößert. — 12. Geflügelte Reblaus; stark vergrößert. — Figur 12 nach Zwirner.
A.J. Nystrom & Co
EDUCATIONAL MAP PUBLIS
CHICAGO, ILLINO

XXIV TYPES

Many educational wall charts cannot be classified into families. These wall charts deal with 'types' – where plants or parts of plants of different species and different families are shown together, in order, perhaps, to compare the morphology of leaves or blossoms among different families or genera; to explore different reproductive mechanisms; or to enlighten students on economically valuable or endangered species in a simple, visual manner.

Like all the wall charts we have looked at in the 'families' section, these 'types' charts offer evidence as to the pedagogical value and artistic beauty of the form. How better to learn about the variations between different kinds of fruit (for example) than by seeing them? And how wonderful that material produced for education should be so pleasing to look at, so exquisitely rendered by the artists who made them.

OPPOSITE:
Title: *Botanical Chart by Japanese Government*
Author: Unknown
Language: Japanese with English botanical terms
Country: Japan
Series/Book: N/A
Plate: N/A
Publisher: N/A
Year: 1873

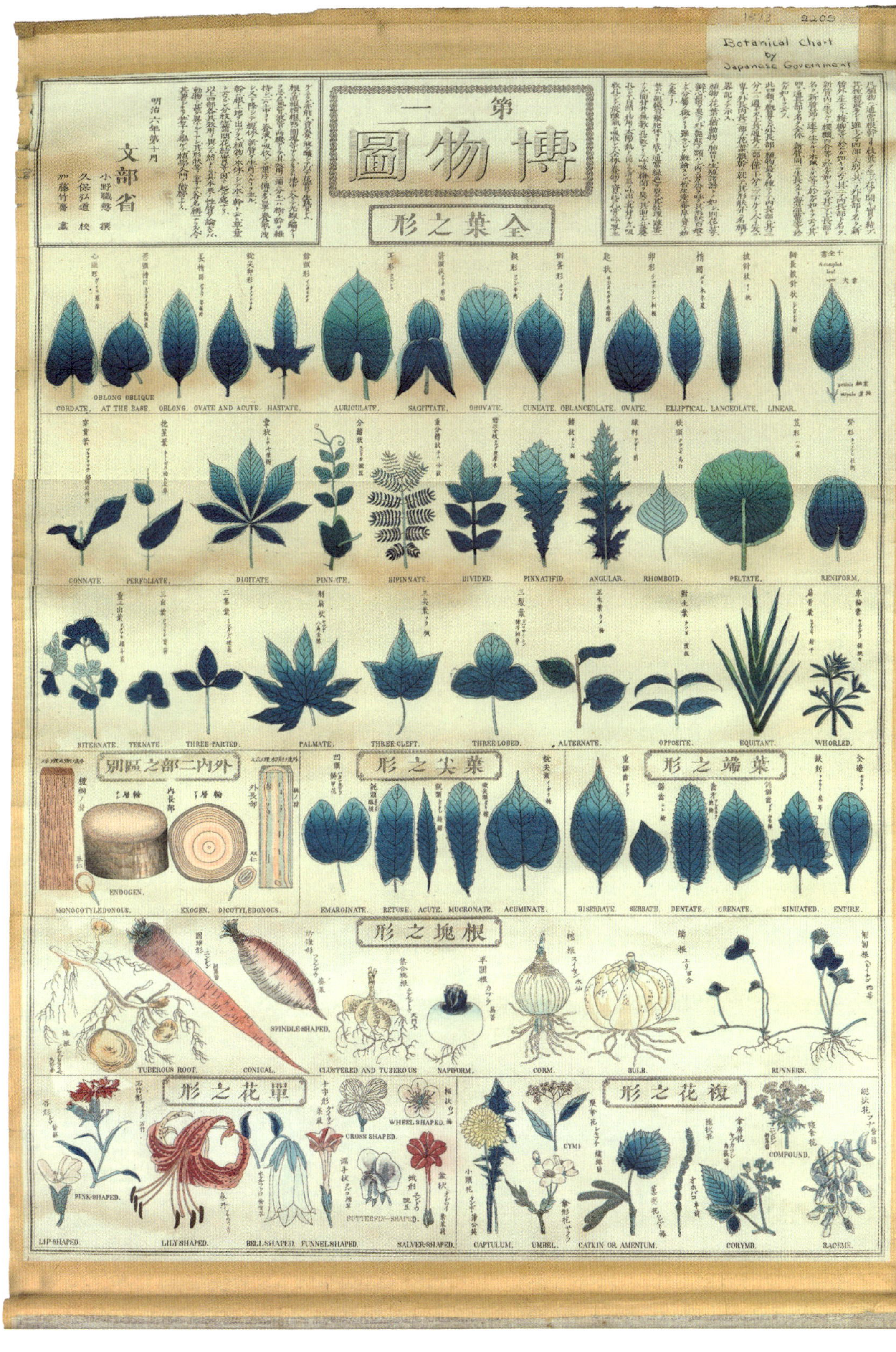

Botanical Chart by Japanese Government
第一
博物圖
明治六年第十月
文部省
小野職愨 撰
久保弘道 校
加藤竹齋 畫
全葉之形
CORDATE.
OBLONG OBLIQUE AT THE BASE.
OBLONG.
OVATE AND ACUTE.
HASTATE.
AURICULATE.
SAGITTATE.
OBOVATE.
CUNEATE.
OBLANCEOLATE.
OVATE.
ELLIPTICAL.
LANCEOLATE.
LINEAR.
CONNATE.
PERFOLIATE.
DIGITATE.
PINNATE.
BIPINNATE.
DIVIDED.
PINNATIFID.
ANGULAR.
RHOMBOID.
PELTATE.
RENIFORM.
BITERNATE.
TERNATE.
THREE-PARTED.
PALMATE.
THREE-CLEFT.
THREE-LOBED.
ALTERNATE.
OPPOSITE.
EQUITANT.
WHORLED.
外内二部之區別
ENDOGEN.
MONOCOTYLEDONOUS.
EXOGEN.
DICOTYLEDONOUS.
葉尖之形
EMARGINATE.
RETUSE.
ACUTE.
MUCRONATE.
ACUMINATE.
葉端之形
BISERRATE.
SERRATE.
DENTATE.
CRENATE.
SINUATED.
ENTIRE.
根塊之形
TUBEROUS ROOT.
CONICAL.
SPINDLE SHAPED.
CLUSTERED AND TUBEROUS.
NAPIFORM.
CORM.
BULB.
RUNNERS.
單花之形
PINK-SHAPED.
CROSS SHAPED.
WHEEL SHAPED.
BUTTERFLY-SHAPED.
LIP SHAPED.
LILY SHAPED.
BELL SHAPED.
FUNNEL SHAPED.
SALVER-SHAPED.
複花之形
CYME.
COMPOUND.
CAPITULUM.
UMBEL.
CATKIN OR AMENTUM.
CORYMB.
RACEME.

OPENING PAGE: Do you know your leaf-forms? After studying this comprehensive Japanese chart you will, as well as flowers, roots and wood growth.

RIGHT: Faded and watermarked, this chart is a sad example of the fate of many beautiful old wall charts. Yet it also serves as a reminder that these were not the kind of works to be hung in hallways or protected behind glass, but practical items subject to the wear and tear of their use in the classroom or the lecture hall.

This chart is one of nine, illustrated by Walter Hood Fitch, based on sketches made by Professor John Stevens Henslow and his daughter, Anne Barnard. Don't be misled by its shabby appearance; a closer inspection reveals how meticulously detailed the chart is, with a key to the many abbreviated labels on the left-hand side, and a guide to the most important characters on the right.

Henslow himself was a pioneer and stalwart of the new pedagogical method; he organized field trips for his students so that they could examine plant species in their natural habitat, and brought his diagrams to the lecture hall, encouraging his students to critically evaluate the specimens represented therein. But probably, he is best known as a mentor to Charles Darwin, whom he taught at Cambridge where he was professor of botany. In 1831, dissuaded by his wife from going himself, Henslow recommended Darwin as his replacement on the voyage of the HMS *Beagle* – the five-year expedition that would be so significant in shaping Darwin's theories of evolution. Throughout the voyage, Darwin corresponded with Henslow and sent him specimens.

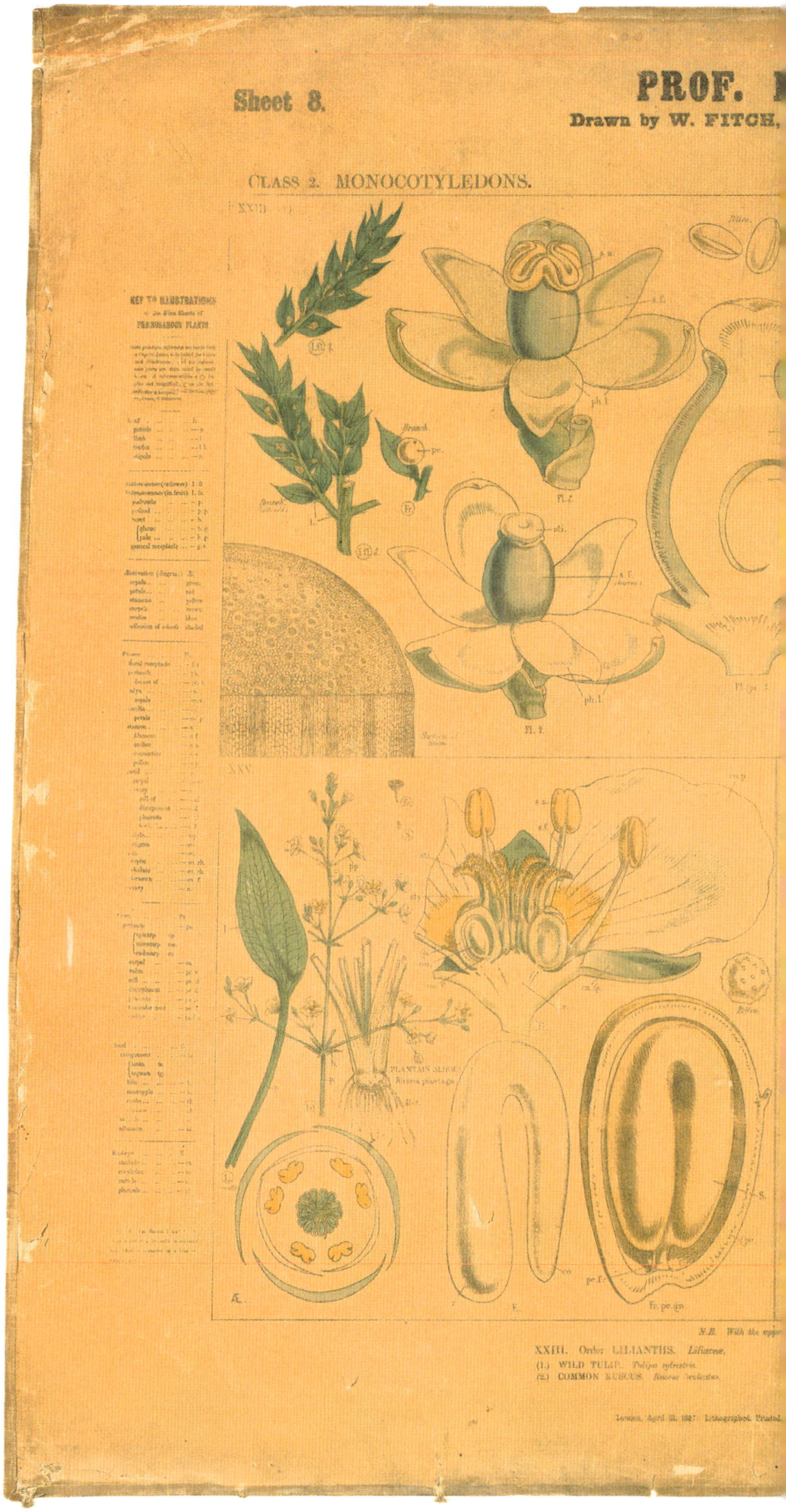

RIGHT:
Title: *Phænogamous Plants*
Author: Walter Hood Fitch
Language: English
Country: UK
Series/Book: *Prof. Henslow's Botanical Diagrams. Drawn by W. Fitch, for the Committee of Council on Education: Department of Science and Art*
Plate: 8
Publisher: Day & Son (London, UK)
Year: 1857

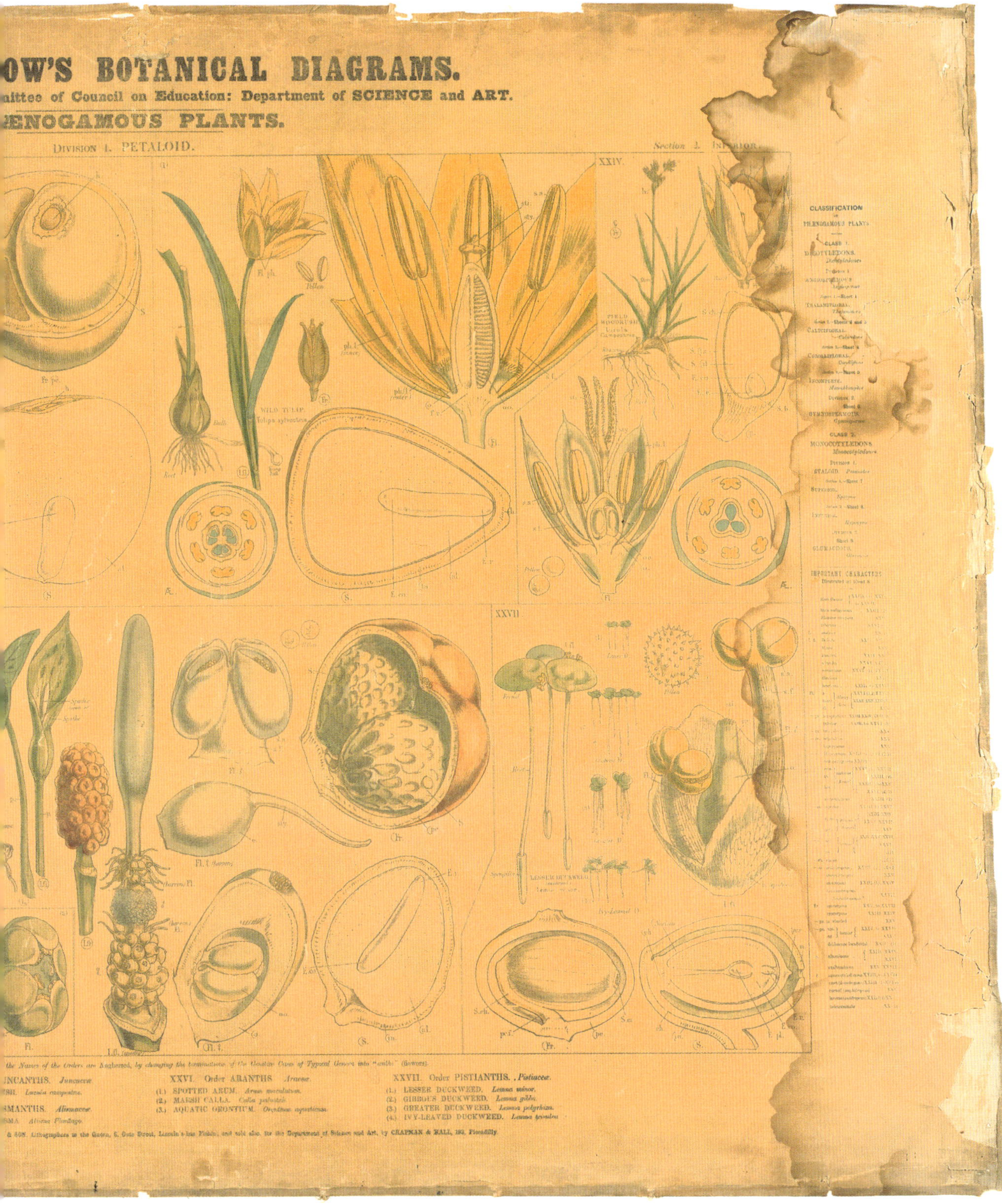
OW'S BOTANICAL DIAGRAMS.
aittee of Council on Education: Department of SCIENCE and ART.
ENOGAMOUS PLANTS.
Division I. PETALOID.
Section 1. INFERIOR.
XXIV.
XXVII
WILD TULIP.
Tulipa sylvestris.
FIELD WOODRUSH
LESSER DUCKWEED
CLASSIFICATION
of
PHÆNOGAMOUS PLANTS
the Names of the Orders are Anglicized, by changing the terminations of the Genitive Cases of Typical Genera into "anths" (flowers).
INCANTHS. Juncaceæ.
SH. Luzula campestris.
SMANTHS. Alismaceæ.
SMA. Alisma Plantago.
XXVI. Order ARANTHS. Araceæ.
(1.) SPOTTED ARUM. Arum maculatum.
(2.) MARSH CALLA. Calla palustris.
(3.) AQUATIC ORONTIUM. Orontium aquaticum.
XXVII. Order PISTIANTHS. Pistiaceæ.
(1.) LESSER DUCKWEED. Lemna minor.
(2.) GIBBOUS DUCKWEED. Lemna gibba.
(3.) GREATER DUCKWEED. Lemna polyrhiza.
(4.) IVY-LEAVED DUCKWEED. Lemna trisulca.
& SON. Lithographers to the Queen, 5, Gate Street, Lincoln's-Inn Fields; and sold also, for the Department of Science and Art, by CHAPMAN & HALL, 193, Piccadilly.

OPPOSITE: This lovely Swedish chart, prepared 'under the supervision of Gertrud Carlberg, teacher of home economics in the trade school in Uppsala,' depicts 'useful plants.' Each of the species has been drawn like a full wall chart in miniature – small, but wonderfully detailed, so that Ms. Carlberg's students at the trade school ought to have had no difficulty in recognizing the plants from which their spices came. Clearly, though, the chart was not intended for domestic use alone, for the text appears with English, French, Spanish and German translations, besides the original Swedish – as well as the Latin botanical names.

Let's take a look at spices included. In the top tow, from left to right: hops (*Humulus lupulus*), aniseed (*Pimpinella anisum*), caraway (*Carum carvi*) and fennel (*Foeniculum vulgare*); middle row: poppy (*Papaver somniferum*), mustard (*Sinapis alba*), coriander (*Coriandrum sativum*), paprika (*Capsicum annuum*); bottom row: cardamom (*Elettaria cardamomum*), capers (*Capparis spinosa*), bay (*Laurus nobilis*) and saffron (*Crocus sativus*).

OPPOSITE:
Title: *Nyttoväxter: Kryddor 1 (Useful Plants: Spices 1)*
Author: Gertrud Carlsberg, prepared by Nils Karlson and M. Richter
Language: Swedish, English, French, Spanish, German
Country: Sweden
Series/Book: N/A
Plate: N/A
Publisher: Gunnar Saietz A. B. (Stockholm, Sweden)
Year: Unknown

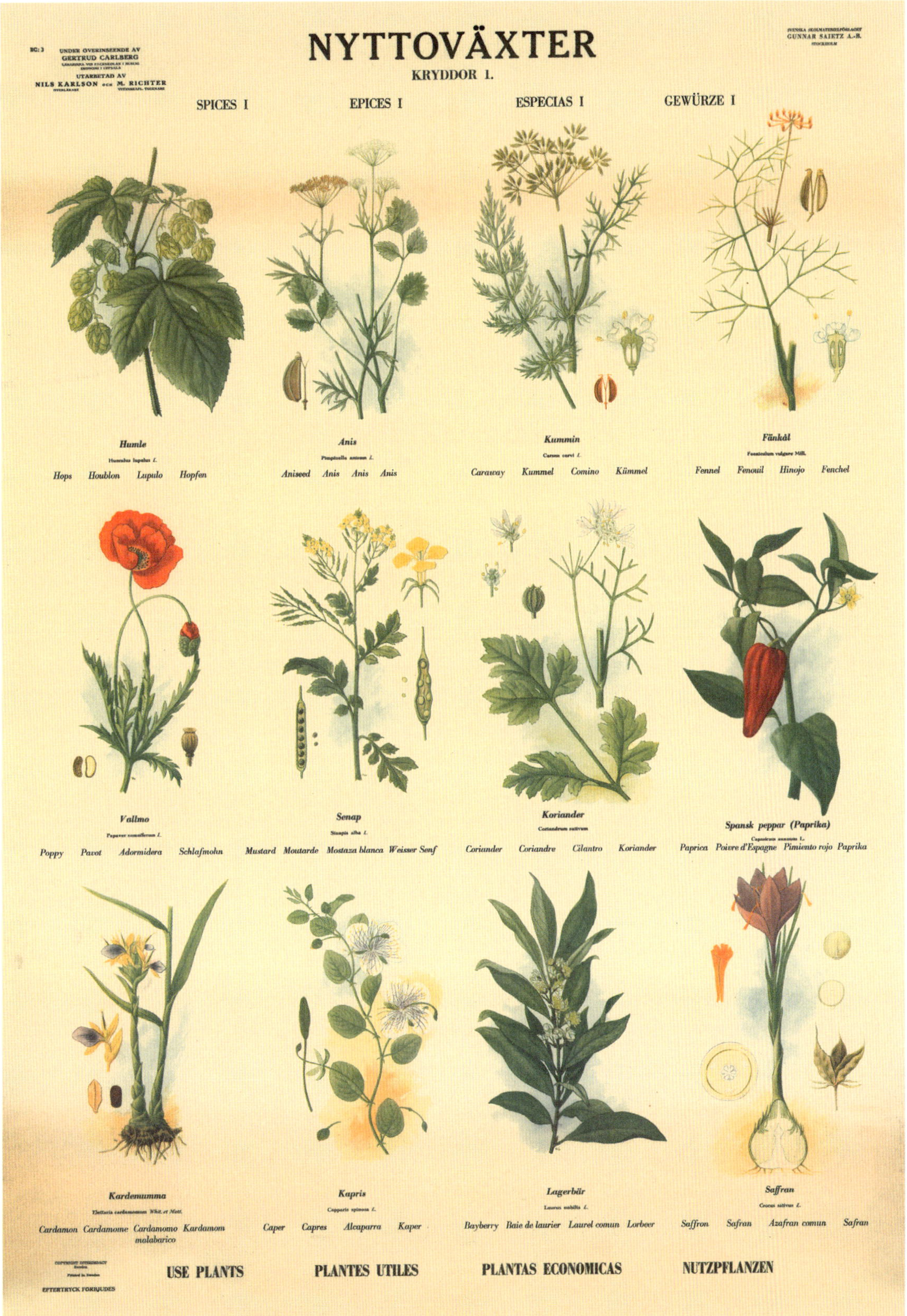
BG: 3
UNDER ÖVERINSEENDE AV
GERTRUD CARLBERG
UTARBETAD AV
NILS KARLSON och M. RICHTER
NYTTOVÄXTER
KRYDDOR I.
GUNNAR SAIETZ A.-B.
STOCKHOLM
SPICES I
EPICES I
ESPECIAS I
GEWÜRZE I
Humle
Humulus lupulus L.
Hops Houblon Lupulo Hopfen
Anis
Pimpinella anisum L.
Aniseed Anis Anis Anis
Kummin
Carum carvi L.
Caraway Kummel Comino Kümmel
Fänkål
Foeniculum vulgare Mill.
Fennel Fenouil Hinojo Fenchel
Vallmo
Papaver somniferum L.
Poppy Pavot Adormidera Schlafmohn
Senap
Sinapis alba L.
Mustard Moutarde Mostaza blanca Weisser Senf
Koriander
Coriandrum sativum
Coriander Coriandre Cilantro Koriander
Spansk peppar (Paprika)
Capsicum annuum L.
Paprica Poivre d'Espagne Pimiento rojo Paprika
Kardemumma
Elettaria cardamomum Whit. et Matt.
Cardamon Cardamome Cardamomo malabarico Kardamom
Kapris
Capparis spinosa L.
Caper Capres Alcaparra Kaper
Lagerbär
Laurus nobilis L.
Bayberry Baie de laurier Laurel comun Lorbeer
Saffran
Crocus sativus L.
Saffron Safran Azafran comun Safran
Printed in Sweden
EFTERTRYCK FÖRBJUDES
USE PLANTS
PLANTES UTILES
PLANTAS ECONOMICAS
NUTZPFLANZEN

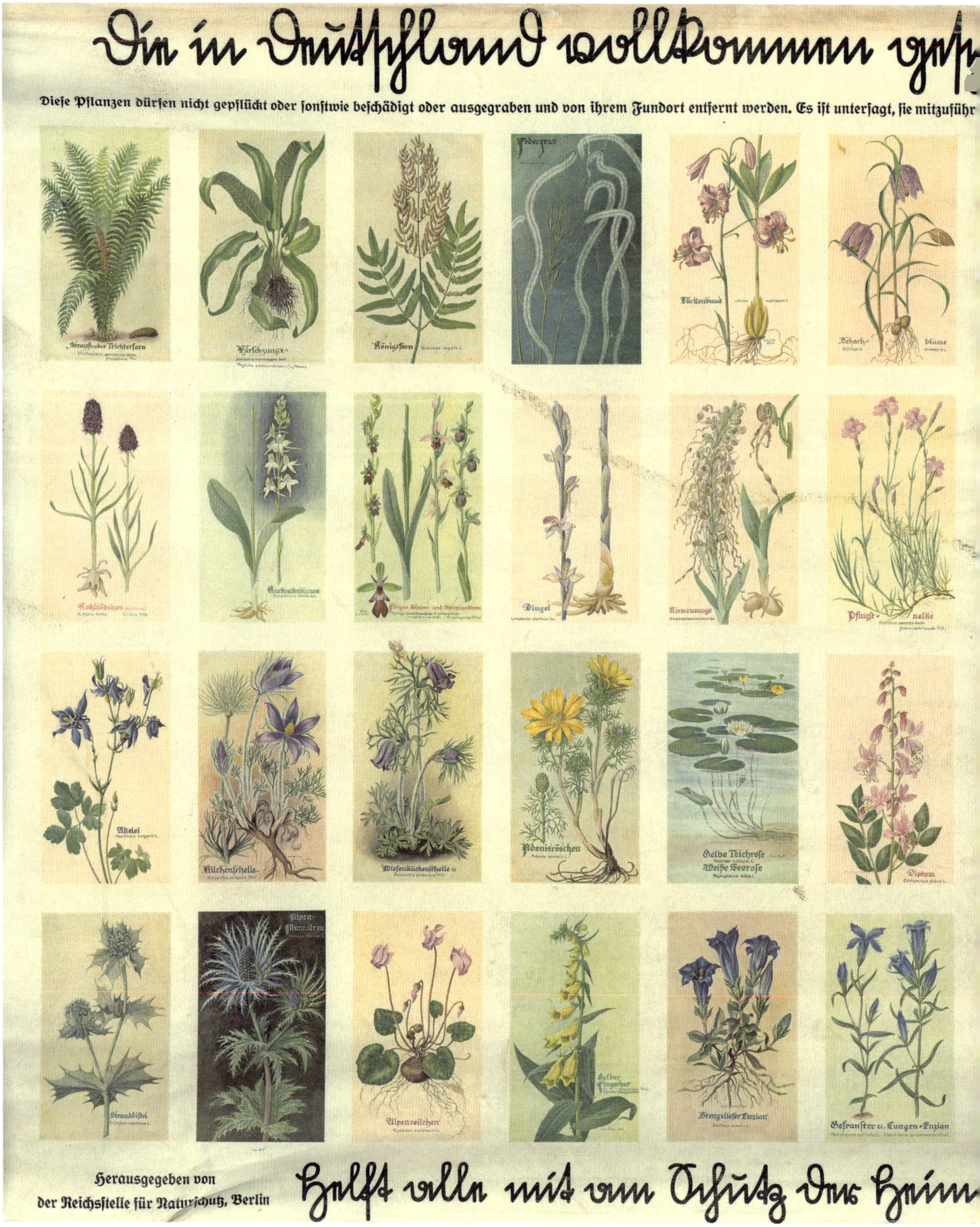
Die in Deutschland vollkommen
Diese Pflanzen dürfen nicht gepflückt oder sonstwie beschädigt oder ausgegraben und von ihrem Fundort entfernt werden. Es ist untersagt, sie
Strauß- oder Trichterfarn
Hirschzunge
Königsfarn
Federgras
Türkenbund
Schachblume
Kohlröschen
Kuckucksblume
Fliegen-, Bienen- und Spinnenblume
Dingel
Riemenzunge
Pfingstnelke
Akelei
Küchenschelle
Wiesenküchenschelle
Adonisröschen
Gelbe Teichrose
Weiße Seerose
Diptam
Stranddistel
Alpen-Mannstreu
Alpenveilchen
Gelber Fingerhut
Stengelloser Enzian
Gefranster u. Lungen-Enzian
Herausgegeben von
der Reichsstelle für Naturschutz, Berlin
Helft alle mit am Schutz der

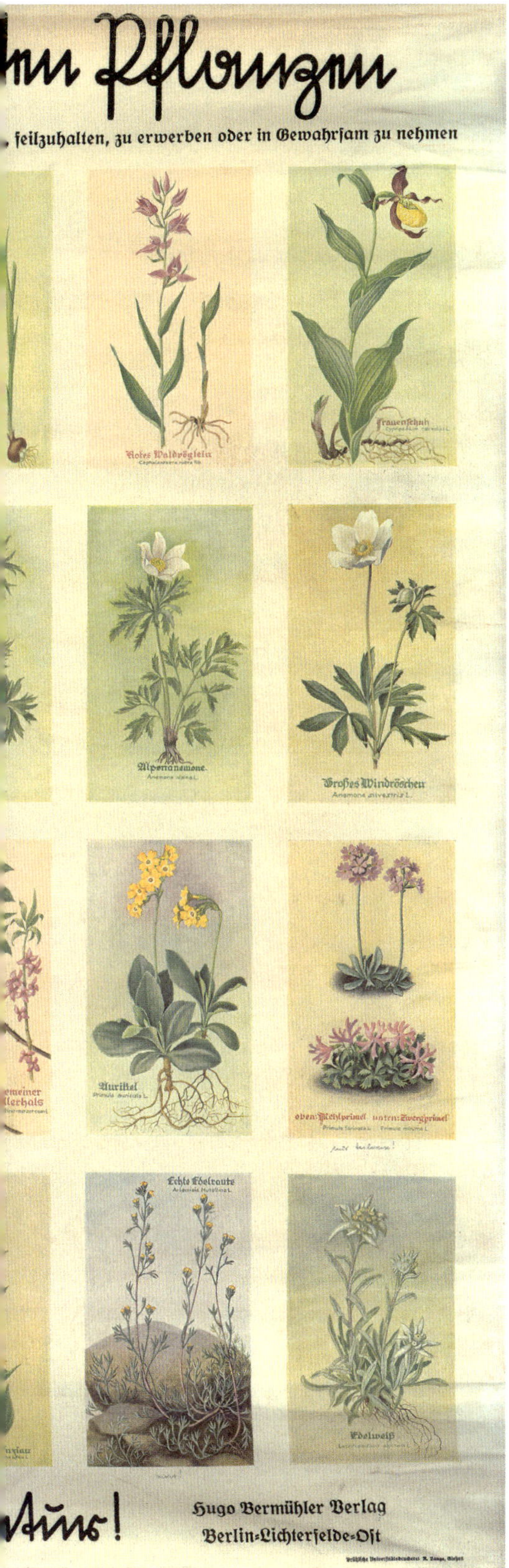

LEFT: You might be surprised to learn that in the years leading up to the Second World War, the Nazi party in Germany should be preoccupied by ecological concerns. But this plate, published in 1936 by Hugo Bermühler in partnership with the Reich Agency for Nature Conservation, carries the entreaty, 'Help all in the protection of our native nature!' – as well as the sterner interdiction, 'These plants may not be picked or otherwise damaged, or removed from their location. It is forbidden to send them, offer them for sale, to purchase, or to take custody.'

Sinister Nazi overtones aside, it is a beautiful chart, depicting 37 endangered plant species with their German and Latin names. What makes this chart exemplary, however, is that you can see, if you look very closely, where a student has crossed out the Latin name and written its new classification underneath. It's impossible to tell when this helpful graffiti might have been done, but it bears testament to the ever-shifting nature of taxonomy, as well as evidencing the educational environment, which was these boards' first use.

LEFT:
Title: *Completely Protected Plants*
Author: Schröder
Language: German
Country: Germany
Series/Book: N/A
Plate: N/A
Publisher: Hugo Bermühler, with the Reich Agency for Nature Conservation (Berlin, Germany)
Year: 1936

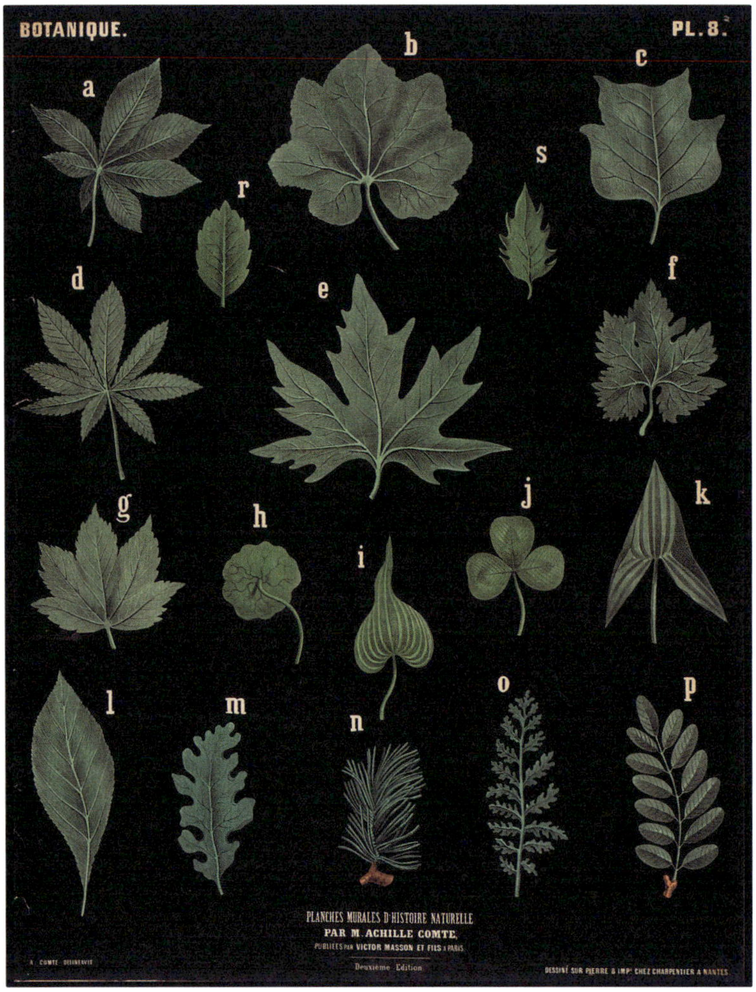

ABOVE:
Title: *Diverses Formes des Feuilles (Diverse Forms of Leaves)*
Author: Achille Comte
Language: French
Country: France
Series/Book: *Planches Murales d'Histoire Naturelle*
Plate: 8
Publisher: Victor Masson et Fils (Paris, France)
Year: 1869

OPPOSITE:
Title: *Bekers (Cups)*
Author: Nelly Bodenheim
Language: Dutch
Country: The Netherlands
Series/Book: Unknown
Plate: Unknown
Publisher: Unknown
Year: 1899

THIS SPREAD: Here we have two different treatments of similar subjects. On the left, Achille Comte's *Diverses Formes des Feuilles* (*Diverse Forms of Leaves*), and opposite, Nelly Bodenheim's *Bekers*, or '*Cups*' – representing cup-shaped leaves. While Comte's subjects stand out boldly on their black background, all rigid lines and compositional symmetry, Bodenheim exercises a freer touch, with samples taking their places in the spaces allotted them on the pale canvas almost as if at random, as if they have simply fallen there.

In the context of their authors' backgrounds and intentions, the contrast between these charts makes perfect sense. Comte was a certified academic, having left a successful career as a medical doctor to take a post as professor of natural history at the Royal College of Charlemagne, and he remained in teaching for the rest of his life. Bodenheim, on the other hand, was first and foremost an artist. Early in her career, at the turn of the twentieth century, she worked in partnership with Hugo de Vries, illustrating many botanical boards for him. But botany was just one subject among many for her, and her artistic ambitions were doubtless greater. She trained at the State Academy of Fine Arts in Amsterdam and was a member of the Amsterdamse Joffers – a group of woman artists, mostly painters, stylistically similar to the Dutch Impressionists.

We can understand, then, that Comte's orderly boards, accompanied by a detailed text, are designed for a pedagogical role. Bodenheim's charts do not lack educational value, but it takes second place to her first priority as an artist.

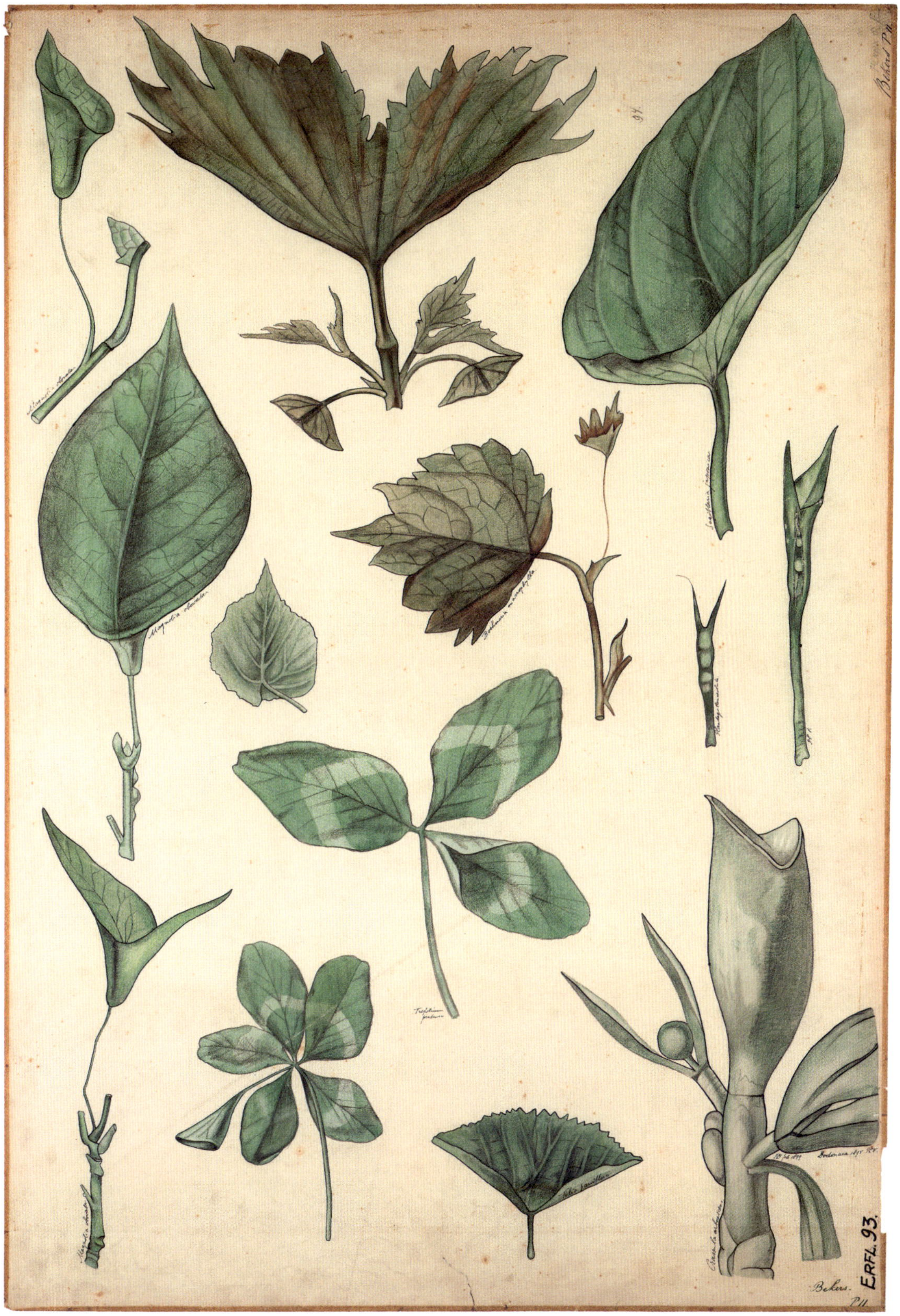
E.RFL. 93.
Bekers.

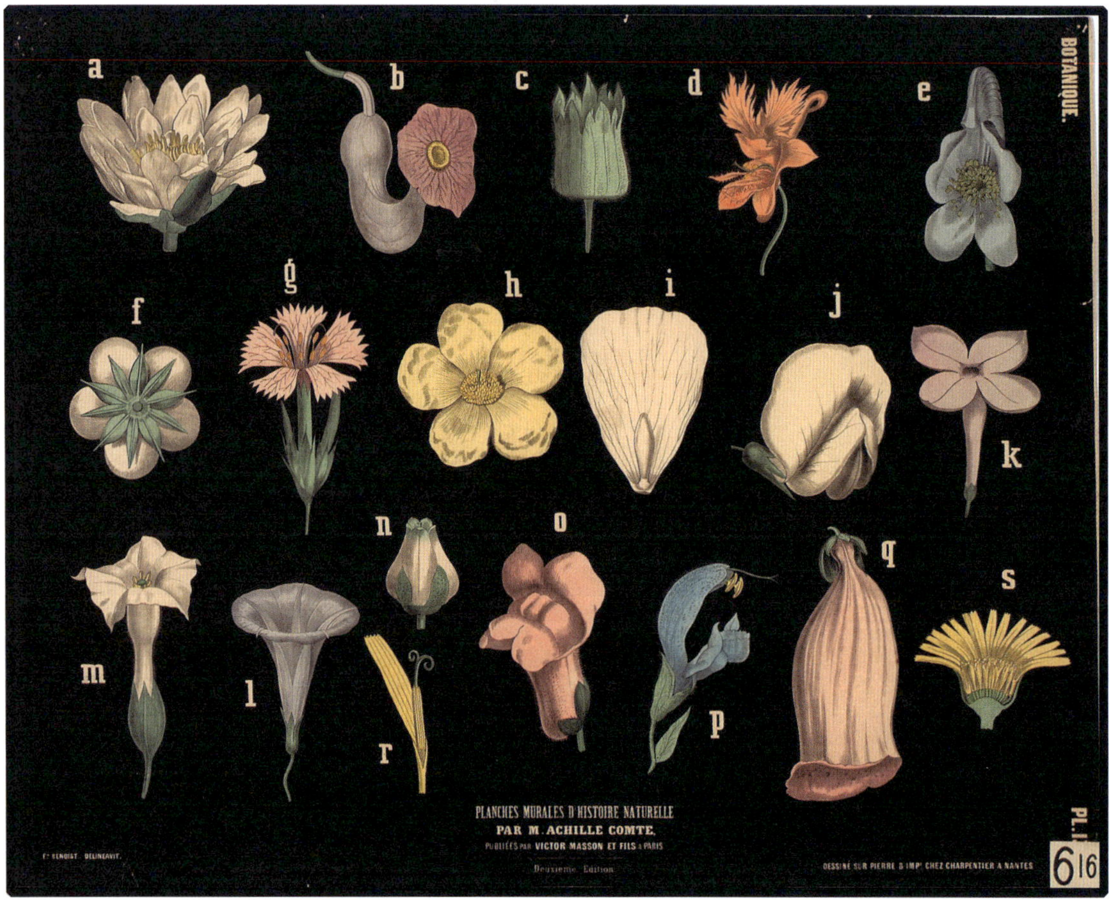

ABOVE & OPPOSITE:
Title, above: *Différentes formes du périanthe (Different forms of perianths)*; Opposite: *Formes diverses des étamines, des pistils et des disques ou nectaires (Various forms of stamens, pistils and discs or nectaries)*
Author: Achille Comte
Language: French
Country: France
Series/Book: *Planches Murales d'Histoire Naturelle*
Plates: 13; 14
Publisher: Victor Masson et Fils (Paris, France)
Year: 1869

THIS SPREAD: Achille Comte's botanical charts are as striking as they are thorough. The chart above is beautiful with its electric-bright colours that pop out against the pitch black of the canvas; appropriate for glorying in the (usually) most extravagant part of the plant. The chart opposite is in one sense more restrained – fixed with a bold colour scheme of black, white, green, and yellow – but apparently delighting in representing as great a variety as possible of the different parts of a plant's sexual organs. In this way, this chart is exemplary in demonstrating the benefits of the 'type' chart. It's also exemplary as a model of the botanical wall chart as a fusion of the scientific and the aesthetic; it appears that Comte has fit in many different samples as he humanly could, but they are neither haphazardly nor dryly arranged.

From the geometric forms of pollen grains at the bottom of the chart (figs. s, t, u); the transitioning petals of the white waterlily (fig. o) making a crown above a cross-section of the anther of a pumpkin flower (fig. p), bordered on either side by different forms of pollen sacs (figs. r, s); also in the bottom section of this chart Comte gives us a vertical cross-section of a primrose ovary (fig. v), stigmas (fig. x, 1), two different kinds of style (figs. y, z), disc (fig. 2), and nectary (fig. 3); and above all of this a diverse show of stamens and anthers – Comte has given us a veritable A–Z (and then some) of various permutations of the reproductive organs of plants – although curiously, it seems he has forgotten the letter 'W.'

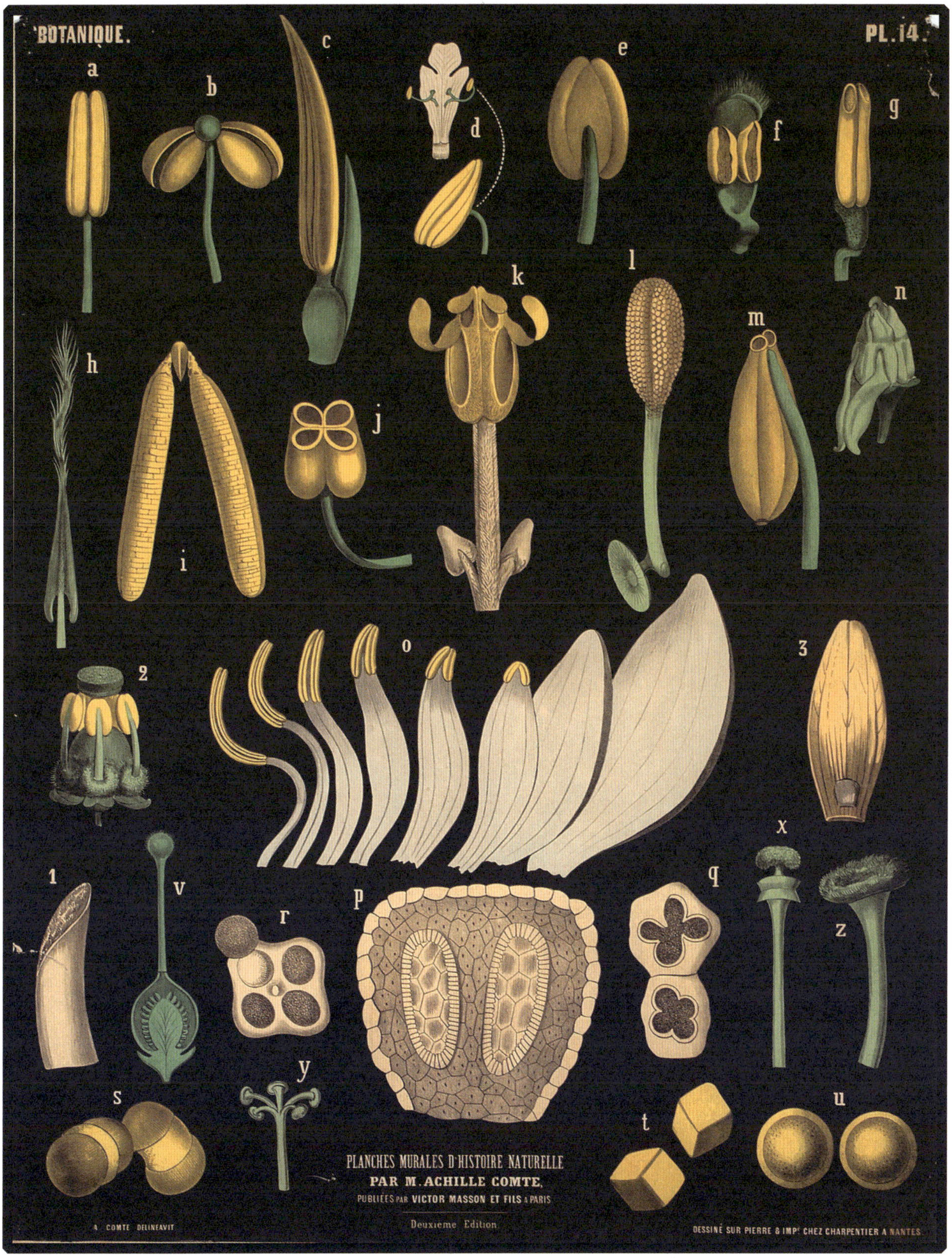
BOTANIQUE.
PL. 14.
a
b
c
d
e
f
g
h
i
j
k
l
m
n
o
2
3
1
v
r
p
q
x
z
s
y
t
u
PLANCHES MURALES D'HISTOIRE NATURELLE
PAR M. ACHILLE COMTE,
PUBLIÉES PAR VICTOR MASSON ET FILS A PARIS
Deuxieme Edition
A. COMTE DELINEAVIT
DESSINÉ SUR PIERRE & IMPr CHEZ CHARPENTIER A NANTES.

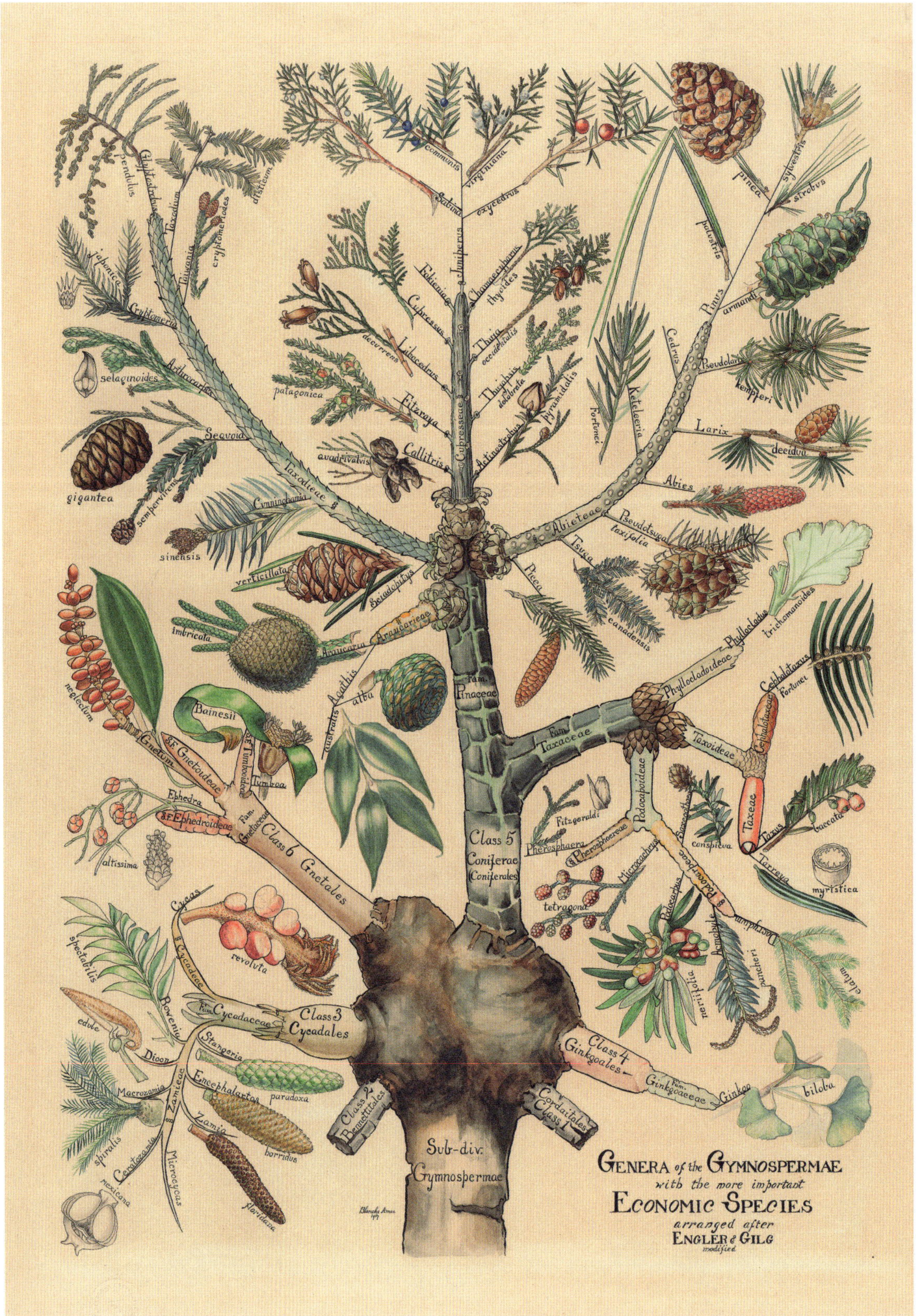
GENERA of the GYMNOSPERMAE
with the more important
ECONOMIC SPECIES
arranged after
ENGLER & GILG
modified
Sub-div. Gymnospermae
Class 1 Cordaitales
Class 2 Bennettitales
Class 3 Cycadales
Class 4 Ginkgoales
Class 5 Coniferae (Coniferales)
Class 6 Gnetales
Fam. Cycadaceae
Fam. Ginkgoaceae
Ginkgo
biloba
Fam. Pinaceae
Fam. Taxaceae
Fam. Gnetaceae
Abieteae
Taxodieae
Cupresseae
Araucarieae
Podocarpoideae
Phyllocladoideae
Taxoideae
Taxeae
Pinus
Larix
Abies
Tsuga
Picea
Cedrus
Juniperus
Sequoia
gigantea
Taxus
baccata
Torreya
myristica
Ephedra
Gnetum
Tumboa
Cycas
revoluta
Zamia
floridana
Encephalartos
horridus
Stangeria
paradoxa
Macrozamia
spiralis
Microcycas

OPPOSITE & OVERLEAF: What better way to end our exploration through the world of botanical wall charts than with these gorgeous illustrations by Blanche Ames. A naturalist in her own right, Ames produced numerous illustrations for her husband, Oakes Ames's Economic Botany courses at Harvard University in the early 1900s. Blanche often accompanied Oakes, on naturalist expeditions, illustrating the species he collected (when he donated his herbarium to Harvard, it included 64,000 specimens). Her sketches and watercolours accompany her husband's herbarium archive, including several she illustrated from a Berlin Collection that was destroyed in a bombing raid in 1943.

The two charts seen here, opposite and overleaf, are like none we have looked at before. Rather than representing the morphology of different species, or comparing the various parts of different species, they deal in the phylogeny – the family tree (so to speak) – of plant forms. Since we are speaking of taxonomy, we should already be cautious that there is the risk of spending days or years in the creation of a magnificent chart, only to find that the following week, classifications have changed and our chart is wrong. Having noted this warning, we can also look at these charts and recognize the educational, historical, and purely aesthetic value of Blanche Ames's charts.

The chart on the page opposite explores the phylogenetic relationships of economically valuable gymnosperms, dividing the group first into classes – including the extinct *Bennettitales* and *Cordaitales*, represented as grey, truncated stumps – then families, then genera, and species. Such a representation offers a wonderfully direct comparison, allowing us to understand, in a purely visual way, what common ground different genera share to place them in the same family – or what separates them.

Overleaf, a chart of the same series, representing the economically valuable plants of the *Metachlamydeae* or *Sympetalae* (meaning 'with fused petals,' in contrast to the free-petalled *Choripetalae*) sub-class; their higher class, the *Dicotyledons*, can be seen at the left-hand side of the class as the bough from which this sub-class branches. In talking of *Sympetalae* and *Choripetalae*, the shifting sands of taxonomy come into play, firstly because it they are no longer favoured classifications, and secondly, because in terms of evolutionary taxonomy, they have little value. The terms are simply descriptive, based on the understanding that such distinct features must relate to a common ancestor, whereas in fact they do not. But this does not mean that they, and this chart, are not useful. Historically, it evidences the evolution of our understanding as our knowledge has grown. Pedagogically, it offers a way to compare similar floral structures across varying groups and families. And aesthetically – certainly, no one can argue with the skill and beauty of Ames's work.

OPPOSITE:
Title, opposite: ***Genera of the Gymnospermae with the more important Economic Species arranged after Engler and Gilg***
Author: **Blanche Ames**
Language: **English**
Country: **US**
Series/Book: ***Ames Charts***
Plate: **N/A**
Publisher: **N/A**
Year: **1917**

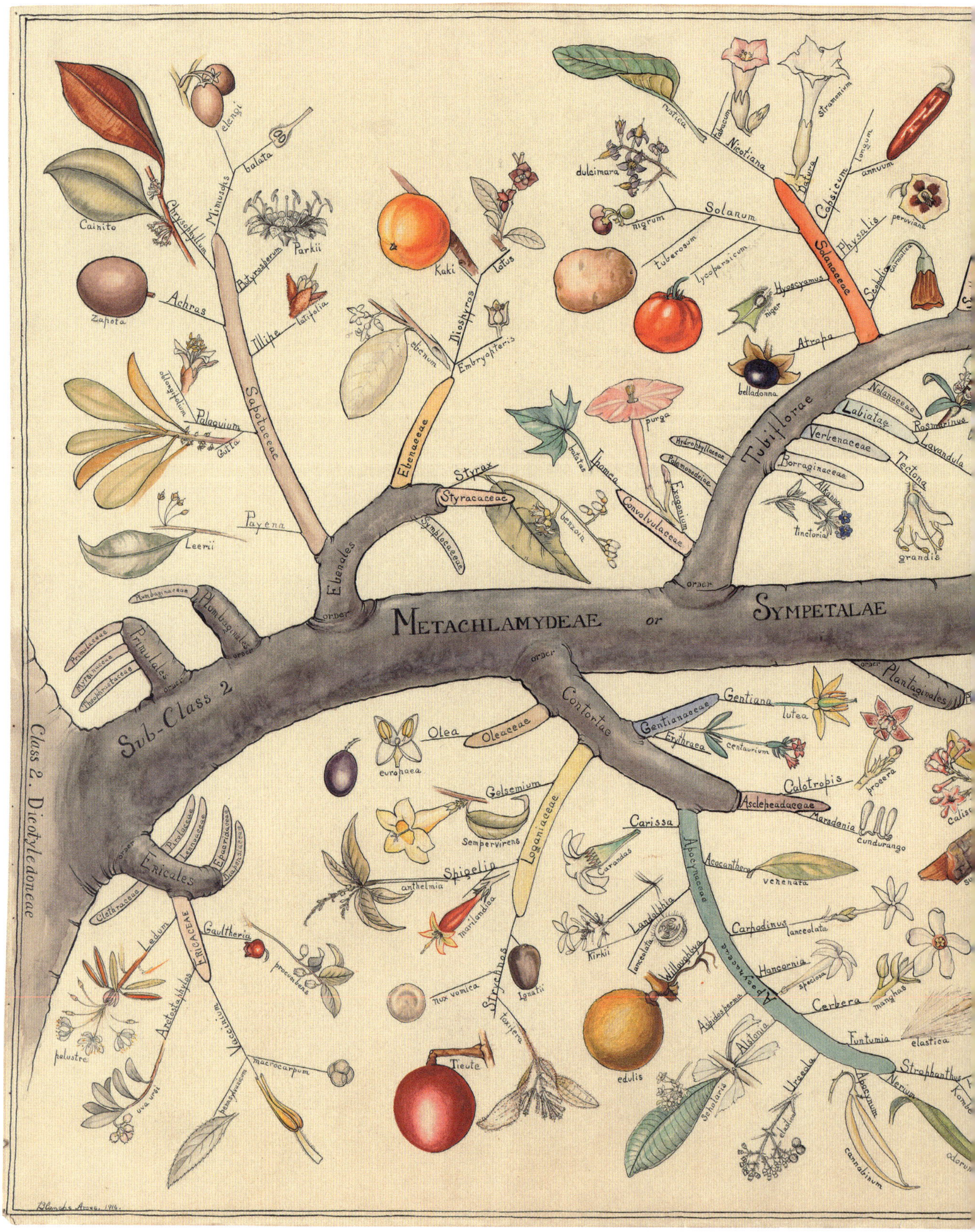

Class 2. Dicotyledoneae
Sub-Class 2
METACHLAMYDEAE or SYMPETALAE
Sapotaceae
Chrysophyllum
Cainito
Mimusops
elengi
balata
Butyrospermum
Parkii
Achras
Zapota
Illipe
latifolia
Palaquium
Gutta
Payena
Leerii
Ebenales
Ebenaceae
Diospyros
Kaki
Lotus
ebenum
Embryopteris
Styrax
Styracaceae
Symplocaceae
benzoin
Plumbaginales
Plumbaginaceae
Primulales
Primulaceae
Myrsinaceae
Theophrastaceae
Ericales
Clethraceae
Pirolaceae
Lennoaceae
Epacridaceae
Diapensiaceae
Ericaceae
Ledum
palustre
Arctostaphylos
uva ursi
Vaccinium
macrocarpum
pennsylvanicum
Gaultheria
procumbens
Tubiflorae
Convolvulaceae
Ipomoea
batatas
purga
Exogonium
Polemoniaceae
Hydrophyllaceae
Borraginaceae
Alkanna
tinctoria
Verbenaceae
Tectona
grandis
Labiatae
Rosmarinus
Lavandula
Nolanaceae
Solanaceae
Solanum
dulcamara
nigrum
tuberosum
Lycopersicum
Nicotiana
rustica
Tabacum
Datura
Stramonium
Capsicum
annuum
Physalis
peruviana
Hyoscyamus
niger
Atropa
belladonna
Scopolia
Contortae
Oleaceae
Olea
europaea
Loganiaceae
Gelsemium
Sempervirens
Spigelia
anthelmia
marilandica
Strychnos
nux vomica
Ignatii
toxifera
Tieute
Gentianaceae
Gentiana
lutea
Erythraea
centaurium
Asclepiadaceae
Calotropis
procera
Marsdenia
cundurango
Apocynaceae
Carissa
Carandas
Acocanthera
venenata
Landolphia
Kirkii
Willughbeia
edulis
Carpodinus
lanceolata
Hancornia
speciosa
Cerbera
manghas
Aspidosperma
Alstonia
scholaris
Urceola
elastica
Funtumia
elastica
Strophanthus
Apocynum
cannabinum
Nerium
Plantaginales
order
Blanche Ames. 1916.

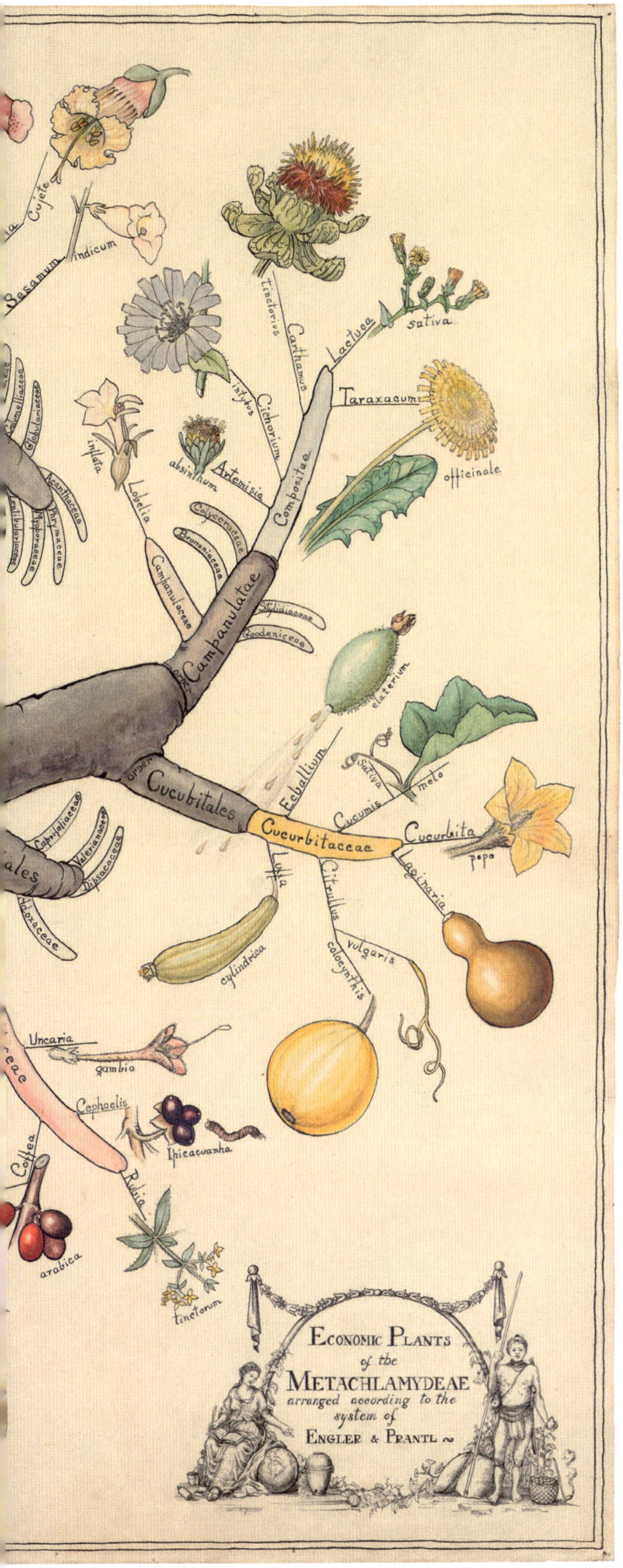

LEFT:

Title: *Economic Plants of the Metachlamydeae arranged according to the system of Engler and Pranti*
Author: Blanche Ames
Language: English
Country: US
Series/Book: Ames Charts
Plate: N/A
Publisher: N/A
Year: 1917

BIBLIOGRAPHY

Ames, Oakes. 'Pollination of Orchids through Pseudocopulation' *Botanical Museum Leaflets*, Harvard University Vol. 5, No. 1, pp 1–27 (July 1, 1937), accessed November 2015. https://www.jstor.org/stable/41762973

Bailey, Liberty Hyde. *The Standard Cyclopedia of Horticulture*. London: Macmillan & Co, 1917.

Benedictine University. 'History of the College of Science: Hilary Jurica.' Accessed December 2015. http://www.ben.edu/college-of-science/about/ history-of-the-sciences.cfm

Brickell, Christopher (ed.). *The Royal Horticultural Society A–Z Encyclopedia of Plants*, 3rd edition. London: Dorling Kindersley, 2008

Butt, Len P. *An Introduction to the Genus Cycas in Australia*. Milton: Palm and Cycad Societies of Australia, 1990.

Caruel, Teodoro. *Storia illustrata dei tre regni della natura del dott. Aloisio Pokorny: Storia illustrata del regno vegetale* (*Dr. Alois Pokorny's Illustrated History of the Three Kingdoms of Nature: Illustrated History of the Plant Kingdom*), Volume 2. Rome: Ermanno Loescher, 1871.

Comte, Achille. *Planches murales d'histoire naturelle: Zoologie, Botanique, Géologie* (*Natural History Wall Charts: Zoology, Botany, Geology*), 2nd edition. Paris: Victor Masson et Fils, 1869.

Cullen, James, Sabina G. Knees and H. Suzanne Cubey. *The European Garden Flora: A Manual for the Identification of Plants Cultivated in Europe, Both Out-of-doors and Under Glass*, 2nd Edition, Volumes 1–5. Cambridge: Cambridge University Press, 2011

Davis, J. R. Ainsworth. *Vegetable Morphology and Physiology*. London: Charles Griffin & Company, Limited, 1893.

Detmer, Dr. Wilhelm. *Practical Plant Physiology: An Introduction to Original Research for Students and Teachers of Natural Science, Medicine, Agriculture and Forestry*. London: Swan Sonnenschein & Co; New York: The MacMillan Company, 1898.

Dittes, Friedrich, Albert Richter, and Heinrich Scherer. *Pädagogischer Jahresbericht* (*Educational Annual Report*), Volume 46. Leipzig: Friedrich Biandstetter, 1894.

Dodel-Port, Dr Arnold and Carolina. *Erläuternder Text zum anatomisch-physiologischen Atlas der Botanik* (*Explanatory Text for the Anatomical and Physiological Atlas of Botany*). Esslingen: J.F. Schreiber, 1883.

Esser, Dr. Peter. *Die Giftpflanzen Deutschlands* (*The Poisonous Plants of Germany*). Braunschweig: Friedrich Vieweg & Sohn, 1910.

Forster, Beat and Dagmar Nierhaus-Wunderwald. 'The Silver Fir Woolly Aphid (*Dreyfusia nordmannianae*),' waldwissen.net (online version: September 4 2006), accessed November 2015. http://www.waldwissen.net/waldwirtschaft/schaden/insekten/wsl_weisstannentrieblaus/index_EN

Fritsch, Karl. *Pokornys Naturgeschichte des Pflanzenreiches für höhere Lehranstalten* (*Pokorny's Natural History of the Plant Kingdom for Higher Educational Establishments*) Volume 2. Leipzig: G. Freytag, 1903.

Gerard, John. *A Catalogue of Plants Cultivated in the Garden of John Gerard in the Years 1596–1599*. London: Privately printed, 1876.

Grey-Wilson, Christopher. *Poppies: The Poppy Family in the Wild and in Cultivation* (revised and updated edition). London: B.T. Batsford Ltd, 2000

Guttstadt, Albert. *Die Anstalten der Stadt Berlin für die öffentliche Gesundheitspflege und für den naturwissenschaftlichen Unterricht* (*The Institutions of the City of Berlin for Public Healthcare and for Science Lessons*). Berlin: Stuhr, 1886.

Härtel, W. and K. Schenkl. *Zeitschrift fur die osterreichischen gymnasien* (*Journal for the Austrian High Schools*). Vienna: Carl Gerold's Sohn, 1882.

Holzner, W. and M. Numata (Eds.) *Biology and Ecology of Weeds*. New York: Springer; The Hague: vccc W. Junk Publishers, 1982.

Howell, Catherine Herbert. Flora Mirabilis: *How Plants Have Shaped World Knowledge, Health, Wealth, and Beauty*. Washington, D.C: National Geographic Society, 2009.

Klein, Aldo Luiz. *Eugen Warming e o cerrado brasileiro: um século depois* (*Eugen Warming and the Brazilian Cerrado: a Century Later*). Sao Paulo: Universidade Estadual Paulista, 2002.

Kny, Leopold. Botanische wandtafeln mit erläuterndem text (Botanical Wall Charts with Explanatory Text) Parts I, II, III. Berlin: Wiegandt, Hempel & Parey, 1874.

Kraus, Gregor. *Geschichte der Pflanzeneinführungen in die europäischen botanischen Gärten* (*History of Plant Introductions into European Botanical Gardens*) Leipzig: Engelmann, 1894.

Lanini, W. T, et al. 'Pest Notes: Dodder,' for UC Statewide Integrated Pest Management Program, University of California, Davis (March 2010), accessed September 2016. http://www.ipm.ucdavis.edu/PMG/PESTNOTES/pn7496.html#REFERENCE

Levy, Clifford J. 'Seeking Purification at Russia's Melon Stands,' *New York Times*, September 21, 2009.

Loudon, Jane. *Botany for Ladies; or, a Popular Introduction*. London: Bradbury and Evans, 1842.
Marinelli, Janet (ed.). *Plant: The Ultimate Visual Reference to Plants and Flowers of the World*. London: Dorling Kindersley, 2004.

Phillips, Roger and Martyn Rix. *The Botanical Garden* (2 vols.), London: Macmillan, 2002.

Parbery, Douglas G. *Daniel McAlpine and The Bitter Pit*. New York: Springer, 2015.

Porter, Duncan and Peter Graham. *Darwin's Sciences*. Hoboken NJ: John Wiley & Sons, 2015.

Riley, Charles Valentine. 'The Grape Phylloxera' *Popular Science Monthly* (May 5, 1874), accessed December 2015. https://en.wikisource.org/wiki/Popular_Science_Monthly/Volume_5/May_1874/ The_Grape_Phylloxera

Rosen, Felix. *Anatomische Wandtafeln der vegetabilischen Nahrungs- und Genussmittel: Text (Anatomical Blackboards of Vegetables for Food and Beverages: Text*), Volume 1. Breslau: J.U. Kern's (Max. Müller), 1904.

Royal Botanic Gardens, Kew, Missouri Botanical Garden, and others. *The Plant List: A Working List of All Plant Species*, Version 1.1 (September 2013), accessed November–December 2015. http://www.theplantlist.org

Schmeil, Otto. *Lehrbuch der Botanik für höhere Lehranstalten und die Hand des Lehrers* (*Textbook of Botany for Higher Educational Institutions and the Hand of the Teacher*). Leipzig: Erwin Nägele, 1904.

Schmeil, Otto. *Pflanzenkunde, unter besonderer berücksichtigung der beziehungen zwischen bau und lebensweise der pflanzen* (*Botany, with Special Consideration of the Relationship between the Form* 2171907.

Schwartz, James. *In Pursuit of the Gene: From Darwin to DNA*. Cambridge MA: Harvard University Press, 2008.

Sibley, David Allen. *The Sibley Guide to Trees*. New York: Alfred A. Knopf, 2009.

Smith, James Edward, George Shaw, and James Sowerby. *English Botany; Or, Coloured Figures of British Plants, with Their Essential Characters, Synonyms, and Places of Growth: To Which Will Be Added, Occasional Remarks*. London: R. Taylor (printed by), 1812.

Stern, William Louis, Kenneth J. Curry, and Alec M. Pridgeon. 'Osmophores of Stanhopea (Orchidaceae),' *American Journal of Botany*, Vol. 74, No. 9, pp 1323–1331 (September 1987), accessed November 2015. http://www.jstor.org/stable/2444310

Tutin, T. G. et al. (eds.) *Flora Europaea Volume 1: Psilotaceae to Platanaceae*, 2nd (revised) paperback edition. Cambridge: Cambridge University Press, 2010
Stern, William Louis, Kenneth J. Curry, and Alec M. Pridgeon. *Flora Europaea Volume 2: Rosaceae to Umbelliferae*, 2nd (revised) paperback edition. Cambridge: Cambridge University Press.

Stern, William Louis, Kenneth J. Curry, and Alec M. Pridgeon. *Flora Europaea Volume 3: Diapensiaceae to Myoporaceae*, 2nd (revised) paperback edition. Cambridge: Cambridge University Press, 2010

Stern, William Louis, Kenneth J. Curry, and Alec M. Pridgeon. *Flora Europaea Volume 4: Plantaginaceae to Compositae* (*and Rubiaceae*), 2nd (revised) paperback edition. Cambridge: Cambridge University Press, 2010.

von Ahles, Wilhelm Elias. *Vier Feinde der Landwirtschaft: das Mutterkorn und der Rost des Getreides, die Kartoffel- und Traubenkrankhet* (*Mehlthau, Honigthau, Russthau etc.*)*: zugleich als Erläuterung der vier Wandtafeln der Pflanzenkrankheiten* (*Four Enemies of Agriculture: Ergot and the Rust of Corn, Potato and Grape Diseases* (*mildew, honey-dew, Russthau* [*soot-dew*], *etc.*)*: With Explanations of the Four Wall Panels of Plant Diseases*). Ravensburg: Eugen Ulmer, 1874.

Webb, D. A. (eds: T. G. Tutin et al.) *Flora Europaea Volume 5: Alismataceae to Orchidaceae*, 2nd (revised) paperback edition. Cambridge: Cambridge University Press, 2010.

Witte, H. *Sieboldia, weekblad voor den tuinbouw in Nederland* (*Sieboldia, Weekly for Horticulture in the Netherlands*), Volume 5. Leiden: E. J. Brill, 1879.

Zippel, Hermann. *Repräsentanten einheimischer Pflanzenfamilien in farbigen Wandtafeln mit erläuterndem text: Kryptogamen* (*Representatives of Indigenous Plant Families with Explanatory Text: Cryptogams*). Braunschweig: Friedrich Vieweg & Sohn. 1880.

Zippel, Hermann. *Ausländische handels- und Nährpflanzen zur Belehrung für das Haus und zum Selbstunterrichte* (*Foreign Trade and Host Plants, for Instruction in the Home and as Lessons*). Braunschweig: Friedrich Vieweg & Sohn, 1885.

Zippel, Hermann, Otto Wilhelm Thomé, and Carl Bollmann. *Ausländische Kulturpflanzen in farbigen Wandtafeln mit erläuterndem text* (*Foreign Crops in Coloured Wall Panels, with Explanatory Text*), Volume 1. Braunschweig: Friedrich Vieweg & Sohn, 1899.

GLOSSARY

Achene: one-seeded, *indehiscent* dry fruit, formed from *ovary* with single *carpel.*

Actinomorphic: describing flowers with radial symmetry, so that two halves cut through the centre would be almost identical; such flowers usually have *perianth* segments of similar size and shape (see also Zygomorphic).

Androecium: male reproductive part of flower, made of *stamens* (see also Gynoecium).

Angiosperm: flowering plant that develops seed inside *ovary* once fertilization has occurred in the *ovule.*

Annual: Plant that lives out its life cycle within one growing season.

Antepetallous: of *stamens*, each aligned in front of one *petal.*

Anther: part of *stamen* that produces *pollen*, usually at tip of *filament.*

Anthocyanin: water-soluble red, blue,or purple pigment in plant tissue, such as leaves, stems, fruits and flowers.

Aril: fleshy or hairy, usually brightly coloured outgrowth from seed stalk that surrounds seed.

Asexual vegetative reproduction: where one plant reproduces from a vegetative part – such as a bud, *bulb*, *rhizome* or stem – of itself rather than seed or *spores*; new organisms are genetically identical to the parent.

Awn: sharp, stiff bristle, as on the grass *inflorescences* and some fruits.

Axil: upper angle formed between stem and leaf, *bract* or lateral branch.

Axillary: referring to organ in *axil*, such as bud or *cyme.*

Basal rosette: circular arrangement of leaves around stem at soil level.

Berry: simple, *indehiscent*, *succulent* fruit, usually with many seeds and often brightly coloured.

Biennial: Plant that lives out its life cycle over the course of two growing seasons.

Bipinnate: of leaves, where leaflets are also divided into leaflets.

Blade: see *Lamina.*

Boss: Protruding part of flower, usually in centre.

Bract: modified leaf-like organ at the base of a flower or *inflorescence.*

Bracteole: modified leaf-like organ at the base of flower, where *inflorescence* is itself subtended by another *bract.*

Bulb (adj. bulbous)**:** shoot modified to fleshy underground storage organ, consisting of scale leaves, apical bud and basal plate.

Bulbil: small *bulb*, formed from bud in leaf *axil*, on stem or in flower head.

Calyx: collective name for *sepals*, which comprise outer whorl of *perianth.*

Cambium: layer of meristematic (growth) tissue responsible for increasing stem and root girths.

Campanulate: of *corolla*, bell-shaped.

Capitulum/a: densely packed *inflorescence* of small, stalkless flowers or florets on *involucre* of *bracts*, so resembling a larger, single flower.

Capsule: dehiscent, dry fruit formed from at least two *carpels.*

Carpel: part of *gynoecium* that encloses *ovules*, usually consisting of *ovary*, *style*, and *stigma* (see also Pistil).

Carpellary: possessing a *carpel.*

Catkin: *raceme-* or *spike*-like *inflorescence* adapted for wind pollination, usually many sessile, unisexual flowers with no *petals.*

Caudex/ices: 1) swollen persistent basal stem of some herbaceous *perennials*; 2) woody stem of cycad, palm or tree fern.

Caudiciform: having a swollen trunk or stem (see also Caudex).

Cell vacuole: fluid-filled sac in plant cell, containing nutrients, waste products, and other substances, that also contributes to rigidity of the plant by means of hydrostatic pressure.

Chalice: a *calyx*, often cup-shaped.

Chlorophyll grain: particles of cell protoplasm colored green by chlorophyll, the pigment involved in photosynthesis.

Circinate: leaf or stem, rolled up so tip is in centre of coil.

Cladode: modified leaf-like stem, usually in plant with reduced true leaves.

Cluster: *inflorescences* that arise, or seem to arise, from a single point.

Compound: 1) of leaf, referring to leaf made of smaller leaflets; 2) referring to flower composed of many smaller flowers or florets.

Cone: seed-bearing structure typical of *gymnosperms*, often composed of woody *scales* arranged spirally.

Corm (adj. cormous)**:** stem modified to fleshy underground storage organ, consisting of starchy tissue and papery outer tunic.

Corolla: collective name for *petals*, which comprise inner whorl of *perianth.*

Corona: crown- or trumpet-shaped outgrowth from *corolla*, formed of fused *filaments* or *petals.*

Cortex: tissue between *epidermis* and *vascular tissue* in stem.

Corymb (adj. corymbose)**:** *inflorescence* with flower stalks arising at different points around the stem, but of different lengths, to end in a flattish or domed flower head.

Cretaceous Period: last period of Mesozoic era, between 146 and 65 million years ago.

Cruciform: describing four-petalled flowers oriented perpendicularly in the shape of a cross.

Cruciferous: of the plant family *Cruciferae*, now called *Brassicaceae.*

Cupule: cup-shaped structure; in *Caytoniales*, refers to hollow structure containing *ovules*, with pore at base to allow entry of *pollen.*

Cyathium: cup-shaped *involucre* of one female and several male flowers that looks like larger, single flower.

Cyme (adj. cymose)**:** branching *inflorescence* in which all axes end in a flower, with the one at the main stem tip opening first.

Cypsela: one-seeded, *indehiscent* dry fruit, formed from inferior *ovary* of single *carpel.*

Daughter bulb: offset, formed at base of *bulb* by *asexual vegetative reproduction.*

Dehiscent: of a fruit or *anther*, which splits along precise lines to release seeds or *pollen* (see also Indehiscent).

Dichasial: of *inflorescence*, two-branched.

Disc: centre of compound flower head, formed from the *receptacle.*

Disc floret: tiny flower in *capitulum* formed from toothed *corolla* tube, sometimes surrounded by outer array of *ray florets.*

Dispersal wing: extension of seed membrane to aid wind dispersal of seed (see also Wing).

Dissected: of leaves, deeply or repeatedly cut into lobes.

Dioecious: where male and female reproductive organs occur on separate plants (see also Monoecious).

Drupe: simple, *indehiscent*, succulent fruit with one or more hard-coated seeds (stones), often brightly coloured, for example a cherry.

Drupelet: small *drupe*, usually one of several in *compound* fruit, as in a blackberry.

Endosperm: nutrient storage tissue in seed.

Ephemeral: short-lived.

Epidermis: outermost layer of cells of plant body, usually only one cell thick.

Epiphyte (adj. epiphytic): plant that does not root in soil, but lives on other plants or surfaces and takes in nutrients from the air, rain water, and organic debris.

Exstipulate: without *stipules* (see also Stipulate).

Filament: stalk of the *stamen*, having the *anther* at its apex.

Filiform: thread-like.

Gall: abnormal, localized swelling or growth of plant tissue, formed in reaction to parasitical attack.

Glabrous: smooth, lacking hairs.

Growth ring: also called annual ring; circle of secondary *xylem* visible in tranverse section of woody plant, showing one year's growth.

Gymnosperm: plant that develops seed from unenclosed or naked *ovules*, usually borne on *cones*.

Gynoecium: female reproductive part of flower, made of one or more carpels (see also Androecium, Pistil).

Haustorium/a: organ of a parasitic plant that penetrates a host plant's tissue in order to absorb nutrients.

Hypanthium: tubular or cup-like structure composed from fused *petal*, *sepal*, and *stamen* bases.

Indehiscent: fruit that does not split to release seeds (see also Dehiscent).

Inflexed: curving downward toward axis.

Inflorescence: flower head of more than one flower.

Inflorescence meristem: undifferentiated growth tissue (*meristem*) at stage between apical meristem, where shoot tip produces vegetative growth, and the floral meristem, which produces flowers.

Infundibuliform: funnel-shaped.

Infructescence: fruiting stage of *inflorescence*.

Interpetiolar: between *petioles*.

Involucre: in flowering plants, whorl of *bracts* at base of compact *inflorescence*, such as *capitulum* or *umbel*.

Keel: of flowers in *Fabaceae*, two lower *petals* fused at their lower edges into a boat-like shape.

Labellum: lip formed from fused *petals* or *sepals*, or lowest petal, in flower; often to form a landing platform for pollinators.

Lamina: typically flattened blade of leaf.

Lanceolate: longer than wide in shape, being broader in the lower part and tapered at the tip.

Legume (adj. leguminous)**:** 1) *dehiscent*, dry fruit with one or more large seeds, formed from a single *carpel*; 2) plant from *Fabaceae*.

Medullary ray: in woody stems, extension of the medulla (pith) in sheets or ribbons, radiating outward to cortex; transports sap.

Meristem: tissue containing dividing cells for growth, sometimes undifferentiated and sometimes differentiated into, for example, leaf or flower cells.

Microsporangium/a: sac in which microspores develop in *spore*-bearing plant; akin to *pollen sac* in flowering plants (see also Microsporophyll).

Microsporophyll: modified leaf that bears *microsporangia* in *spore*-bearing plant (see also Pollen cone).

Molecular phylogenetics: study of how an organism has evolved, by analysis of molecular structures, especially DNA.

Monocarpic: plant that flowers and fruits only once before dying; plants may live for several years without flowering.

Monoecious: where male and female reproductive organs occur in separate flowers on the same plant (see also Dioecious).

Monotypic: taxon with only one member of lower rank, for example a genus with one species.

Morphology (adj. morphological)**:** in botany, study of form and structure of organisms.

Mucilage (adj. mucilaginous)**:** organic substance that absorbs water to forms a viscous, sticky solution.

Nectary: gland that secretes nectar.

Needle: long, needle-like leaf, formed to reduce water loss, particularly in conifers.

Node: point on stem from which leaves, shoots or flowers arise.

Nutlet: one-seeded, *indehiscent* dry fruit, formed from more than one *carpel*, although only one carpel develops into a woody-coated, small nut.

Nymph: juvenile form of some insects, as in dragonflies.

Ovary: swollen base of *carpel*, or several fused carpels, containing *ovules*.

Ovate: of leaf or fruit about 1.5 times long as it is broad, egg-shaped in outline and broadest below the centre.

Ovule: female organ in the *ovary* that becomes seed after pollination and fertilization.

Palmate (adj. palmately)**:** 1) *compound* leaf with at least four leaflets, arising from same point; 2) pattern of veins where several conspicuous veins radiate from the base of the *lamina*.

Panicle: branching *inflorescence*, often made of *racemes* arising from the main axis, as in grasses.

Pangenesis: hypothesis that all cells in an organism shed hereditary particles into the blood, which then accumulate in reproductive cells.

Papilionaceous: 1); flowers shaped like those of the *Papilionoideae*, i.e. *zygomorphic* flowers with *petals* differentiated into large, upper *standard*, two lateral wings and *keel* of two fused lower petals.

Papillate: covered in minute, rounded protuberances.

Pappus: tuft or whorl of hairs, bristles or *scales* formed from modified *calyx*, to aid wind dispersal of fruit or seed.

Parthenogenesis (adv. parthenogenetically)**:** type of reproduction where an unfertilized ovum or egg develops into an embryo.

Pedicel: stalk of single flower.

Peduncle: main stem of an *inflorescence*.

Peltate: structure such as leaf that is attached to stalk at the centre of its lower surface.

Pepo: modified fleshy *berry*, with hard skin.

Perennial: Plant that may live over many seasons, often partially dying back over winter and regrowing in the spring.

Perianth: collective name for *calyx* and *corolla*.

Petal: modified leaf and part of *corolla*, often conspicuous in flowers that depend on pollinators.

Petiole: leaf stalk, attaching *lamina* to stem.

Phloem: transport or *vascular tissue* mainly responsible for carrying nutrients around a plant (see also Vascular system, Xylem).

Phylogenesis (also phylogeny; adj. phylogenetic)**:** the evolutionary history of a group of organisms, often depicted as a family tree.

Pinnate (adj. pinnately)**:** describes *compound* leaf that has alternate or opposite leaflets arranged along its central axis.

Pinnatisect: of leaf, with deeply cut lobes.

Pistil (adj. pistillate)**:** female reproductive part of flower, made of single or several fused or separate *carpels* (see also Gynoecium).

Placenta: tissue within *ovary*, to which may be attached *ovules*, *spores*, or *sporangia*.

Plasma: substances within cells.

Plicate: of leaf, folded lengthways like a concertina.

Polar nucleus: female nucleus, one of a pair, present in the *ovule*; during fertilization, the two polar nuclei combine with one sperm cell (male gamete) to form endosperm nucleus (see also Pollen tube, Ovary).

Pollen: singular or collective name for *pollen grains*.

Pollen cone: male *cone* in conifers, with *microsporophylls* extending from central axis; each microsporophyll shields one or several *microsporangia*.

Pollen grain: fine, powdery microspore of seed plants that contains and protects the sperm cells (male gametes); also called *pollen* (see also Pollen sac).

Pollen-mass: bundle of *pollen grains* held together by fine filaments or waxy substance; also called pollinium in *Orchidaceae*.

Pollen sac: pouch where *pollen* forms; in *angiosperms*, pollen sacs – usually four – are located in the *anther*, in conifers, in *axils* of the *pollen cone*.

Pollen tube: outgrowth of *pollen grain* that forms after pollination and grows towards *ovule* in *angiosperms*, in order to bring male gametes to ovum or egg.

Polycephalous: bearing many heads.

Pome: firm, fleshy false fruit formed from the hypanthium with the true fruit at the centre, or core, such as apple and pear.

Poricidal capsule: *capsule* that opens pores to distribute seeds (see also Dehiscent).

Prickle: thorn-like outgrowth from plant *epidermis*, as in rose (see also Thorn).

Primary wood: primary *xylem*, formed during first growth of plant (see also Secondary wood).

Pseudobulb: swollen, *bulb*-like stem base used as water storage organ in some orchids.

Raceme: unbranched, long-stemmed *inflorescence* with stalked flowers that open first at the base.

Rachis: main axis of *inflorescence* or *pinnate* leaf.

Ray floret: tiny flower in *capitulum*, formed from strap-shaped *corolla* tube, sometimes around an inner array of *disc florets*.

Receptacle: 1) enlarged tip of *pedicel* bearing simple flower; 2) enlarged, shortened tip of *peduncle*, usually convex but sometimes flattened, bearing *capitulum*.

Resin canal/duct: intercellular channel that secrets resin, such as in leaves of many *gymnosperms*.

Reticulate: with a netted pattern, such as the veins of a leaf.

Rhizome (adj. rhizomatous)**:** branching underground stem, capable of *asexual vegetative reproduction*, that grows horizontally; some are fleshy storage organs, others are rope-like.

Rootstock: 1) underground part of plant; 2) crown and roots of herbaceous *perennial*; 3) part of grafted plant, onto which scion (top growth) is grafted.

Runner: trailing stem capable of rooting and developing plantlets from nodes.

Samara: type of *achene* – one-seeded, *indehiscent* dry fruit, formed from *ovary* with single *carpel* – with wing to aid wind dispersal of seed.

Saprophyte (adj. saprophytic)**:** plant, usually lacking chlorophyll for photosynthesis, which feeds off decaying organic matter.

Sarcotesta: fleshy outer coat of seed.

Scale: 1) reduced, modified leaf, often *sessile* and lacking chlorophyll; 2) flattish outgrowth from *epidermis*, or membrane, such as on leaf or stem surface; 3) modified, woody leaf in *cone*.

Scape: leafless stem of solitary flower or *inflorescence*.

Schizocarp: dry fruit, formed from at least two *carpels*, that splits when mature into one-seeded units; one-seeded unit may be *achene*, *berry*, *nutlet* or *samara*.

Secondary wood: secondary *xylem*, responsible mainly for increase in girth of woody plant (see also Growth ring, Primary wood).

Sepal (adj. sepaloid)**:** constituent part of *calyx*, usually green and smaller than a *petal*, but sometimes petaloid.

Serrated: finely toothed, as in leaf margin.

Sessile: lacking a stalk or *petiole*.

Silicula: *dehiscent* dry fruit formed from two *carpels* into flat pod divided by a membrane; seeds are exposed as the pod peels away from dividing membrane; silicula is less than three times longer than it is broad.

Siliqua: similar to silicula, but three times or more longer than it is broad.

Spadix: fleshy, spike-like *inflorescence*, bearing tiny, *sessile* flowers, usually sheathed or surrounded by a *spathe*.

Spathe: Large, conspicuous *bract* enclosing a *spadix*, often brightly colored to attract pollinating insects.

Spike: unbranched *inflorescence* with *sessile* flowers.

Spine: sharp, pointed, modified leaf that provides plant with means of defence and reducing water loss.

Sporangium/a: sac in which *spores* develop in spore-bearing plant.

Spore: basic unit in asexual stage of reproduction in non-flowering plants.

Sporophyll: modified leaf bearing *sporangia*.

Stamen: male part of flower, comprising *filament* and *anther* (see also Androecium, Pistil).

Staminate: of flower, lacking female parts.

Standard: large, uppermost, often upright *petal* typical of flower of plants in *Fabaceae*.

Stigma: tip of *pistil*, adapted to receive *pollen*; site of *pollen-grain* germination.

Stipule: protective, leaf-like structure or *scale*, usually borne in pairs at base of leaf or *petiole* (see also Exstipulate, Stipulate).

Stipulate: possessing *stipules* (see also Exstipulate).

Stoma: pore in the *epidermis* of the aerial part of plant such as a leaf that allows gaseous exchange between atmosphere and plant tissue.

Style: part of *carpel* or *pistil* that connects *stigma* to *ovary*.

Succulent: plant that occurs in arid habitats and stores water in fleshy, swollen stems or leaves; succulents sometimes have greatly reduced leaves to mimimize water loss; one group of succulents are cacti.

Syncarpous (adj.)**:** of *gynoecium*, having united *carpels*.

Taproot: single primary root that grows vertically downward, sometimes also a fleshy storage organ.

Taxonomy: study of classification of living organisms.

Tepal: collective name for undifferentiated *petals* and *sepals* (see also Calyx, Corolla, Perianth).

Thallus: vegetative parts of algae, lichen or fungi that have no differentiation between stems, leaves or roots.

Thorn: sharp, pointed outgrowth from stem, as in hawthorn (see also Prickle).

Triassic: first period of Mesozoic era, between 252 and 201 million years ago.

Trichome: outgrowth from epidermal cell in plant.

Tuber (adj. tuberous)**:** stem or root modified to fleshy storage organ; stem tubers have buds, or 'eyes', whereas root tubers do not.

Umbel: flat-topped *inflorescence* in which each flower stalk or *pedicel* arises from a single point on the main stem or *peduncle*, with the oldest flowers at the margins.

Umbellate: describing *inflorescence* arranged in *umbel*.

Unilocular: possessing one chamber or locular, for example of *carpel*, *ovary* or *sporangium*.

Vascular bundle: collection of strands of primary *vascular tissue* in a plant (see also Vascular system).

Vascular system: network that transports *vascular tissue*, such as *phloem* and *xylem*, around a plant.

Vascular tissue: transport tissue, such as *phloem* and *xylem*, in a plant.

Viscidium/a: specialized part of some orchid flowers – sticky part of one *stigma* to which adhers waxy *pollen-mass*, or pollinium, so that it is transferred more easily to pollinating insects.

Wing: 1) one of two lateral *petals* typical of flower of plants in *Fabaceae* (see also Keel, Standard); 2) membraneous extension of fruit, to aid dispersal by wind, for example in *samara*.

Xylem: *vascular tissue* mainly responsible for carrying water and soluble minerals from the roots to plant (see also Phloem, Vascular system).

Xylem fibre: support cells in *xylem* tissue that have lignified or woody cell walls.

Zygomorphic: describing flowers that are bilaterally symmetrical, i.e. are symmetrical if cut along only one plane; such flowers usually have *perianth* segments of varying size and shape (see also Actinomorphic).

Zygote: single fertilized cell after fusion of male (sperm) cell and female (ovum) cells, before cell division occurs.

INDEX

AUTHOR BIOGRAPHY

Anna Laurent (1979–2024) dedicated her life to showcasing the beauty and functionality of plant forms and bringing them to a wider audience. An artist in her own right, her work is incorporated in the permanent collections of the Royal Botanical Gardens at Kew, and in the Los Angeles Museum of Natural History, California, as well as being exhibited in botanical gardens around the world and at the Linnean Society, London. Dispersal, a photographic project devoted to the beauty of seedpods that originated in a column for Print magazine, won prizes at the RHS Chelsea Flower Show, the International Photography Awards, and the International Garden Photography Awards. Originally from Boston, Massachusetts, Anna's interest in botany developed upon moving to Los Angeles in 2008 to work in documentary film production. What began as a hobby became all consuming and within a short time she quit her job to devote herself full time to writing and creating art about plants. She found further success on relocating to Europe in 2017 where she split her time between London and the French Basque Country. Anna passed away from cancer at the age of 45 while working on a book on the history of human interactions with the botanical world.

PICTURE CREDITS

Ilex Press would like to acknowledge and thank the following sources, collections and their staff for their kind help in putting this book together, including Jens Jan Andersen, Marie-Laure Baudement, Jos Beerens, Lisa DeCesare, Anita Dijkstra, Eline Delftstra, Frans van den Hoven and Monique Jaspars, Debbie Gale, Jan Waling Huisman and Ciska Ackermann, Dr Norbert Kilian, Tom Lamb, Sytse van der Leest, Saskia Speur and Yvonne de Wit, Sara Miori, David Murray, Margaret Pezalla-Granlund, Elisabeth Price, Dr Lesley Robertson FRSB, Lydia Rosenberg,Philippe Rossignol, Carrie Roy, Stephen Sinon, Milan Skalický, Ruud Wilstra, and Eleonora Zen.

Allard Pierson, University of Amsterdam 28–9, 38–9, 55, 66–7, 76, 90–1, 92, 112–13, 128, 146, 149, 154, 156–7, 164, 165, 170, 171, 183, 189, 193, 198, 209

AU Library, Emdrup – Royal Danish Library 36, 142, 143, 187

Botanic Garden and Botanical Museum Berlin, Freie Universität Berlin 18–19, 62–3, 70, 205, 206–7

Botanical Collections, University Museum, Groningen, Netherlands 12, 13, 56, 57, 64, 65, 67–8, 136, 208, 210, 211

Collection UMU, University Museum Utrecht 30, 82

Department of Botany and Plant Physiology, FAFNR, Czech University of Life Sciences, Prague 37, 42–3, 46–7, 74–5, 78–9, 120–1, 130–1, 151, 178–9, 190

De Kantlijn 11

Delft School of Microbiology Archives at Delft University of Technology 9(a & b), 15, 41, 59, 60, 86, 98, 104, 114, 127, 137, 144, 148, 158, 196

Economic Botany Archives, Harvard University, Cambridge, MA 2, 49, 81, 108, 133, 139, 172, 175, 191, 195, 199, 212, 214–15

Editions Rossignol 176–7

Fondazione Museo Civico di Rovereto 16, 25, 52, 95, 140–1, 155, 160, 161, 184–5

Gould Library, Carleton College, Northfield 119

The Hocken Collections, Uare Taoka o Hakena, University of Otago, Dunedin 22–3, 45

Hunt Institute for Botanical Documentation, Carnegie Mellon University, Pittsburgh 202–3

Jos Beerens Handelsonderneming 85, 163

The LuEsther T. Mertz Library and Archives of The New York Botanical Garden 166, 201

Randolph College, Lynchburg 7, 122

Special Collections, Wageningen UR Library 35, 72, 100, 103, 107, 152, 167

University of Burgundy, Dijon 20–1, 50–1, 88–9, 96–7, 111, 116–17, 134–5, 180